Supercomputing

NATO ASI Series

Advanced Science Institutes Series

A series presenting the results of activities sponsored by the NATO Science Committee, which aims at the dissemination of advanced scientific and technological knowledge, with a view to strengthening links between scientific communities.

The Series is published by an international board of publishers in conjunction with the NATO Scientific Affairs Division

A Life Sciences	Plenum Publishing Corporation
B Physics	London and New York
C Mathematical and Physical Sciences	Kluwer Academic Publishers Dordrecht, Boston and London
D Behavioural and Social Sciences	
E Applied Sciences	
F Computer and Systems Sciences	Springer-Verlag Berlin Heidelberg New York
G Ecological Sciences	London Paris Tokyo Hong Kong
H Cell Biology	Barcelona

Series F: Computer and Systems Sciences Vol. 62

Supercomputing

Edited by

Janusz S. Kowalik

Boeing Computer Services, Engineering Computing and Analysis
P.O. Box 24346, Seattle, WA 98124, USA

and

Department of Computer Science and Engineering
University of Washington, Seattle, WA 98195, USA

Springer-Verlag Berlin Heidelberg New York
London Paris Tokyo Hong Kong Barcelona
Published in cooperation with NATO Scientific Affairs Division

Proceedings of the NATO Advanced Research Workshop on Supercomputing, held in Trondheim, Norway, June 19–23, 1989.

ISBN-13:978-3-642-75773-0   e-ISBN-13:978-3-642-75771-6
DOI: 10.1007/978-3-642-75771-6

2145/3140-543210 – Printed on acid-free-paper

Preface

This book contains papers presented at the NATO sponsored Advanced Research Workshop on Supercomputing held in Trondheim, Norway, from June 19 to June 23, 1989. The purpose of the Workshop was to assess the state of the art and the key trends in supercomputing technology, and identify the most pressing needs and trends in the field.

Supercomputing is an important science and technology that enables the scientist or the engineer to simulate numerically very complex physical phenomena related to large-scale scientific, industrial and military applications. In many industrialized countries the value of supercomputing is fully appreciated and several national programs have been funded to foster and commercialize high performance computing technologies. We expect that the impact of supercomputing will be significant and wide in scope; from better cars, planes and manufacturing processes to solving the societal and scientific grand challenges such as mapping the human genome, global environmental issues, quantum chromodynamics, design of drugs, origin and structure of the universe, and so on.

Since the first NATO Workshop on High-Speed Computation held in Julich in 1983, the science and the technology of supercomputing have made considerable progress: new architectures and programming tools have been designed and implemented, we have learned more about networking and visualization, and we have constructed new parallel and vectorizable algorithms. At the same time we have generated new questions and issues that require individual research and coordinated team efforts. They include:

1. Theory and Algorithms

Parallel computing is one of the principal sources of opportunity to improve computational performance. Unfortunately there exist large areas of human activity related to resource allocations and decision making that give raise to intrinsically difficult combinational optimization problems that cannot be solved much faster by parallel computers. We need to learn how to design heuristic algorithms for these problems. This may involve coupling numerical and symbolic computation.

Data structures are key to efficiency of algorithms. Research is needed in the areas of structuring spatial data for large-scale numerical simulations and "parallel data structures" (suitable data structures for parallel computation) for nonnumerical problems.

2. Computer Architecture

A vast variety of computer architectures is currently available. It is unclear which of these architectures are scalable or general purpose, and which will remain suitable for problems of limited size or in narrow domains.

3. Compilers and Parallel Program Development Tools

Sophisticated restructuring and compiling systems are required for parallel computers. It is believed that the detection of parallelism will remain the programmer's responsibility and the architectural mapping will be automated. Can we do better?

4. Supporting Technologies

Three most important supporting technologies are networking, visualization and mass storage. Supercomputers generate massive amounts of data which need to be transmitted, displayed and archived. Intelligent ways of reducing the amount of raw data and enhancing the semantic value of the data transmitted or stored should be among our research goals.

5. System Integration

The number of supercomputer users who are not programmers is increasing. For them the major obstacle to effective use of computers is the difficulty of setting up, controlling and analysing computation without the help of a software expert. Heterogeneous computing environments with several types of computers and workstations exacerbate this situation. Improved user interfaces and utilities combined with integrated seamless environments will be required to make the use of supercomputers easier.

The collection of papers presented in this volume addresses some of these issues.

We are indebted to NATO, the US Air Force Europeon Office of Aerospace Research and Development (EOARD), Office of Naval Research London (ONRL), US Army Research Standardization Group (USARSG), The University of Trondheim, SINTEF and the Boeing Company for the generous financial support.

Individuals who deserve special thanks include the Workshop co-director Dr. Bjornar Pettersen, Dr. Johannes Moe and Dr. Al Erisman. Thanks are also due to those who served with me on the Organizing Committee: B. Buzbee, NCAR, U.S.A., L. Dadda, Politecnico di Milano, Italy, J. Dongarra, ANL, U.S.A., I. Duff, AERE, United Kingdom, S. Fernbach, Consultant, U.S.A., R. Hockney, University of Reading, United Kingdom, F. Hossfeld, KFA, Federal Republic of Germany, E. Levin, RIACS, U.S.A., G. Michael, LLNL, U.S.A., K. Neves, Boeing, U.S.A. and S. Norsett, NTH, Norway.

Mrs. Cynthia Actis has been extremely helpful in all phases of the Workshop preparation and activities. Her meticulous assistance contributed greatly to the success of the Workshop.

Seattle, January 1990   Janusz S. Kowalik

Table of Contents

Part 3 Supercomputer Architecture

Part 4 Programming Tools and Scheduling

Part 5 User Environments and Visualization

Part 6 Requirements and Performance

Part 7 Methods and Algorithms

Part 8 Networking

Part 9 Storage

Appendix

PART 1

INTRODUCTION

SUPERCOMPUTING: KEY ISSUES AND CHALLENGES

Kenneth W. Neves and Janusz S. Kowalik

Boeing Computer Services
P. O. Box 24346 M/S 7L-22
Seattle, Washington 98124-0346

1.0 INTRODUCTION

In this treatise research issues related to high speed computing will be explored. Important elements of this topic include hardware, algorithms, software and their interrelationships. One quickly can observe that, as in the past, the nature of computing will change dramatically with technological developments. The continued growth in computer power/capacity, workstation capabilities, animation, and host of new technologies such as machine vision, natural language interfaces, can all have a great impact on research requirements and thrusts in scientific computing. To better appreciate what might be needed in future systems, one must explore those likely "drivers" of computation -- important science that cannot be done or understood today because of limited computing technology. In this brief introduction, drivers of computation will be explored, and in subsequent sections, the research challenges they suggest will be elaborated.

Traditional Drivers of Computation

Most of today's programs that push computer capacity (in scientific computing) often have some relation to the solution of (perhaps time dependent) partial differential equations. Continuous models are discretized in time and/or space and approximated at grid points. Other approaches include the use of finite elements, integral equation approaches and so on. Many of these models have been either simplified (such as two-dimensions instead of three) or could benefit in accuracy (fidelity to the continuous model) by employing more refined grids in time and/or space. To achieve the desired results in a reasonable amount of time, there usually is a need for larger memory in machines that can produce more computations per second than current technology.

NATO ASI Series, Vol. F 62
Supercomputing
Edited by J. S. Kowalik
© Springer-Verlag Berlin Heidelberg 1990

In many areas, the continuous process being modeled discretely, is itself an approximation to a more realistic model. For example, today we model commercial aircraft with linear inviscid methods for complicated geometries, and perhaps can handle simpler geometries with Euler Equations or Navier-Stokes models. The driver for using the more complex equations with viscosity and turbulence is clear. Such considerations can produce more realistic, fuel efficient designs and dramatically shorten the product design cycle. Similar examples of the need for more complex models abound in fields such as molecular modeling, petrochemical problems, structural analysis, and electromagnetics, to mention a few.

Higher dimensional models and more sophisticated models have been traditional drivers for computation since the inception of scientific computing. However, while these traditional drivers of computation continue, far greater computer requirements will be generated by the following areas.

Solving Inverse Problems and Optimization

Today large supercomputer applications are often analysis programs such as those alluded to above. A specific process or physical phenomenon is modeled in as much detail as possible. The results are then reviewed, new input parameters and data is readied, and another analysis is attempted to gain more insight. Over time this iterative process, human to program, yields some approximation to desired result. This process is depicted in Figure 1.1.

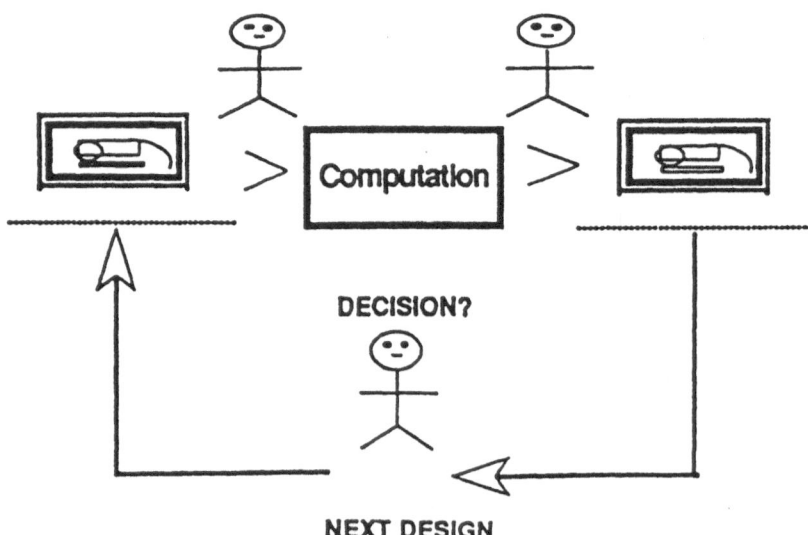

Fig. 1.1 Person-In-The-Loop

Moreover, the loop depicted above has subloops. Many large complex scientific programs, in academe and industry, are so complex that developing the input (e.g. grids for a PDE) can be difficult. That is, the modeling of the input is sometimes a difficult problem in itself. Also, interpretation of the output can be complex, even with graphical tools. It is not uncommon for the output to dictate that more resolution was required in some area of interest dictating a change in the input solely due to the interest of the viewer, and not because the model was incorrect or that the accuracy was insufficient. These human interface difficulties can cause as significant a loss of productivity as the lack of necessary computational power for more sophisticated models or refined analysis.

In the 1990s we expect to see more analysis programs being turned into design systems. One approach could be through large-scale optimization algorithms. This discipline of numerical analysis has been active for years. Recent techniques in non-linear optimization (constrained and unconstrained) offer new opportunities for complex optimization. Techniques that permit reasonable convergence with "noisy" or erratic functions offer the hope of substituting large analysis programs as the "function" in the optimization process. One can easily envision today's supercomputer programs being tomorrow's inner-loops.

Armed with these techniques, scientists can begin to solve the "inverse" problems of engineering. For example, today we analyze an airfoil by using complex computational fluid dynamics techniques. Quite often the objective is to create a particular pressure gradient over the surface of the wing. Aerodynamicists have learned a number of heuristics about the desired pressure gradient profiles over the wing, and iterate with the wing design to produce the desired result (again in the style of Fig. 1.1). With today's algorithm technology, it is possible to "suggest" the desired pressure gradient, and ask a design program (using advanced optimization techniques) to drive the analysis program as an inner loop and "solve" the problem. To some extent, this process is taking the "man out of the loop." (See Fig. 1.2.) Several technical hurdles still exist of course. It is probable that the optimization technique will need to be "steered" by heuristics. The enormity of evaluation of a large number of CFD problems suggests several orders of magnitude more computer power may be required. As more sophisticated problems can be addressed, there likely will be a role for expert systems technology in the solution process.

Optimization will likely drive other techniques. For example, structural analysis offers many opportunities in this arena. Build a structure that is "so" strong and yet is of minimal weight. Build a structure that has certain restrictions on the modes of vibration. These are two types of problems that are solved today with the engineer in the loop, often using cut and try techniques, and convergence arrives through engineering intuition, if at all.

Casting the full design problem mathematically offers the potential for improved productivity and better design quality, but is only feasible with much more computer power than available today.

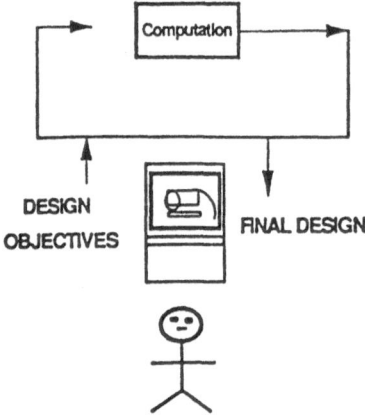

Figure 1.2. Tomorrow's Design Systems: Person-out-of-the-Loop

Cross Discipline Analysis

Another area that will drive the need for increased computation will be the requirement to mix traditionally separate disciplines during analysis and design. Science and engineering, from a computational perspective, has been compartmentalized. Classic examples in the aircraft industry include structural analysis, computational fluid dynamics, acoustics, control, thermal analysis, propulsion. The creation of an advanced aircraft, of course, requires all of the disciplines. Effective designs can only be achieved by the integration of these types of analyses in the final design. However, computational programs in the above areas have often been so computationally intense, that they are often run separately. The integration of the disciplines is done at the human or department level. Competitive pressures are causing industries to re-examine the process of design.

As an example, newer airplanes are using composite materials more and more to improve efficiency, reduce weight. These structures are often more flexible. By modeling the flex in flight greater efficiency can be achieved by improving lift in various flight regimes. This combination of control, structural analysis, and fluid dynamics poses a very complex and computation intensive model. Virtually every industry has multidisciplinary science that could provide improved product design or process if combined in this fashion. The major inhibitor for this integration has been lack of computational power.

New Modes of Interaction

One of the most rapidly changing fields in computation today is workstation and graphics technology. As the power of these tools increases, the nature of computing is being redefined. The nature of computing has always involved gaining insight to complex physical phenomenon. An important component of this process has been the human. Computation was used much like a tool to help gather facts for the human to synthesize and postulate the behavior of the phenomenon being studied. In this process, one of the greatest strengths of the human brain, was often not utilized -- the human capabilities in pattern recognition and visual inference capabilities which have long been a goal of other disciplines in AI and now, neural network theory. The advent of advanced graphics has given the human a new window on the world of physics, chemistry and engineering in general. The animation of computed results, interactive exploration of volumetric data, systematic display techniques for multi-dimensional data with dimensions greater than three, and host of new techniques have opened the whole field of computing to realistic simulation.

In time, computer users will be able to interact with simulation, direct computation in process to explore more meaningful areas (interactive animation), or design by hands-on parameterization of unknown variables. This process could bring the computer into the world of discovery, much like a surgeon in exploratory surgery.

IMPLICATIONS:

The implications of the above scenarios are by no means trivial. Even if some of what is postulated does not happen, it is clear there will continue to be a need for more computational power. Electronics forecasters tell us this power cannot be realized through circuitry alone, and there will be an increased use of parallelism. In Section 3, we will explore trends in machine organization. Clearly, there will be an ever increasing need to develop new algorithms (discussed in Section 2). The manner in which we deal with computing will clearly have to support the changes in hardware and algorithms. The tools for implementing algorithms on hardware begin with computer languages and other optimization tools in scientific computation. These will be discussed in Section 4. Finally, in Section 5 we will examine the technologies that support computing. Network technology, visualization, systems integration, data storage and management, languages and performance issues will be discussed. While detailed discussion of these topics is beyond the scope of this work, and indeed this workshop, their influence and importance cannot be ignored.

2.0 ALGORITHMS

Cocke in his 1987 A.M. Turning Award acceptance lecture stated that the main contributors to performance in scientific processors included the machine organization, the algorithm and the compiler (Cocke [1988]). Among these three factors, the algorithmic improvements can be most effective since they offer the potential for speedups nonlinear in terms of the problem size, that is, speed improvements which increase as we solve larger problems. If, for example, we replace a polynomial algorithm whose average time complexity is $T=0(N^2)$, where N is the problem size, by a better algorithm whose complexity is $T=0(N \log N)$, we can achieve under ideal conditions a speedup proportional to $N/\log N$. This can be far superior to the speedups due to parallel machine organization that at best can be constants determined by the number of CPU's, functional units or pipeline segments. Furthermore, properly designed algorithms can contribute to the total performance by minimizing processor to processor and processor to memory communication, as well as other overheads encountered in parallel processing. In vectorized and parallel algorithms, the distribution of, and the access to data have decisive impact on computational performance. This is true for both the machines with shared and distributed memories.

These observations suggest that the study of vectorized and parallel algorithms in conjunction with access to data is fundamental and crucial to the future use of supercomputers. It should include the entire life cycle of the parallel algorithms, from initial design to their long term maintenance in software libraries.

One of the authors of this paper participated in the workshop entitled "Parallel Algorithms and Architectures" conducted for the Supercomputing Research Center by the University of Maryland Institute for Advanced Studies during July and August 1985.

Subsequently, a brief version of the workshop report was published in the open literature, see Buell et al [1988]. Since the content of this report remains timely and relevant to our discussion of algorithms, we bring here some of the main points made in the report and offer an updated elaboration of these suggestions.

We begin with the observation that many theoretically and practically significant algorithms have complex and often lengthy life cycles. These cycles typically include:

1.   Selection of a generic algorithm,

2.   Architecture-dependent algorithm development and coding,

3.   Execution on one or more parallel computers, or a simulation on sequential computers,

4.   Measurement of performance that includes theoretical and specific explanation for the observed performance,

5.   Documentation and long term maintenance of the algorithm in software libraries which may be experimental, public or commercial.

The life cycle of an algorithm often begins with the initial selection of a generic algorithm which defines the computational steps to be performed but without the mapping of these steps to a specific parallel architecture. At this stage of the design, we decide to use a generic computational method such as Gaussian elimination, FFT, conjugate gradient and so on. This is followed by the development of the architecture-dependent algorithms which takes into account all relevant features of the specific supercomputer, such as: the number of processors, the memory type and hierarchy, data distribution and access, and inter-processor communication. A significant obstacle in our ability to develop architecture-dependent parallel algorithms for supercomputers is the current lack of a theory of computational and communication complexity based on sufficiently high level and yet realistic models of parallel computation. Such a theory will have to include several critical aspects of parallel computation and machine organization such as: algorithm granularity, load balancing, processor-to-memory and processor-to-processor communication, and accuracy of numerical results. The main advantage of such a theory would be a better understanding of the limits to parallel processing and an accurate assessment of the proximity to these limits exhibited by the existing algorithms. Our search for superior algorithms would be better focused and directed toward achievable goals.

Tarjan [1987] provides an excellent illustration of the quest for algorithmic efficiency and the importance of theoretic investigations. Typically, algorithms superior theoretically, that is, minimizing the total amount of computation and using data structures that correctly represent the information needed for the problem solution, are also superior in practice. Since parallel computation extends the range and size of solvable problems, the significance of asymptotic complexity analysis will be increasing. What has to change is the model of computation and the scope of issues that will have to be considered in parallel computation.

Several elements of a useful theory of parallel algorithm complexity have been already proposed. They are extensions of the current theory of algorithm complexity and account for the time penalties involved in communication. Levitan [1987] has attempted to characterize the communication structures of common parallel architectures and the communication needs of parallel algorithms. A set of measures useful in this characterization has been suggested and used to predict algorithm performance. Jamieson [1987] has researched the mapping of algorithms to parallel architectures. She has focused her attention on the transition process from generic algorithms to architecture-dependent algorithms. She has made general assumptions about the target architecture, characterized parallel algorithms along a number of attributes and has attempted to measure the degree of fit between algorithms and architectures. These and similar contributions bring us closer to a practical theory of parallel computation but until such a theory is established and practically used, the design of algorithms for supercomputers will be guided by the accumulated experience and commonly applied techniques developed in the past. In this respect, our empirical knowledge and understanding of algorithmic issues are far more advanced for vector machines than for parallel machines. Also, an argument can be made (see, for example Duff [1986]), that the influence of vectorization on algorithms has not been profound, in contrast to the influence of parallel computers and hybrid systems that combine vector and parallel capabilities. Vectorization deals primarily with reorganizing computation within inner-loops, and its objective is to obtain the longest possible vectors eligible for pipelining. Parallel computation, on the other hand, introduces entirely new degrees of freedom for algorithm design. In its most general form it allows parallel execution of any set of independent tasks whose granularity can vary from a single operation to large subproblems requiring many thousands of operations. Various parallelizing techniques can be used to develop parallel algorithms, but roughly speaking, there exist two major styles of computational parallelism: (1) MIMD or control parallelism, and (2) SIMD or data parallelism (see section 3 for definitions).

The MIMD style requires a programmer to use various decomposition methods, some of which have been known in science and engineering as means for solving large problems. This style of parallelism is coarse-grained oriented to minimize the cost of communication and requires a global, top-down analysis currently performed by humans. Two crucial tasks here are to identify properly the processes which are indivisible units of computation (such as independently executable FORTRAN subroutines), and to construct the precedence graph which determines the allowable sequences of the process execution (see paper by Kumar and Kowalik in this volume). Once this is done, one can use a tool such as SCHEDULE (Dongarra and Sorenson [1986]) to schedule the execution of processes on a system of parallel processors. The use of SCHEDULE requires that the large grain parallelism is understood by the programmer who is fully responsible for (1) the programming partitioning and (2) construction of the process precedence graph. The second step can be automated by a program capable of data dependency analysis. Partial automation of the first step is currently one of the high priority research goals in the field. By partial automation, we mean a program development environment which would allow rapid testing of different versions of algorithm decompositions defined by the programmer or suggested by a heuristics based programmer-assistant system. Eventually, we may be able to construct programming tools that can automate to a great extent, the process of extracting parallelism from high level language description of algorithms and mapping these algorithms to specific parallel architectures.

The SIMD style of parallelism, also called data parallelism, exploits simultaneous operations across large sets of data and is appropriate for massively parallel machines with thousands or even millions of processors, exemplified by the Connection Machine or the DAP machine. The data parallel approach (algorithms and computers) are most effective for problems with large amounts of data and small amounts of code; the reverse is true for the MIMD computation. At first sight, the data parallelism seems to be limited to just a few application domains. Contrary to this intuition, the practical use of the Connection Machine has shown that it can be applied well to many domains: molecular dynamics, VLSI circuit simulation, computer vision, finite element methods, atmospheric dynamics, document retrieval, memory-based reasoning, to mention some. These applications include both numerical and symbolic processing applications (Waltz [1987]). More surprisingly, it is possible to construct data parallel algorithms utilizing O(N) processors to solve problems of size N, often in O(log N) time (Hillis and Steele [1986]). A well know case of such a problem is the summation of N numbers. A less known problem in this category is that of finding the end of a serially linked list.

In the area of numerical analysis, McBryan [1988] has reported an impressive performance of the Connection Machine, handling a parallel preconditioned conjugate gradient method applied to elliptic partial differential equations reduced to sets of sparse linear equations by finite element or finite difference method. The 65,536 processor CM-2 has attained a truly "macho performance" of 3.8 GIGAFLOPS. In another application also related to PDE equations, the CM-2 outperformed the CRAY X-MP4/8 with 4 processors by factor of 3 and with a single processor by a factor of 11. Moreover, the CM-2 could handle larger problems than the CRAY without reducing the speed.

These applications also have shown some disadvantages of data parallel algorithms. For instance, it was necessary to keep most of the Connection Machine idle while processing the boundary points which require different formulas than the non-boundary points. Clearly, any computational heterogeneity of the computation can result in extreme inefficiencies.

An example of such heterogeneity would be a discretization problem using multigrid technique with several grid levels. An example cited by McBryan in the same study shows the rate dropping from 1800 MFLOPS for the fine grid phase to 200 MFLOPS for the coarse grid phase of the total solution. This severely obviates the advantage of multigrid techniques. The problem would not have occurred on a sequential computer for which the multigrid methods have been originally designed. To overcome the difficulty, McBryan has proposed a new algorithm that allows use of all CM-2 processors at all times. The new algorithm does not have a sensible sequential version. Here, the worlds of sequential and parallel algorithms have separated; there is no transition from a parallel algorithm to a corresponding serial version.

Another application domain that seems to be uniquely suitable for data parallel algorithms, is that of simulating cellular automaton on large lattices for obtaining solutions to partial differential equations such as incompressible Navier-Stokes equations (Boghosian and

Levermore [1987]). Such simulations may, in the future, replace solutions of PDE's together with the related mathematical machinery of linear algebra.

Given the expected broad applicability of the massively parallel SIMD computers, the relative simplicity of the programming style and the impressive performance levels, three general questions relevant to SIMD algorithms arise:

1. What are suitable classes of problems and algorithms (numerical and symbolic) for and the limitations of these architectures?

2. How to handle sequential parts of the algorithms or parts that have limited parallelism?

3. How to exploit the unique advantage for graphics in the algorithm development and problem solving of the massively parallel machines such as CM-2?

3.0 MACHINE ORGANIZATION

3.1 GENERAL TRENDS IN PARALLELISM

There seems to be little doubt the parallel architectures will dominate the field of computing. It is also quite likely that the major force in parallelism will be highly parallel systems with distributed memories. The key issues often overlooked are when and how computing hardware and software will migrate to such systems. Certainly, machines like the Connection Machine (65 thousand CPUs with distributed memory) indicate a future of massive parallelism might not be far away. Nevertheless, the bulk of high-end computation today is done on vector supercomputers with a modest amount of parallelism. What can be expected of parallelism in the 3-7 year range?

Categories of Parallelism

To examine supercomputer architecture trends, the following three categories of parallelism will be used:

- Shared memory
- Pseudo-shared memory
- Distributed memory

Figure 3.1 displays a typical shared memory system. Separate CPUs share a main memory through an interface, e.g., cache, bus, registers. Conceptually, this type of parallel architecture is easy to design. There a numerous examples of this type of architecture, but the most visible supercomputer examples, are the Cray X-MP, Y-MP, and the Cray 2, 3. On most applications, practical and theoretical results show that the ultimate efficacy of this type of parallelism, when all CPUs are applied to a single job, begins to fade quickly after 20 CPUs, even using the most clever algorithmic approaches.

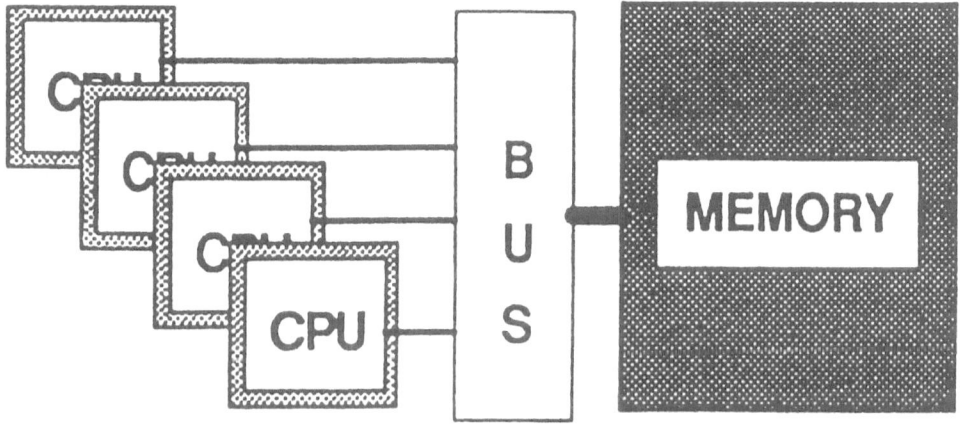

Fig. 3.1 Shared Memory

To overcome the obvious bottlenecks due to memory contention associated with shared memory systems, many computer designs provide memory local to each CPU, while still allowing for a large shared global main memory. An example of this in supercomputing was the CDC/ETA Systems supercomputer. This was one form of compromise between purely distributed memory and purely shared memory, which we shall call pseudo-shared memory design. The variety of pseudo-shared memory designs, if one includes the mid-range, is large. The BBN Butterfly, for example, really is a distributed memory machine, but it is made to behave more like a shared memory device by its "mode" of operation. The memories are treated as one pool of memories which any CPU can access via a high speed switch. The degradation in going to a distant memory versus a local memory is only a factor of 2 or 3. (See Figure 3.2.) Another important sub-class in this category is clustered memory- machines. This is the true hybrid of distributed and shared architectures. The idea is usually to begin with a shared memory design (e.g. 16 CPUs sharing a main memory), then replicate the shared memory and CPUs in clusters.

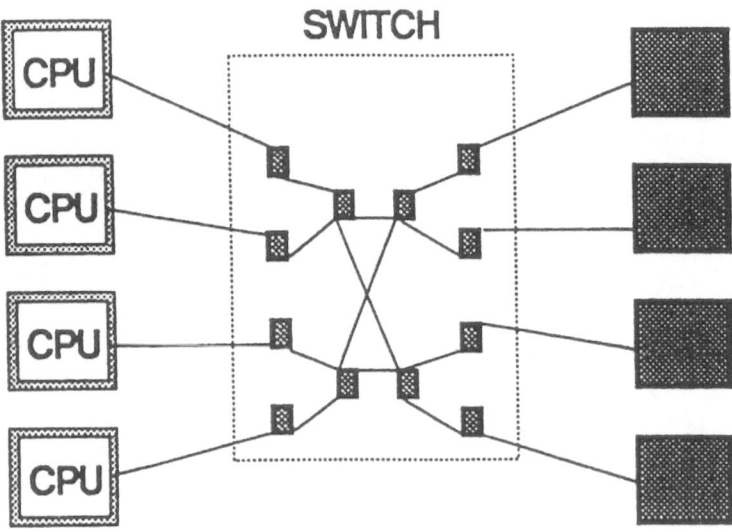

Fig. 3.2 Pseudo-Shared Memory

The third category is that of distributed memory systems. Today the largest class of such machines are the hypercubes. The issue relative to purely distributed memory parallel systems, is the topology of connectivity among the CPUs. Hypercubes are so named because the connectivity topology is that of the N-cube hypercube. A 3-cube is pictured in Figure 3.3.

Grid or net topologies, ring topologies, and various combinations abound. As both the power and number of the nodal CPUs increase, the total power of the system can rival high-end (and more expensive) designs. For example, the Connection Machine uses relatively low-power CPUs in enormous numbers to claim a very high potential compute power.

Algorithm design to achieve optimal performance depends critically on various hardware features. Variance in any one of the following factors can seriously impact algorithm performance.

1. The computational capacity of each CPU.

2. The amount of memory of each nodal CPU.

3. The number of CPUs and memories.

4. The connect topology of the CPUs/memories.

5. The communication speed among CPUs

6. The mode of operation (discussed below).

7. System overhead to synchronize CPUs.

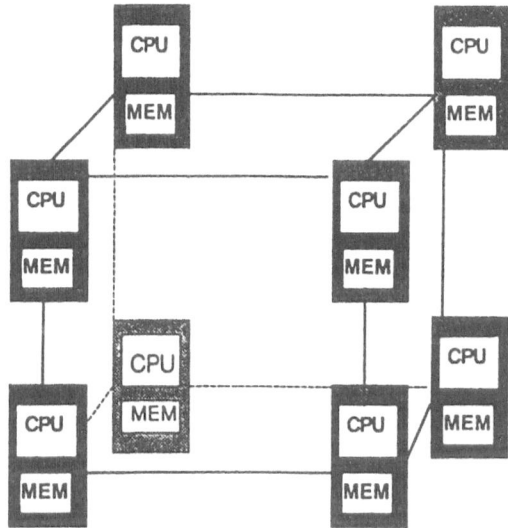

Fig. 3.3 Distributed Memory Configuration

One should not be surprised that even though two machines may come from the same loosely defined categories above, that they, in fact, are quite different machines to operate. The reason for this is that there can be fundamental differences among machines of identical hardware design in the manner in which they allow the user to compute. Probably the most fundamental differences lie in the allowable mode of operation. Three major approaches are discussed next, Flynn [1972].

SISD (single instruction stream/single data stream) - this is the conventional computing process, today often called scalar computing. The concept is that instructions are processed sequentially and result in a flow of data from memory to functional units and back to memory (perhaps through caches or registers). The data flow is equally sequential. This does not exclude the use of pipelining in the hardware, but does exclude manipulation of vector data types explicitly. The key is that operands are scalar objects and their manipulation is paced by the instruction processing rate.

SIMD (single instruction stream/multiple data stream) - the processing of instructions remains the same, but the data manipulated (i.e., the operands of the operations) can be explicitly "vectors" or arrays of data. In parallel machines, the data could come from separate memories simultaneously, but all CPUs perform the same set of instructions simultaneously. In either case, the floating point operations are "paced" by functional unit throughput as opposed to instruction processing rates. (Thus, it is no longer considered meaningful to rate vector/parallel computers in MIPs, millions of instructions per second.)

MIMD (multiple instruction stream/multiple data stream) - this process implies the ability to simultaneously process instructions. These instructions are then free to process independent data streams. Of course, the variety of MIMD computing paradigms is enormous: data flow architectures, message passing approaches, parallel functional units of various varieties. (Note: each instruction stream may be allowed to issue SIMD instructions to vector units or synchronous sub-processors allowing a mixtures of SIMD/MIMD modes.)

MIMD architectures seem to be preferable since they can allow for SIMD or synchronous processing as a subcase. Two drawbacks are often the case with such architectures: 1) the overhead of synchronization is often higher, requiring large grained subtasks, and 2) the management of very large number of CPUs in MIMD mode is very difficult. Later it will be observed that the combination of asynchronous and synchronous processing may very well be the ideal compromise in supercomputing for the next few years as an evolution to highly parallel systems is taking place.

3.2   PARALLELISM IN SUPERCOMPUTER DESIGN: TODAY

Having discussed parallelism in general terms, it is instructive to return to the domain of today's supercomputer architectures and observe the current state of parallelism. While many industrial research projects, commercial companies, and academic research projects are heavily involved with distributed memory architectures, supercomputer vendors are currently exploiting a modest amount of parallelism in order to continue to capitalize on the enormously powerful vector CPUs. Their purpose is to achieve the greatest power. One might argue that this is not the best approach, but it is "evolutionary" and in the same that it has the least impact on the large body of applications software so important to users and vendors alike. This approach appears to be the trend well into the 1990s. Cray Research, IBM, Fujitsu, HITACHI and NEC offer multiple CPU supercomputers. The ETA system notably was the only one, so far, to use large local (hence distributed) memories while still sharing a large main (global memory). However, the clustering of CPUs about local memories is inevitable as the number of CPUs grows beyond 8 or 16.

The initial foray into multiple CPUs in mainframes, minicomputers, and supercomputers has been largely motivated by job stream throughput considerations. In optimizing a single job, two or four CPUs, at best, can only increase performance by a factor of two or four, respectively. This can only be accomplished through some modifications to the application at hand, usually by the user. This requires time and effort, and usually, is somewhat deficient in producing theoretical linear speed-up, for a variety of reasons. On the other hand, providing two CPUs, in proper coordination through the operating system, can easily, and without user intervention, nearly double the throughput of a system. For this purpose, DEC, IBM and its plug compatible competitors have been offering multiple CPU mainframes for some time in their scalar computers. The multiple CPUs of Cray X-MPs, the most successful supercomputer in terms of sales, are largely used in this fashion at most user sites today. When the number of CPUs exceeds 4, the gains from improving job stream throughput begin to wane. Eight job streams would tend to randomly compete for critical resources such as memory and I/O ports to the detriment of both individual job performance and throughput. This random contention of resources, from a large number of job streams, would likely exceed the well organized utilization of resources resulting from carefully parallelized programs in a single job stream.

To effectively use higher degrees of parallelism, the power of multiple CPUs have to be brought to bear on a single job. In order for this to take place, the application program must be redesigne and the operating system has to provide the mechanisms for higher level languages (such as Fortran) to create and coordinate tasks within a single job. Through this coordination (synchronization) the programmer can perhaps eliminate random contention and create much greater "job" efficiency. Ideally, if N processors are brought to bear on a single job, a performance improvement of a factor of N in real time is attained. Invariably, not all the contention for resources can be eliminated. In addition, the CPU coordination itself introduces overhead that can degrade potential performance. Of course, not all applications can be 100% parallelized.

It is instructive to observe how commercially available machines have implemented the same degree of parallelism differently -- the Cray X-MP/4-CPU, the ETA-10 (the 4-CPU version), the NEC SX-2 (single CPU). All three of these machines have the same amount of computational parallelism in the following sense: they each have 4-add pipelines and 4-multiply pipelines which yield their respective peak performance rates by allowing simultaneous operation to produce one result per pipe per cycle when performing 64-bit arithmetic. We make the following observations:

- The Cray has 4-CPUs that must contend for main memory. Each CPU is independently programmable.
- The ETA-10 had 4-CPUs each endowed with its own memory sharing a secondary memory. Each CPU was independently programmable.
- The NEC SX-2 has a single CPU with 4 pairs of pipelines and no contention for memory among CPUs. This is an entirely vector-technology with respect to floating point operations.

At this point, there is no reason to suggest that any of the three designs is superior. ETA had substituted the memory contention problems of the X-MP with communication overheads for its distributed memory processors. The NEC design avoids resource contention altogether, but requires very long vector operations to ameliorate the start-up times of 4-pipes. The ramifications of the three approaches on compiler design is significant. It is not at all clear that one single application program could run well on all three designs. Compiler considerations will be explored in Section 4.

3.3   PARALLELISM IN SUPERCOMPUTERS: THE NEXT DECADE

There is no question that the quest for more computer power drives supercomputer design. Yet, the fuel for that quest is more than technical, its financial. For every new and innovative supercomputer, there must be at least one very committed target customer, or a general market. This aspect of supercomputing is often overlooked. In the past, government agencies have spurred the creation of new and more powerful supercomputers. Today, the entire industrial complex drives the need for these types of machines. For more than a decade, supercomputer users have developed a vast body of software that new supercomputers must run to be marketable. Tempered by this reality, the 6-10 major supercomputer companies seek to tap the promise of parallelism in an "evolutionary" rather than "revolutionary" manner. In the years to come, there will be several exceptions, particularly from brand new companies who can survive the long haul of developing a new community of users. The new supercomputers for the next three years are already in prototype, and those of the next 5 to 7 years are predictable from an architecture perspective.

In this setting, it is likely that supercomputers through the mid to late 1990s will utilize relatively small numbers of very powerful (probably vector) CPUs. By the mid-1990s these machines will probably have on the order of 100 to 200 CPUs. It is likely that at any given time the range of parallelism among vendors will vary, as it does today with the Japanese having just begun to opt for parallel CPUs (over a single CPU with replicated floating point pipelines). Leading manufacturers have already indicated that a likely evolutionary path to this level of parallelism will be that of the pseudo-shared memory approach, particularly that of clustering of CPUs about small numbers of distributed memories. With this scenario, it is useful to examine the ways in which a relatively small number of CPUs can be used to improve performance on a single job. Two alternate, yet perhaps, complementary directions have emerged in commercially available systems. One approach is based on asynchronous (i.e., MIMD) utilization of parallel CPUs. With this approach even inherently scalar (sequential) processes can be parallelized. This can be characterized as top-down parallelization. The other approach is a more bottom-up approach, which can capitalize on highly vectorized code to orchestrate automatic parallelization at the inner DO-loop level. This will be examined in more detail in Section 4.

The two techniques previously discussed are not mutually exclusive and could be combined to achieve greater performance, particularly in clustered architectures likely in future supercomputers. As a hypothetical example, imagine a 16-CPU machine, with each CPU, itself, a vector architecture (e.g., a CRAY-3). One could employ a strategy of decomposition of an algorithm into four main synchronized subtasks. Quite often it is easy to recognized three or four relatively independent (and equally complex) tasks high in the program structure. Assume each subtask is generally utilizing long vector operations. One could cluster 4-CPUs to each major task using a top-down approach. Within the four clusters of four CPUs, one could employ a bottom-up approach breaking the long vector processes into four identical vector process spread across the four CPUs as illustrated in Figure 3.4. The vector lengths would be one quarter the original length. (With some luck there might be a local nest of do-loops allowing a group of vector operations to be spread among the four CPUs with no vector length degradation.) This hypothetical situation is probably quite likely to be utilized on the CRAY line. In fact, CRAY users are already experimenting with macrotasking (top-down multitasking) and microtasking (bottom-up multitasking) interaction. (CRAY has implemented the microtasking automatically as CPU's are available. Priority allocation of CPU's is given to macrotasking.) The advantages and disadvantages of two approaches will be discussed further in Section 4.0.

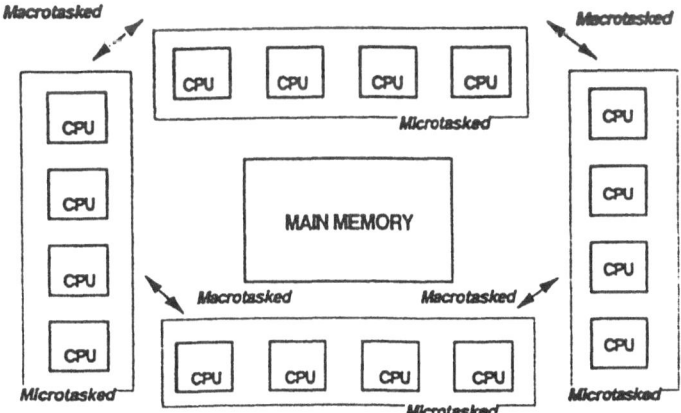

Fig. 3.4 Macro- and Micro-Tasking - A Hypothetical Example

In the previous example, there was one memory. Clearly, memory contention is the largest potential bottleneck. It is questionable that 16 CPUs can really productively share a single memory. The design is simple, but the implementation is critical. To increase parallelism further, say to 64 CPUs, it seems apparent that the individual clusters of CPUs will require some measure of local memory. It is likely that designs would have 4, 8 or 16 CPUs clustered about a shared memory, then each cluster replicated. Figure 3.5 illustrates one possible topologies for clustering of 64 CPUs, 4 local memories, and one global memory.

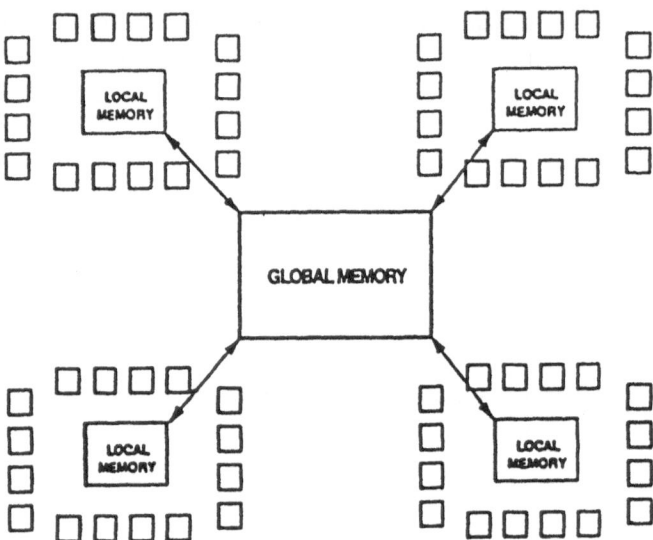

Fig. 3.5 Clustered Architecture

As simple as the above scenario is, several complex decisions must be considered. First Amdahl's law must be considered. The fundamental penalty for parallelism of the variety being discussed, is overhead for synchronization and "wait time" due to memory contention and/or communication. When these overheads are large, it becomes hard to capitalize on performance improvement unless the "granularity" is also large. The term granularity refers to the "size" (measured in time) of the subtasks. Since supercomputer designs will likely employ vector CPUs, the overhead of a vector operation must also be considered. This overhead is the start-up time for the pipelined process. Vector length, in fact, is analogous to granularity. The longer the vector, in a sense, the greater the vector task's granularity; consequently, the more productive the vector operation, since the overhead of start-up time is amortized over a larger number of operations. In a parallel environment, the overhead for task starts is constant. Therefore, if a computational task is being performed by 50 subtasks, the overhead is 50 times a single task start. Thus, one would strive to have the computational task being performed large enough to absorb this overhead. The primary issues that arise, deal with exploiting the parallel structure of the hardware through the design of the software. This question is explored in the next section.

4.0 COMPILERS AND PROGRAMMING TOOLS

4.1 HISTORICAL SETTING

Generally, vector compiler technology has been successful in achieving reasonable peak efficiency on "good" code. Moreover, the scientific community's ability to generate "good" vector code has improved dramatically throughout the 1980s. As we move into the era of parallelism, particularly in supercomputing, we can observe certain trends among the leaders in compiler technology. The basic techniques of computer optimization are extensions of strategies for vector machines, but, so far, have had limited effectiveness in a parallel environment.

In this introductory paper, current techniques for automatic parallelism are examined with respect to trends in hardware design outlined in the previous section. Elsewhere in this volume, the current state of the art of compiler technology is reviewed in detail by eminent researchers in the field. Our concern here, in looking into the next decade of computing, is to examine the ability of optimizing compilers to be as effective on parallel machines as they have been on vector machines. Technical reasons for this concern are presented. Human intervention, prior to, or during compilation, to achieve peak optimization seems to be inevitable, even in vector technology. Reasons for this abound. Even the most sophisticated dependence analysis techniques cannot avail themselves of knowledge possessed by the program designer or user. New approaches will require sophisticated run time analyses and, perhaps, expert system technology to aid human intervention.

To appreciate the magnitude of the challenges ahead, it is critical to understand the growing disparity between compiler technology and hardware design. This disparity will be described here, but it in no way meant to be a condemnation of current techniques. It is simply a realization that computer architecture is, perhaps, changing faster than compiler technology can assimilate. The impact of this growing gap is to be felt by the user of computers, who for the most part have relied on compiler technology to achieve optimal performance. This gap, in fact, is a major topic of discussion at this workshop, and hopefully the summary report (in this volume) will offer some fruitful research directions for the next decade.

Historically, the high level language of science and engineers has been Fortran. (Note: the likelihood that Fortran will continue to dominate scientific computing is debated heatedly. Our perspective is to consider the relationship between architecture and compilers, predominantly Fortran. Whether Fortran is enhanced to accommodate the new technology, or replaced by other languages is really secondary to understanding more fundamental barriers to efficiency. For this reason we shall avoid the debate of Fortran, good or evil.) During the era of "scalar" computers the gigantic strides in computer performance were achieved by circuitry. The basic design of computers, of course, changed, but not to the extent that the compiler writers could no longer accommodate "performance" within the Fortran compiler. Indeed, the sophistication of compiler writers in optimizing performance through judicious instruction scheduling, register and stack optimization, and memory and I/O management far exceeded that of the vast majority of users. Moreover these techniques had little impact on program design (with the possible exception of memory growth which often allowed for more sophisticated models.)

4.2   NEAR-TERM COMPILER AND SOFTWARE CONSIDERATIONS

With the introduction of single CPU vector computers, it was discovered that through minor changes in algorithms and, often through inner-loop resequencing, better performance could be achieved. Techniques to improve performance in vector compiler technology include:

- Removing unallowable vector forms: e.g., on some machines (ETA 10, Fujitsu VP-200) a degradation occurs in vector performance when vectors are not stored contiguously in memory. This can sometimes be obviated by altering the sequence of inner-loops nest. From an algorithmic perspective this can occur when the row of a column-stored matrix is required. This often can be avoided.

- Judicious use of registers, caches, or local memories: e.g., one can alter the sequence of nested loops to reduce the need to store and fetch vectors from memory. This is particularly important when memory access is restricted by memory paths (ports) causing a bottleneck. See Dembart and Neves [1977]. A similar approach can be used when excessive memory fetching and storing causes potential "cache" misses. See Samukawa [1988].

In many cases, the above techniques quite often can be handled by compilers. Minor algorithm improvements are often effective in this regard. The definition of the level 3 BLAS, which are matrix oriented linear algebra building blocks, was motivated by the increased efficiency that can be obtained on most vector computers by careful coding of these routines which have enough structure to minimize memory references as suggested above. See Dongarra [1988b].

There are, of course, algorithm changes that cannot be accomplished readily by compilers. For example, replacing a Gauss-Seidel iteration scheme with a red/black formulation, or by a direct method to improve efficiency, would be difficult and dangerous to automate in the compile process. In the parallel environment, algorithmic considerations will be even more important in achieving high performance.

The highest stages of the modeling process are often filled with parallelism. Many applications dealing with physical models have obvious parallel decompositions related to the geometry of the model itself. This "model" parallelism cannot generally be exploited at the inner do-loop level. Moving to a vector or parallel computer can have, and has had, an impact in all stages of software design. Even if the physical model isn't changed, the algorithms used, the mathematical models, and range of the physical model parameters are often impacted. The technological challenge comes from the increasing requirement to obtain greater single job performance, particularly when a factor of 8 or 16 can result. In fact, in an 8-CPU vector computer, the combined performance improvement from exploiting both parallelization and vectorization will have a potential of improving performance two orders of magnitude.

With this in mind, we focus on the process of parallelization of a single job. An interesting fact that is exploited in today's compiler technology is that "a single vector operation or a loop of vector operations can be trivially parallelized." It is also a fact, however, that parallel operations cannot all be vectorized; i.e., the converse is not necessarily true. It is therefore clear that focusing only on vector processes, may overlook some obvious parallel processes, particularly processes high in the tree structure of a program. Several other facts related to the parallelization of vector loops need observation.

1. Parallelizing a single vector operation of length N into M parallel operations creates M vector operations of length N/M. This introduces more overhead. Instead of one vector start-up being amortized over a vector operation of length N, there are M vector start-ups being amortized over vectors of length N/M. Thus, there is M times as much overhead on vector operations of much shorter length. Parallelism is achieved at the expense of vector inefficiency. This can be a good trade-off provided the vector length N is very long. Thus, what may be considered a long vector in the single CPU vector world, may not be long enough for the parallel vector world.

2. While shipping off a loop of vector operations to independent CPUs may be an obvious approach to parallelism, the frequency of occurrence of such operations may be too limited to payoff over the entire application.

With every process that is being carried out in a parallel environment, there is an associated overhead for synchronization. This overhead varies among hardware designs. A parallel task is said to be of fine *granularity* when the overhead is small compared to the work performed in the task. The requirement in (2) above, that vectors be "long," more precisely stated that the vector loop must be of high granularity. The length is for this condition will vary among parallel computer designs.

Figure 4.1 displays a typical scientific application. The above discussion suggests that focusing on the obvious parallelism found at the loop level near the "leaves" of the tree may be an ineffective approach to parallelism.

With this in mind, let us examine the state of parallelizing compiler technology. Two approaches are available today. The first approach is to exploit the advanced compiler technology to achieve parallel vector optimization. This approach may be called "bottom-up" parallelization for it is applied to the loops (one, two, or perhaps three nests of loops) at the end of the branches of the program tree structure. (See Figure 4.1). This technique has the advantage of being amenable to automatic use of parallel CPUs through preprocessors, the compiler and/or other software tools. For a highly vectorized program, where a high percentage of the CPU time is spent processing long vector operations (i.e. large granularity), this technique can be efficient for small numbers of CPUs. Theoretically, this approach can be applied to any Fortran program without user intervention, and when effective can be a very painless way to migrate to a parallel environment. In addition, parallelization tools of this kind often produce reports of barriers to parallelization indicating possible user intervention.

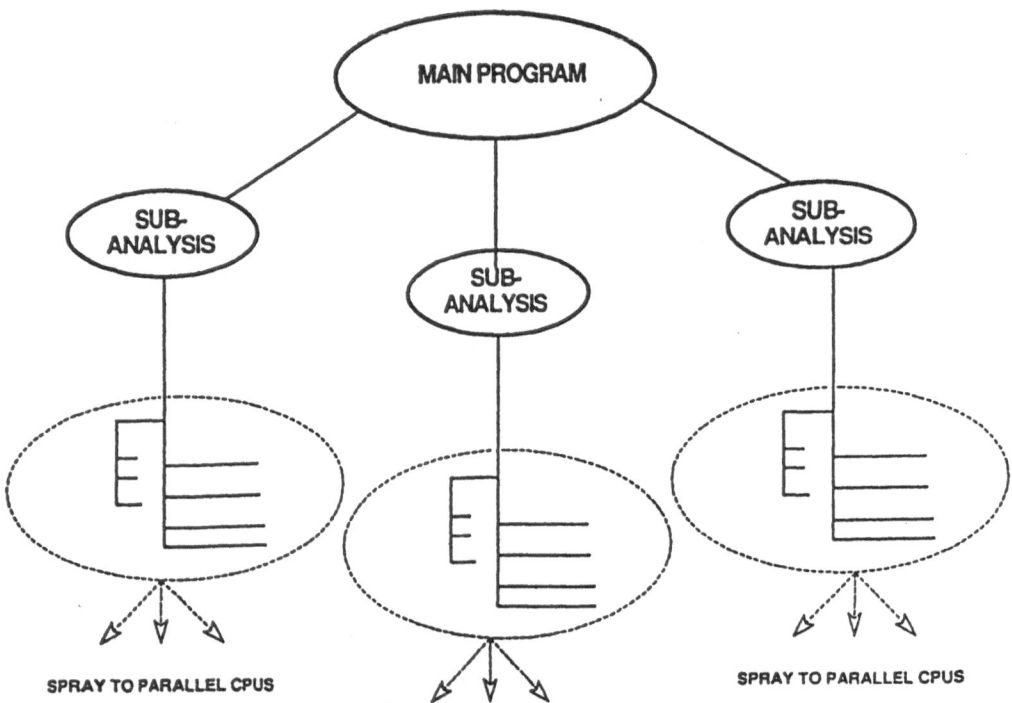

Fig. 4.1 Bottom-up Parallelism

A second approach to parallelization is to extend the Fortran language by added task creation and synchronization commands that allow the user (programmer) to structure the computation to take advantage of parallel execution. This approach, which we shall call the "top-down" approach requires human intervention, at least in today's compiler technology. The granularity of the bottom-up approach is limited to, at best, several nested do-loops. The top-down approached mentioned above has the advantage of potential granularity comparable to the total computation time of the program divided by the number of useable CPUs. This is illustrated by Figure 4.2.

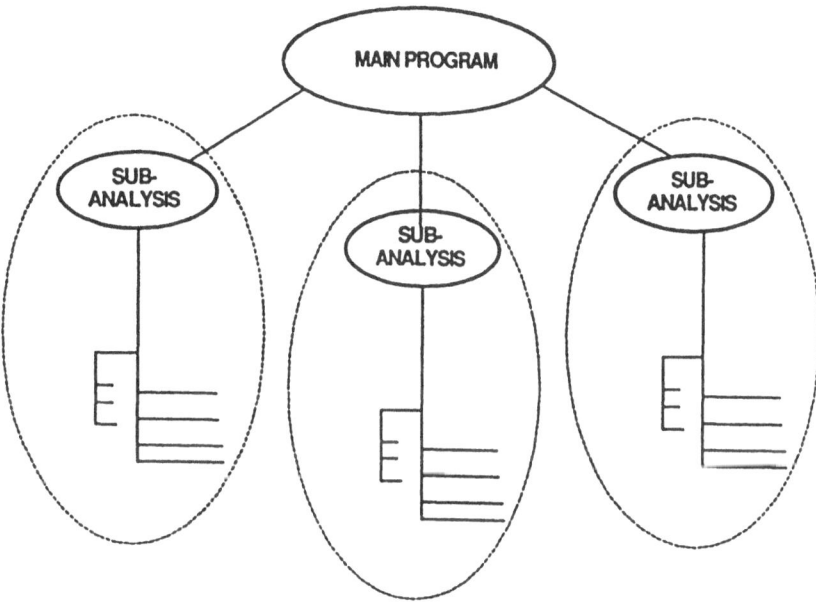

Fig. 4.2 Top-Down Parallelism

Important questions remain, however, as to the overall efficacy of either the bottom-up or top-down parallelization approaches. As hardware evolves, we anticipate more and more parallelism. Techniques that work well on today's supercomputers (that possess relatively small numbers of very powerful vector CPUs) may not be techniques that will fare well over the next decade. It is not really known whether the paradigm for parallel computing will be SIMD or MIMD, nor whether memory will be shared, clustered, or distributed. The long term trend of distributed memory is predictable, but when it will be the "bread and butter" architecture for scientific processing is difficult to state with any certainty. With these unknowns in hardware architecture, what is the best area to spend our research efforts in exploring parallelization tools? Many questions can be posed. The answer to these questions will depend greatly on the degree of parallelism, the overhead for parallel synchronization, and memory structure that future hardware designs will impose.

- Can top-down parallelism be automated or assisted by deterministic tools delineating program structure?
- Should a new programming language or technique be explored now, or should we continue to bend Fortran to fit new architectures?
- Can expert systems technology be employed to advantage in top-down parallelization?
- Can bottom-up automatic parallelization techniques achieve high efficiency on a broad enough spectrum of applications to handle the mainstay of production software for the next 5 to 10 years?
- Is the bottom-up approach too ineffectual to spend time and effort nurturing this approach over the next 5 years?
- What are the evolutionary possibilities of parallelization? Can compilers be linked to run-time statistics and "learn" to be more optimal?
- Can the bottom-up approach and the top-down approach find a middle ground where both techniques are integrated (first with, and eventually without, user intervention)?

Hopefully, these and other questions will be explored in this volume and as a result of this workshop some focus of effort for the next several years can be established. To set the stage, a further examination of today's technology is useful.

4.3 CURRENT COMPILER TECHNOLOGY: A BRIEF SUMMARY

By and large today's compiler technology, on most, if not all, advanced scientific computers, begins with a "vector" or array dependency analysis. The vectors or arrays are the results of loops common in scientific processing. The objective is to identify operations on arrays such that the input arrays (e.g., the two vector operands) are independent of the result array. This will insure there are no backward array referneries (recursions) that would invalidate vectorization or parallelization. Often loop sequencing or nesting is examined and altered with some eye toward creating parallel vector processes. In fact, one can view the process as being one of mapping Fortran source to an intermediate set of known constructs such as popular vector instructions. The known operations include strided-vector-dyad and triad operations, dot products, and sparse vector (gather/scatter/list) operations. More recently the direct recognition of matrix operations is being employed as well. The dependences-analysis techniques employed have become more sophisticated over the years, but quite often the dependences can not be determined by the compiler *a priori*. This can often require user intervention or "run-time" checking.

In practice the actual implementations on a specific machine can be difficult because of the interplay of vector or parallel technology. To illustrate we examine a typical situation.

```
DO 10 I = 1, N
DO 10 J = 1, M
     A(I,J) =  A(I,J) + B(I,J)
10   CONTINUE
```

In a parallel vector CPUs, a number of options exist. First, the loop could be executed as follows:

```
DO 10 I = 1, N*M
A(I,J) =  A(I,J) + B(I,J)
10   CONTINUE
```

In fact, a good vector compiler will make this optimization. The parallelization of the first loop could be accomplished by spreading the outer-loop to the available processors each of which would have the inner-loop available for vectorization. Several questions come to mind:

What if the overhead for so many processes is too great?

What if both microtasking and macrotasking are available? Which should be used? Or should both be used?

What if M is very small and N is very large?

Some of the answers are fairly obvious once overhead characteristics are known for both vector, macro-, and micro- processes.

The vector instructions are then decomposed into parallel processes and shared among the available and/or appropriate CPUs. This is often done through SIMD utilization of the CPUs. That is, the various CPUs execute identical instructions on sub-vectors of the original vector operation. This overall approach is often related to parallelization through data partitioning. (We have already commented that if this were the only way a multiple CPU system were to utilize its parallel capability, then in would probably be better to imbed the parallelism within a single CPU! Refer to the comparison of the NEC SX-2 and its 4-CPU competitors earlier.) The data partitioning approach is in sharp contrast to the flow control approach of the top-down parallelization technique.

Clearly, the automated approach gives performance improvement without undue effort on the part of the user. In fact, a user knowledgeable of vector computing can employ the same modeling techniques creating code ideal for this compiler approach. Unfortunately, there is no guarantee that the frequent stopping and starting of inner-loop parallel tasks created in this fashion will be truly efficient on a parallel system. For example, the automated approach of the Alliant FX-80 (4-CPU) yields only 5.8 MFLOPs performance on the Argonne Linpack benchmark, Dongarra [1988a], out of a peak potential of 94 MFLOPs for all 4 CPUs.

This is 6.2% of peak. Using the same reference, we observe that vector machines such as the Cray X-MP, Convex C-2, and the SCS-40, achieve 20% of peak on this small problem. On a larger problem using Fortran more amenable to optimization (Table 5 of Dongarra [1988]), one can observe that the vector machines all exceed 80% of peak, while no parallel machine exceeds 20% of peak. One could argue that the solution of dense linear equations should offer as rich an opportunity to exploit bottom-up parallelization as one is likely to find in most scientific programs, with the possible exception of large iterative solutions found in partial differential equations.

In the previous Section we discussed the natural evolution of the next generation of supercomputer architectures. There it was observed that the clustering of CPUs about shared memories which, in turn, are clustered about a more global memory, is apparently the trend of the next 5 years. This compromise of distributed and shared memory parallelism offers a natural platform to investigate a compromise of "top-down" and "bottom-up" parallelism. This compromise could also find applicability in systems with a very large degree of parallelism.

4.4   CONCLUSIONS

Compiler technology is seriously lagging behind changes in architecture. Current efforts in compiler technology seem to be concentrated on parallelism based on bottom-up, inner-loop techniques. This approach is important, but suffers from

- not exploiting multiple CPUs at the highest levels of granularity, important in supercomputer designs into the mid 1990s;
- being dependent on data not always available at compile time;
- difficult to manage on a distributed memory system (where this form of data partitioning can cause needless memory transfers.

In contrast, hardware technology is definitely moving toward parallel CPU architectures. The complacency of users and software tool developers to generate approaches to parallel programs could stand as an international research priority.

No commercial products are available today that will automatically decompose Fortran applications (from the top) into balanced multitasked processes. This will probably be the domain of human expert intervention for some time to come. Program analysis tools are sorely needed. Standardized Fortran extensions to provide task creation and synchronization are also imperative if portable and distributed processing is to be supported. Currently, *de facto* standards seem to be evolving based on the large community of Cray users exposed to tools which that company has provided. Several *ad hoc* committees have been formed among parallel CPU architecture vendors. Software tools that can perform static analysis of program flow exist, but are not being integrated into vendor supplied Fortran development environments, and no vendor is seriously investigating expert systems technology which may be an aid in this area. Modern workstations permit the visualization of program structure and offer the possibility of a much different form of interactive compilation process. Through these techniques Fortran could, perhaps, evolve to be an after-thought behind the person-machine interface, and in time, replaced by the interface itself as a software development language.

As a matter of observation, it seems likely that the gap between compiler technology and hardware architecture is enlarging. The ability to use new technology computers to achieve performance near their potential is diminishing. While some researchers may claim that 8-CPU shared memory systems are old hat, it is probably more accurate to say that the practical use of such machines on large-scale scientific applications is a first rate technological problem yet to be addressed. In the meantime, users will probably require support tools for top-down parallelization, and whenever possible take advantage of efficiency gains from bottom-up compiler technology and sophisticated dependence analysis they provide.

5.0 OTHER ISSUES

We examine here several computer technology areas that will make significant contributions to high speed computation. Most of them have already been recognized as crucial elements of supercomputing, others begin to play increasingly important roles. We begin with an area which is futuristic in nature but has promising potential for increasing levels of machine assistance in supercomputer application: the fusion of knowledge-based systems with conventional numerical number crunching.

5.1   COUPLING NUMERICAL AND SYMBOLIC COMPUTATION

The current process of solving large scale scientific/engineering problems has two distinct parts: (1) numerical computation executed by a supercomputer, and (2) reasoning about the process and results done by an analyst. With the emerging expert system technology, it is tempting to consider and develop computer programs that would assume some responsibilities exercised by the analysts.

Such programs would have to combine numerical and symbolic computation - and we call them the coupled systems. Coupled numerical-symbolic software systems, with the increasing power of automatic reasoning and explanation capability, have the potential for extending the usefulness of supercomputers and for solving new classes of problems we discussed in Section 1. They differ from conventional software in a number of ways. They access, use and maintain knowledge, specifically heuristic and experiential data. They also interact intelligently with users and other systems. For examples of prototypical coupled systems, the reader is referred to Kowalik and Kitzmiller [1988]. Erman et al. [1988] described an architecture for developing coupled systems. Abelson et al. [1989] discuss the development of intelligent techniques appropriate for the automatic preparation, execution, and control of numerical experiments.

Any serious attempt to design, implement and use coupled systems will create new technological challenges for software and hardware developers. One of them might be a need for a new supercomputer architecture capable of efficient numerical and symbolic processing. We believe that without computers, that have very large memories and are highly parallel, coupled systems for advanced applications involving human-like knowledge will not be possible.

Unfortunately, a general purpose commercially parallel computer for knowledge based systems that involve heavy numerical computation has not yet been built or even designed. No general approach to parallelizing symbolic computation has been discovered. Thus, in the area of building and processing coupled systems, we face formidable theoretical and engineering challenges.

5.2   VISUALIZATION

In the broadest sense, visualization is the art and science of transforming information to a form "comprehensible" by the sense of sight. In general computing circles, visualization is broadly associated with graphical display in the form of pictures (printed or photo), workstation displays, or video. In scientific computing circles, visualization is taking on a more particular conceptual meaning for many. In the physical sciences, supercomputing is playing an ever greater role as a simulation tool.

Visualization seems to be a natural form for computed results to be analyzed by scientists. Moreover, the massive amount of computed data in large scientific program, as well as complicated data input requirements, in many cases, can be comprehended in a visual fashion. Visual displays and animation allow understanding of interrelationships of phenomena which simply cannot be observed in any other way. Scientists are discovering that visual output, particularly sequenced visual displays (such as animation) can reveal anomalies in data input, unusual physical behavior, and trends in data/phenomena not easily discovered by non-visual techniques. The human brain is very good at pattern recognition, inference, and intuition. Given the visualized phenomenon, the human becomes a natural complement to the scientific computing process. In the future, there is anticipation of greater use of visualization in scientific computing, particularly interactive animation. Visualization techniques can generate their own computational drivers for computer power. For example, if each still frame of an animation requires the computation of a supercomputer class problem, then observing a 1 minute animation, at 15 frames/second, could require 900 supercomputer problems to be solved. Interaction with the animation, requires that these frames be generated from memory digitally at the viewing rate (i.e. in real time). This latter capability is now on the horizon, and requires the 900 solutions being done in "real time". Clearly, technology is not yet there, but the usefulness of this type of animation can be seen in many fields, visualizing fluid flows, exploring parameter sensitivities in many processes, etc.

There is clearly a choice between using the highest computational power to generate one specific, but difficult, model, or using the greatest possible computing capacity to animate a, perhaps, less complicated model. Both approaches require order(s) of magnitude greater power at any point in time. Both give valuable new insight. The former, uncovers new phenomenon through more detail, more valid models, etc. The latter produces new insights through observation of trends, anomalies, and relationships.

Visualization has a multi-disciplinary flavor and interacts with other sciences and computer related technology disciplines: operating systems, computational geometry, numerical analysis, data bases, optics, perception psychology, etc. It also shares with other computer disciplines some common problems, such as lack of standards, limited software portability, near nonexistence of meaningful visualization benchmarks and confusing multiple choices of hardware solutions.

There are two extreme approaches to hardware configuration: (a) calculate on supercomputer and display results remotely, and (b) calculate and display at the workstation. For large scale computation (a) is the only option and the attendant problem of data transfer to distant processors.

An interesting approach to reducing amount of the output data for display would be a "caricature technique" allowing to extract from the visualization data and transmit only the most essential features which are unexpected and representative. Developing a theory and methodology of "data caricaturization" is a challenging task.

Finally, we should realize that visualization techniques may help us analyze and imagine very abstract images and notions that do not exist in the physical reality. We tend to express graphically abstract ideas, use pictorial analogies and metaphores to explain otherwise invisible concepts. Supporting these functions is a great challenge for both computer visualization and imagination in display concepts.

5.3   DATA STORAGE AND MANAGEMENT

Large numerical computations performed on supercomputers, such as aerodynamic simulation, weather prediction and structural design require storing, accessing and manipulating many millions of numbers. Methods have to be developed that take advantage of parallel/multiple memory levels (registers, caches, RAM, SSD, disks, etc.) and multiple processors. The involved data structures frequently have natural spatial organization (related to grids) and are used in the computation according to some regular patterns. To minimize I/O operations, transfer of data between processors and waiting time for data, it is necessary to use data partitioning which naturally supports the flow of execution. The problem of data organization, space partitioning and relationship between numerical analysis and data arrangement is a key issue in supercomputing and parallel processing. It has been discussed in literature (see, for example, Bell and Patterson [1987]) but the involved methodology still remains a heuristic art. Bell and Patterson [1987] report that the effective data partitioning is not highly hardware-dependent for machines in the same architectural class, but some additional tuning may be required to optimize the designed code for a particular computer configuration. Some experimental software aids (expert systems) for assisting in the data design strategy have been built. Clearly, more tools are needed for data partitioning and management. In the past, scientific users ignored commercial database management capabilities and used simple file organization for their data. In fact, no commercial database has been built specifically for efficient handling of spatial data. The tasks of designing such databases, and providing tools for optimizing performance of large scale computation are key issues in supercomputing.

In addition to data partitioning and organization methods, we need to develop techniques for reducing storage of raw data. One such technique, "data caricature" is mentioned in Section 5.2. Other possibilities of data reduction include: curve and surface fitting, and interpolation techniques which would allow to reconstruct data from a reduced set of stored representative information.

5.4   PERFORMANCE

The ability to measure supercomputer performance precisely would be useful not only for users, but also for hardware and algorithm designers; especially if we could determine how hardware, software and applications separately effect the total performance. Typically, these contributions to performance are strongly intertwined and what we measure is a comprehensive value of performance for a specific program or a workload.

A simple way to look at supercomputer performance is to assume that it has two modes of processing, fast at vector/parallel speed and slower at scalar/sequential speed. A sufficiently small program fitting the random access memory runs with a speed whose value is between those two extremes. If s is the fraction of computational work that has to be performed in scalar/sequential mode, then the program speed up due to vectorization is:

$$\text{Speedup} = 1/(s + (1-s)/N)$$

where N is the ratio of vector/parallel speed to scalar/ sequential speed. As N increases, the speedup approaches asymptotically 1/s. This phenomenon, called Amdahl's Law, has been observed and reported very often. It explains why many codes exhibiting vectorization levels (v=1-s) in the range 0.5 - 0.75 run 20 to 10 times slower than the theoretically possible peak processor speed. Amdahl's Law rests on the assumption that s is invariant with respect to the problem size for a given code. If we assume that for large problems the vectorizable/parallelizable portion of the code grows faster than the scalar/sequential portion, the speedup is better since s shrinks. Such performance has also been observed (Gustafson [1988]). Other useful performance parameters which measure the overhead associated with vectors and parallel processing have been proposed by Hockney [1985].

Martin and Mueller-Wichard [1987] define a multilevel set of groups of programs which may be used to assess the supercomputer performance with more fidelity. They are: basic programs (simple computational or I/O operations), program kernels (typically loop structures), basic routines, strip-down versions of major applications, full applications and developmental programs. Furthermore, they propose an approach to performance evaluation that would allow one to predict performance of future systems on specified programs. This may eventually lead to identifying classes of architectures and applications (algorithms) that are mutually compatible. More work has to be done to accomplish this goal.

5.5   NETWORKING

The productivity of the supercomputer user depends, among other factors, upon the computer network performance: the communication speeds between user terminal, processors performing numerical model computation and graphics, and file servers. The network bandwidth is particularly essential for interactive supercomputer environment where large amounts of information are displayed graphically at speeds compatible with the user thought process. The main factors driving network performance in today's supercomputer installations are:

- processor speeds reaching and exceeding 1 billion operations per second.
- main memory capacity getting close to 1 gigaword,
- graphics processors and display engines with increasing speeds and fidelity,
- capabilities and number of workstations,
- storage servers providing capability to store and use billions of words on-line,
- communication lines such as optical fiber links,
- interactive mode of problem solving,
- new large scale applications discussed in Section 1,
- parallelism in algorithms, processing hardware and data storage hardware.

Perdue [1988] discusses these factors and concludes that the near term target for the communication speeds between the supercomputer system nodes is from one to several billion bits per second, and for host channels and external nodes, one billion bits per second. These targets are about one to three orders of magnitude higher than the current capabilities of networks supporting scientific and engineering supercomputer users.

Other major networking issues include:

- development and standardization of efficient protocols to ensure reliable data exchange between diverse processors and permit sharing of network resources,
- open network topology allowing for the network growth, security and compatibility with existing network applications,
- new methods for data reduction which would slow down the growing demand for super fast networks,
- better ways of distributing and sharing data which would minimize the need for moving large amounts of data between the network modes.

While the drivers for more network capacity just listed, will keep ever increasing pressure on non-standard protocols and systems; the key challenge will be to continue to research ways in which the need for transferring and storing data will be reduced. As the computing nodes across the tiers of computing become more powerful, the options to recreate, reconstruct, and abstract data rather than send and store it, will increase. However, if exploring these techniques is not part of the mind set of code and algorithm developers, the opportunity will be lost, for these techniques are not available at the systems level, only at the application level. This ability to abstract or characterize data (as described in the visualization discussion) is peculiar to scientific data processing when compared to say information systems data processing.

5.6 LANGUAGES

Fortran is the traditional language for scientific and engineering computing and it is highly unlikely that the Fortran user community will switch *en masse* to another language. It is more likely that Fortran will continue to evolve and incrementally keep acquiring extensions to accommodate the users interested in parallel processing. There are many examples of such extensions in the current versions of Fortran offered by the computer manufacturers. Unfortunately, every vendor of parallel hardware has provided a different set of extensions to Fortran, influenced by the underlying architecture and operating system. This creates potentially very difficult and expensive conversion tasks for parallel software users who are interested in software portability between different parallel computers. To avoid such conversions, attempts have been made to construct tools (e.g. SCHEDULE, Dongarra and Sorenson [1986]) that allow existing Fortan subroutines to be used without modification across a wide variety of parallel computers. This 'least effort' approach may not have a long term future; but it will help to minimize the immediate inconvenience and cost of distributed processing and conversion.

A futuristic approach is to develop an entirely new language for scientific and engineering parallel computing. To assure its application universality and portability, the language has to be high level above architectural specifics and used to express uniquely and naturally parallel algorithms. Underneath this user-oriented language, there would be other languages (such as Fortran) plus a collection of modules and packages tuned for a specific computer system. A typical user would not be concerned with the lower level implementation details of his parallel algorithms. This ideal long term objective will be hard to achieve, since we are assuming the possibility of designing a super-parallelizer knowledgeable about: target computer architecture and configuration; data distribution, traffic and access; algorithm decomposition including tasks granularity and dependencies; and synchronization and scheduling overheads. It would know as much as a competent analyst. It is safe to assume that in the foreseeable future, the development of efficient parallel codes will require some degree of human assistance. We believe that until this assistance is reduced in technical complexity and cost, the appeal of parallel processing will be limited to the computer users with huge problems, time-critical applications, and research interests.

5.7 INTEGRATION

The supercomputer is not an isolated island and requires many resources and services before it can be functional and fully utilized. A productive supercomputer environment includes:

- facilities: a fully operational and reliable computing environment,
- supporting hardware: computer mainframes and integration software to maximize the efficiency of a supercomputer,
- maintenance: procedures to facilitate maintenance of hardware and software system,
- systems software: an operating system which drives the supercomputer and integration software which links the multiple computers together,
- applications software: to provide the discipline and organization in the acquisition, installation, and consultation of application programs,
- communication network: to provide the data communication network and to support the total supercomputer systems environment,
- user support: to provide efficient data center service and to give training and documentation for successful management and operation of a supercomputer center,
- implementation and operations: to establish a single point of contact in the day-to-day operational management of the center that is trained and equipped with procedures and techniques to work in the new environment,

- transition management: facilities provided by a combination of software tools and skills for analysis and conversion of software, to achieve a higher utilization of the new resource sooner,

- mission management: the binding force which an experienced systems integrator provides in the integration of all the components of a successful supercomputer center,

- "seamless" distributed environment that promotes cooperative processing among all computing tiers on the network.

We need to understand better how to provide the integration function, optimize the total supercomputer environment given its purpose, and how to manage it in the rapidly changing world of technology and user expectations.

Also, we would be remiss if we do not mention the positive trend in the scientific community to support standards in operating systems, network protocols and distributed processing tools. In particular, the emergence of UNIX/POSIX standards is a constructive and unifying force in many of the issues discussed in this section.

6.0 SUMMARY:
RECOMMENDED RESEARCH OBJECTIVES

1. Parallel algorithms should be studied through their entire life cycle. It is essential that the study includes data distribution, access and communication, especially for large-scale problems requiring hierarchical storage and extensive I/O activity.

2. New theories of time and space complexity must be developed for parallel processing. We need the capability of performing mathematical analysis of the designed parallel algorithms which would be independent of many details related to specific implementations, provide a reliable indication of performance on real machines, and help to avoid searching for inachievable speed-ups.

3. Tools for developing, debugging and testing algorithms are important. Such tools will assist the programmer in: algorithm decomposition, automatic load balancing, scheduling, data distribution, error identification including performance bugs and rough performance evaluations. Graphical displays may be required to make these tools effective.

4. As our experience with parallel processing grows, we will attempt to classify parallel algorithms, theoretically and empirically. This classification will help us to identify the best matches between the algorithms (old and new) and computer architectures.

5. We support the idea of the Performance Evaluation Club (Berry et al [1988]) whose goals are:

    •   to develop a methodology for creating a FORTRAN benchmark suite of portable application programs for supercomputing,

    •   to produce a collection of realistic benchmark application codes that have been measured and verified on a variety of supercomputers,

    •   to identify and categorize the types of transformations used in optimizing the suite on each computer.

6. We also need to create public libraries of parallel algorithms in order: to avoid duplication of effort, to teach parallel algorithms, and to accumulate and disseminate the state of the art.

7. We expect that some future scientific and industrial applications will require coupling numerical and symbolic computation (Kowalik and Kitzmiller [1988], Kowalik [1988], Adeli and Balasubramanyan [1988]). This opens up an entirely new opportunity for developing coupled parallel algorithms which would have to include both parallel symbolic and numerical components to avoid detrimental performance bottlenecks.

8. Visualization plays an increasingly important role in scientific and engineering supercomputing and algorithm design. The area of parallel visualization algorithms is currently quite immature. Efficient interactive supercomputing will require parallelizing these algorithms. Existing visualization software is not easily vectorized or ported to parallel computers (McCormick et al [1987]).

9. Algorithms for solving large-scale numerical problems require access to databases containing large amounts of data. We need to research the structure and access mechanisms of these databases to minimize the possibility of I/O operations dominating the computing time. Multiprocessor supercomputers will impact considerably data management and will necessitate appropriate partitioning of data among processors executing parallel algorithms.

10. Seamlessly integrated computing environments that promote cooperative problem solving should have high research and development priority.

7.0 REFERENCES

Abelson, H., M. Eisenberg, M. Halfont, J. Katzenelson, E. Sacks, G. J. Sussman, J. Wisdom and K. Yip, Intelligence in Scientific Computing, CACM, vol. 32, no. 5, 1989.

Adeli, H. and K. V. Balasubramanyan, A Novel Approach to Expert Systems for Design for Large Structures, AI Magazine, vol. 9, no. 4, winter 1988.

Bell, J. L, and G. S. Patterson, Jr., Data Organization in Large Numerical Computation, The Journal of Supercomputing, vol. 1, no. 1, 1987.

Berry, M. et al., The Perfect Club Benchmarks: Effective Performance Evaluation of Supercomputers, CSRD Rep. No. 827, Univ. of Illinois at Urbana-Champaign, Nov. 1988.

Bieterman, M. "Impact of Multitasking on Job Stream Throughput," Proceedings of the Cray User Group, October, 1987, Bologna.

Boghosian, B. and C. D. Levermore, A Cellular Automation for Burgers' Equation, CA87-1, Thinking Machines Corporation, 1987.

Buell, D. A. et al., Parallel Algorithms and Architectures, The Journal of Supercomputing, vol. 1, no. 3, April 1988.

Cocke, J., "The Search for Performance in Scientific Processors: Turning Award Lecture", CACM, vol. 31, no. 3, March 1988.

Dembart, B. and K. Neves, "Sparse Triangular Factorization on Vector Computers," Exploring Applications of Parallel Processing to Power Systems Problems, EPRI EL-566-QR, October 1977.

Duff, I. S., The Influence of Vector and Parallel Processors on Numerical Analysis, AERE R12329, Harwell Lab., September 1986.

Dongarra, J. and D. Sorenson, SCHEDULE: Tools for Developing and Analyzing Parallel Fortran Programs, TM-86, Argonne National Laboratory, Math and Computer Science Division.

Dongarra, J. Performance of Various Computers Using Standard Linear Equations Software in a Fortran Environment, TM-23, Argonne National Laboratory, April 18, 1988a.

Dongarra, J., J. Du Croz, I. Duff and S. Hammarling, A Set Of Level 3 Basic Linear Algebra Subprograms, TM-88 (Rev.1), Argonne National Laboratory, Math and Computer Science Division, May 1988b.

Erman, L. D., J. S. Lark and F. Hayes-Roth, ABE: An Environment for Engineering Intelligent Systems, IEEE Transactions on Software Engineering, vol. 14, no. 12, 1988.

Flynn, M. "Some Computer Organizations and their Effectiveness," IEEE Transactions on Computation, Vol. C-21, 1972.

Gentzsch, W. and K. Neves, Computational Fluid Dynamics: Algorithms and Supercomputers, NATO AGARDograph No. 311, March 1988.

Gustafson, J. L., Reevaluating Amdahl's Law, CACM, vol, 31, no. 5, 1988.

Hillis, W. D. and G. L. Steele, Jr., Data Parallel Algorithms, CACM, vol. 29, no. 12, 1986.

Hockney, R. W., $(r_\infty, n_{1/2}, s_{1/2})$, Measurement on the 2-CPU Cray X-MP, Parallel Computing, vol. 2, no. 1, 1985.

Jamieson, L. H., Characterizing Parallel Algorithms, in The Characteristics of Parallel Algorithms, edited by L. H. Jamieson, D. B. Gannon, and R. J. Douglass, MIT Press, 1987.

Kowalik, J. S., Parallel Processing of Programs Coupling Symbolic and Numerical Computation, International Conference on Parallel Computing, Methods, Algorithms and Applications, Sept., 28-30, 1988, Verona, Italy.

Kowalik, J. S. and C. T. Kitzmiller, (editors), Coupling Symbolic and Numerical Computing in Expert Systems, North-Holland, Amsterdam, 1988.

Levitan, S. P., Measuring Communications Structures in Parallel Architectures and Algorithms, in The Characteristics of Parallel Algorithms, edited by L. H. Jamieson, D. B. Gannon and R. J. Douglass, MIT Press, 1987.

McBryan, O. A., New Architectures: Performance Highlights and New Algorithms, Parallel Computing, 7, 1988.

McCormick, B. H., T. A. DeFanti and M. D. Brown (editors), Visualization in Scientific Computing, Computer Graphics, vol. 21, no. 6, Nov. 1987.

Martin, J. L. and D. Mueller-Wichard, Supercomputer Performance Evaluation: Status and Directions, The Journal of Supercomputing, vol. 1, no. 1, 1987.

Perdue, J. N., Supercomputers Need Super Networks, Proceedings from the Third International Conference on Supercomputing, Boston, 1988.

Samukawa "Programming Style on the IBM 3090/VF," IBM Systems Journal, Vol. 27, No. 4., 1988, pp 453-474.

Tarjan, R. E., Algorithm Design, CACM, vol. 30, no. 3, March 1987.

Waltz, D. L., Applications of the Connection Machine, IEEE Computer, January 1987.

PART 2

SUPERCOMPUTING CENTERS

Supercomputing Facilities for the 1990s

Bill Buzbee
Scientific Computing Division
National Center for Atmospheric Research *
P. O. Box 3000
Boulder, Colorado   80307

Introduction

Historically, the rate of change within computing technology has always been high relative to other technological areas and for the next five years, change within supercomputing technology will be particularly fast paced.   For example, by the mid-'90s:

- We will see an increase in computational processing and memory
  capacity by an order of magnitude,
- Visualization will become an integral part of state-of-the-art
  supercomputer centers,
- The transition to UNIX and related developments will result in a
  common and rich software environment, and
- Very high-speed networks will be available.

While change is healthy and brings with it many opportunities, it also presents a challenge to management, particularly in planning and funding.   This paper will survey those areas where we anticipate significant change in the next three to five years.   In particular, we will discuss:

- Parallel processing via
    Shared-memory supercomputers and
    Massively parallel systems
- Data storage
- Visualization
- Distributed computing on a national scale
- Model development environment

*The National Center for Atmospheric Research is operated by the University Corporation for Atmospheric Research and is sponsored by the National Science Foundation. Any opinions, findings, conclusions, or recommendations expressed in this publication are those of the author(s) and do not necessarily reflect the views of the National Science Foundation.

A summary statement of trends and needs in each of these areas will also be provided.

Finally, we will discuss how all these technologies can be integrated into a supercomputer center typical for the 1990s.

Parallel Processing

Because of the diminishing rate of growth in the speed of a single processor, there is increasing agreement that in the 1990s supercomputing and parallel processing will be synonymous. Two promising architectures for high-performance parallel processing are: (1) shared-memory systems in which a few (less than 100) high-performance processors share a common memory; and, (2) distributed-memory systems in which thousands, even tens of thousands, of low-performance processors with local memory cooperate through a communications network of some sort. Consequently, laboratories such as the National Center for Atmospheric Research (NCAR), are taking steps to encourage scientists to experiment with both types of architectures and, when appropriate, to incorporate parallel processing into their models.

Parallel Processing on Shared-Memory Systems

In 1986, NCAR installed a CRAY X-MP/48 and this machine can function as a shared-memory system for parallel processing. To encourage use of the X-MP/48 as a parallel processor, a Mono-Program (MP) job class with relatively favorable charging rates was created. To optimize system performance, MP jobs are given the entire system and certain network functions are disabled to minimize system interruption. Dedicating the system in this fashion is necessary because operating systems are designed to optimize throughput as opposed to model performance. Two wall-clock hours per day are allocated to the processing of MP jobs and about four models regularly take advantage of this time. Measurements show that the X-MP/48 generally achieves its highest performance when executing MP jobs.

Albert Semtner, Naval Postgraduate School, and Robert Chervin, NCAR, have used the MP job class in their work[1] on eddy resolving, global ocean simulation. To achieve parallel processing, they took an existing model and reconstructed it into independent tasks that process sections in longitude and depth. These independent tasks are parallel processed using the microtasking capabilities of the X-MP/48 system. Peak performance of the X-MP/48 on this model is approximately 450 Mflops. Nevertheless, since a typical simulation uses 0.5 degree grid spacing in the horizontal, and 20 vertical levels, some 200 wall-clock hours are required to complete a 25-year simulation.

Supercomputers that have 16 processors sharing at least 250 million words of memory will become available in the 1990-91 timeframe. These machines will offer an order of magnitude increase in computational power over today's equipment provided we take advantage of their parallel processing capability. The work of Semtner and Chervin demonstrates that by careful formulation and implementation of a model, very high performance can be achieved on a shared-memory parallel processor. As a result, they were able to undertake a simulation that was previously intractable and, in so doing, produce new scientific results. For example, their model is the first to correctly simulate the Antarctic Circum-polar current. This is the primary benefit of parallel processing--expanding the set of tractable simulations. And that is why parallel processing has such an important role in our future.

Massively Parallel Systems

In the spring of 1988, NCAR joined with the University of Colorado at Boulder and the University of Colorado at Denver to form the Center for Applied Parallel Processing (CAPP). CAPP is administered by the University of Colorado at Boulder with Professor Oliver McBryan as the Director. Whereas, research over the past decade has shown the viability of parallel processing, most applications of it have involved model problems and low-performance computers. The objective of CAPP is to solve "real-world" problems using systems with at least one hundred processors and to achieve performance levels that equal or exceed those of supercomputers.

In the fall of 1988, the Defense Advanced Research Projects Agency (DARPA) granted CAPP an 8,000 processor Connection Machine, Model 2 (CM-2) manufactured by Thinking Machines Corporation. As part of its contribution to CAPP, NCAR provides space and operational support for the machine. The CM-2 is a distributed-memory architecture in which each processor has 8,000 bytes of local memory. When configured with 64,000 processors, the machine has achieved from one to five Gflops while executing a variety of applications. More than 40 researchers from at least 10 organizations are using the CAPP machine to develop and test parallel models and algorithms.

NCAR is particularly interested in the potential of the CM-2 to support climate and ocean simulations. Our research plan is to start with a relatively simple model and determine if it can be formulated for the CM-2 such that high performance is achieved. If so, we will then progress through a sequence of increasingly sophisticated models to see if they can be so formulated. The first step was taken in the spring of 1988 when the shallow water model was implemented[2] on a 16,000 processor CM-2. When the performance is extrapolated to a 64,000 processor system and a 4K by 4K grid, we estimate that the model will run at over 1.5 Gflops. We have also implemented a spectral algorithm for solving the shallow water model and achieved an extrapolated performance of 1.13 Gflops on a 2K by 2K grid. The CRAY X-MP/48 achieves about 0.5 Gflops when all four of its processors are applied to these problems.

Distributed-memory machines such as the CM-2 will probably have a relatively limited domain of application as compared to shared-memory supercomputers. But machines like the CM-2 offer the potential of both high performance and low cost. Many people in the high-performance computing community believe that by tightly coupling distributed-memory and shared-memory systems, we will reap the benefits of both.

Some Challenges from Parallel Processing

Achieving high performance on a parallel processor requires careful attention to model formulation and implementation. The first pre-requisite is concurrency in the model. That in turn requires good knowledge of parallel algorithms. When suitable parallel algorithms are not known, they will have to be developed and the associated research requires detailed knowledge of computer architecture. Once sufficient concurrency is available, the next challenge is to implement the model such that there are no significant overheads from task management, e.g., task synchronization, intertask communication, contention for memory, and so forth. Supercomputer users do not normally possess this expertise. So in the '90s, supercomputer centers must provide a critical mass of experts who will provide training in parallel processing and assist with algorithmic research as needed.

Summary of parallel processing

In the 1990s, parallel processing and supercomputing will be synonymous. Shared-memory systems will offer an order of magnitude increase in performance if parallel processing is exploited, while at the same time offering support for general-purpose simulations. Massively-parallel systems may provide more than an order of magnitude increase in performance and better cost performance than supercomputers. However, the application domain of these systems may be relatively narrow.

Data Storage

Many supercomputer centers have concentrated their archival storage into a single node of their network. This concentration makes possible substantial economies in support and expansion of the facility. For example, if all archival storage is concentrated into a network node, then supercomputers do not have to be configured for long-term storage of

Table 1.   New Storage Media

	8MM Video Tape	Digital Audio Tape	Write Once Optical Disk	Rewriteable Optical Disk
Capacity	2.3 GB-4.6 GB	1 GB	800 MB	100 MB
I/O Access Performance	Low	Low	High	High
I/O Device	.25 MBs-.5 MBs	.183 MBs	.312 MBs	.125 MBs
Media Cost	Low	Low	High	High

Source: Compiled by R. Abraham and R. Freeman, Jr., Ninth
IEEE Symposium on MSS, 10-88.

information.   Also, information in the storage network node is accessible by any computer attached to the associated network.

The NCAR network storage node is called the Mass Storage System (MSS) and it allows users to store and access massive amounts of data. Currently, it holds about 380,000 files totaling 80 terabits (Tb) of information.   About 20 terabits are observed data and the remainder is output from large simulation studies.   On a daily basis, the system handles about 3,000 requests involving about 800 gigabits (Gb).   The system consists of a control processor, as well as a tape and a disk farm.

The IBM 3480 tape cartridge is the basic MSS storage medium; our cartridge inventory exceeds 60,000 units.   Because the system has its own control processor, file purging and data compaction on cartridges is performed as background processing.   Also, the control processor keeps frequently used files on disk to reduce manual loading of tapes. Included in the MSS is an "IMPORT/EXPORT" capability whereby information can be transferred from (to) our IBM 3480 cartridges to (from) other media such as those in Table 1.   Thus, field experiments can use whatever media is appropriate and once data has been imported into the MSS, long-term availability is assured.

Summary of data storage systems

Supercomputer centers tend to archive approximately 100 bytes of information for each megaflop of computation. Thus in the 1990s, we will need mass storage systems that have an order of magnitude greater capability than those available today. We will also need the ability to "IMPORT/EXPORT" information between these systems and the new low-cost, high-capacity storage media that will be widely used by experimenters in the '90s.

Visualization

Large time-dependent, three-dimensional simulations can produce an enormous amount of output. For example, a recent ocean basin simulation at NCAR produced 50 Gbytes of output for five years of simulated time, i.e., several variables at each of thousands of mesh points and thousands of time steps. Visualization technology integrates graphics imaging, human interface, and data manipulation technologies to provide an interactive and visual capability for examining large datasets. It greatly leverages the intellect of the scientist by presenting computer-generated information in a readily-comprehensible form. It also greatly accelerates the rate at which scientific progress can be made. Put another way, those organizations that have good visualization capability will have a competitive edge over those that do not.

Summary of visualization technology

Visualization systems will be an integral component of super-computing facilities in the 1990s.

Distributed Computing on a National Scale

Most university scientists must provide computing resources from their research grants. As a result, many university researchers only have minicomputers and desktop systems locally available to them. The power of a supercomputer exceeds these machines by one to two orders of magnitude. A problem requiring an hour of supercomputer time is simply not doable using many university local systems. An immense benefit of the national networks is to make supercomputers accessible to a large portion of the nation's university scientists.

By the early 1990s, the speed of national networks will increase at least by an order of magnitude. Already we have systems whereby scientists can work on their local computer to develop a job and/or data which can then be sent to a supercomputer for processing. Once the processing has been completed, the associated output is automatically returned to the scientist's local system for post analysis. These systems essentially make remote supercomputers appear as an attached processor. Text and graphics metafiles are currently shipped over national networks but, in the 1990s, it will be feasible to routinely ship gigabyte sized files over these networks.

Summary of networking and communications

High-speed national networks will be an order of magnitude faster in the 1990s. This will significantly expand the utility of supercomputers to university researchers.

Model Development Environment

In the 1990s, we will enjoy a much improved software development environment that will be made possible through ease-of-use, software tools, and powerful desktop systems.

Ease-of-use will result from much larger memories, common software, on-line documentation, and windowing. Supercomputers of the 1990s will have hundreds of millions, even billions, of words of directly addressible memory. Thus it will no longer be necessary for three-dimensional simulations to be disk resident. Development and maintenance of associated models will be greatly simplified. The rapid movement toward UNIX as a standard operating system within science and engineering will simplify the transportability of software between systems and thus between scientists. Today, desktop systems have on-line documentation structured and organized such that the user can quickly and efficiently get the detailed information concerning questions of usage. The same will be true of supercomputers in the 1990s. Finally, windowing technology will make it much easier to concurrently utilize local and remote systems and, in particular, to distribute labor between them.

In the area of software tools, one of the most important aids to productivity in the '90s will be the availability of interactive symbolic debuggers on supercomputers. Using these debuggers, not only can software be developed with approximately an order of magnitude less time, it can also be developed in approximately an order of magnitude fewer computer cycles. To facilitate parallel processing, we will need elaborate performance monitors and analyzers, both dynamic and static. Subroutine libraries in mathematics, graphics, and other areas will continue to evolve and will incorporate parallelism in them as appropriate. The Computer Aided Software Engineering (CASE), technology will be available on supercomputers in the '90s and the associated capabilities for version control and model configuration should be particularly valuable. Finally, automatic parallelizing and autotasking compilers will be relatively sophisticated in their ability to recognize and exploit parallelism within models.

In the '90s, powerful workstations will be available at relatively low cost. These machines will offer a friendly-user interface through pull-down menus, pointers, icons, and audio input/output. They will also offer full color and high-resolution graphical displays. They will have relatively large memories and, as today, they will be rich in software. Further, and as noted previously in Table 1, capacious and low-cost storage media will be available for them. Powerful workstations will be the center of the scientist's computational universe during the 1990s. This will be a dramatic change because, for decades, the supercomputer or mainframe system has been the center of that universe. That is, all resources including human time were subservient to the supercomputer. When the workstation of the '90s have access to supercomputers and mass storage systems with high-speed networks, the result will be a new dimension in model development capability.

Using these machines, scientists will be able to use electronic hypermedia research notes that include animation and audio. It will be possible to annotate electronic manuscripts in the margin and, of course, the manuscripts will combine both text and graphics. Shared windows among scientists will be possible, perhaps at great geographic distance.

Summary of software environment

The software development environment of the 1990s will produce a quantum jump in scientific productivity through improved ease-of-use, software tools, and powerful workstations.

Putting It All Together

As discussed previously, the next generation of supercomputer will offer a sustained level of performance that surpasses a gigaflop. These machines will have gigaword memories and they will generate an enormous amount of output that must be stored. Also, visualization workstations and associated technology will be closely coupled to supercomputers in the '90s. All of these capabilities imply, in fact require, a new generation of networking technology. For example, the effective bandwidth connecting supercomputers, mass storage systems, and visualization stations must be in the neighborhood of at least a gigabit per second. Fortunately, that technology is now coming into general use. We will also need standard interfaces for connecting devices to these networks. The HSC proposal from Los Alamos is a good first start for standards in this area and it too is gaining wide acceptance. Storage device I/O rates will have to be much higher than they are today. The Redundant Array Inexpensive Disks (RAID), technology shows a great deal of promise, however, the technology is not yet proven and thus a great deal of work remains to be done. The density on storage media will have to increase by at least an order of magnitude. Video tape technology appears to be the most promising in this area. Again, it is in an embryonic state and much work needs to be done to perfect it. Finally, network security will be essential because of the tremendous economic value of the software and data which is available in the typical supercomputer center.

The networking requirements of the above are formidable and will likely be met via three levels of networking. The cost-effectiveness of 10 megabit per second, ethernet technology will persist well into the '90s. On the other hand, the emerging Fiber Data Distributor Interface (FDDI), will be available in the early '90s and will provide a cost-effective way for connecting large numbers of high-performance equipment. Finally, a gigabit per second or higher networks will be required to processors, mass storage and visualization systems.

Conclusion

The supercomputing community is faced with a formidable rate of change over the next five years that will affect almost every component of a typical center. In particular, we will see supercomputers become available that offer an order of magnitude increase in computational power provided parallel processing is used. Visualization will be an integral part of such a center and will be essential for analyzing the output of large three-dimensional simulations. The emergence of UNIX as a standard supercomputer operating system, combined with the power of desktop systems and the emergence of new software tools, will result in a common and rich software development environment. Finally, all of the above will be interconnected through very high-speed national networks by which these resources will be accessible by most of the nation's scientific and engineering community.

References

1. Semtner, A. and R. Chervin, "A simulation of the global ocean circulation with resolved eddies," to appear in *J. Geophys. Res.*

2. Sato, R. and P. Swarztrauber, "Benchmarking the Connection Machine 2," *Proceedings of the Supercomputing Conference*, Orlando, Florida, Nov. 14-18, 1988, pp. 304-309.

RESEARCH AND DEVELOPMENT IN THE
NUMERICAL AERODYNAMIC SIMULATION PROGRAM

F. Ron Bailey
Mail Stop 258-5
NASA Ames Research Center
Moffett Field, CA 94035

Abstract

NASA's Numerical Aerodynamic Simulation (NAS) Program provides a leading-edge computational capability supporting aerospace research and development applications. The NAS facility, located at NASA's Ames Research Center (ARC), provides supercomputing services to over 1000 researchers from NASA, the Department of Defense, the aerospace industry, and university sites across the United States. In addition to providing advanced computational resources, the NAS Program acts as a pathfinder in the development, integration and testing of new, highly advanced computer systems for computational fluid dynamics and related disciplines. In fulfilling this role, the Program acquired early serial number supercomputers and integrated them into a new system paradigm that stresses networking, interactive processing, and supercomputer-workstation connectivity within a single uniform UNIX® software environment. The NAS system is described; and its pioneering advances in supercomputing operating system and communication software, local and remote networking, scientific workstations, interactive graphics processing and user interfaces are highlighted. Finally, some results of the newly initiated applied research effort are presented.

Introduction

The NAS Program's origins date from 1975 (ref. 1), when ARC researchers in computational fluid dynamics began to investigate the computer technologies needed to support the advances then being made in fluid dynamics modeling. After nine years of program advocacy and several

---

UNIX® is a registered trademark of AT&T Information Systems.

NATO ASI Series, Vol. F 62
Supercomputing
Edited by J. S. Kowalik
© Springer-Verlag Berlin Heidelberg 1990

technical studies, the NAS Program became a NASA New Start Program in Fiscal Year 1984, with the following objectives:

- To provide a national computational capability, available to NASA, DoD, other government agencies, industry, and universities to ensure the nation's continuing leadership in computational fluid dynamics and related disciplines.

- To act as a pathfinder in advanced large-scale computer system capability through systematic integration of state-of-the-art improvements in computer hardware and software technologies.

- To provide a strong research tool for the NASA Office of Aeronautics and Space Technology.

To meet these objectives, Program implementation has stressed the importance of advanced development coupled with a thorough understanding of user requirements. A pioneering network and open systems architecture approach was adopted to facilitate integration of new technologies, and to provide a common operating environment upon which further capabilities could be built. From the beginning, the importance of interactive processing, visualization, and the potential role of scientific workstations in creating a more productive user environment were emphasized. The result of these early efforts was the Initial Operating Configuration (IOC) of the NAS Processing System Network (NPSN), which began full operation in March 1987 (ref. 2). This system pioneered a number of features now widely implemented in state-of-the-art supercomputing facilities.

The NAS IOC introduced the first large- memory supercomputer and the first UNIX-based supercomputer. The Cray-2 system installed in 1985 had 2.048 gigabytes of central memory and represented an increase of two orders of magnitude in memory capacity over previously available supercomputers. The NAS Cray-2 was the first with Cray's UNICOS operating system. Although it started with a limited capability at installation, the operating system was significantly enhanced at NAS. Berkeley-style networking with Department of Defense (DoD) Internet protocols (TCP/IP) was added, so that for the first time a supercomputer could fully participate in an open system network environment. A Network

Queuing System (NQS) was also developed and incorporated into the Cray-2 UNIX environment to provide the capability for batch resource management across the network.

NAS was the first supercomputing facility to install standard operating system and communication software on all processors. Ease of use is enhanced since all NAS computers operate under the UNIX operating system. UNIX offers users the flexibility of both interactive and batch supercomputing. Moreover, the software system presents a common system interface to the users, and provides the same environment (i.e., common utilities and commands) on all user-accessible subsystems. The uniform environment enables NAS users to move easily between processors, allows for easy file access and command initiation across machine boundaries, and enhances user code portability within the NAS processing system. Furthermore, the "open architecture" concept (implementation based on openly available definitions for hardware and software interfaces) used in the design of the NAS system, allows for greater modularity and for easier implementation of new capabilities.

NAS is a pioneer in the direct networking of workstations and supercomputers that has literally transformed the supercomputer into an extension of the workstation. The user "resides" at his workstation where text and small data files are created and modified, and where complex input data and results are displayed and analyzed. Interactive graphical post-processing is an essential aspect of the today's sophisticated numerical simulation process and the NAS Program continues to emphasize the development of graphics workstations as the principal means for researchers to interface with supercomputers (ref. 3).

Since the NAS Program serves a national user community, remote access has been an important consideration. This has led to the development of NASnet, a unique, high-performance nationwide communication network, that links remote user locations to the NAS facility via satellite and land-circuits, with communications speeds up to 1544 kilobits/sec. At the forefront of Ethernet bridging technology, NASnet provides an environment in which researchers at remote locations (Fig. 1) have virtually the same interactive capability as users at the NAS Facility. One example involves graphical co-processing using the Ames-developed Remote Interactive Particle-Tracing (RIP) application (ref. 4, 5). With this application, the

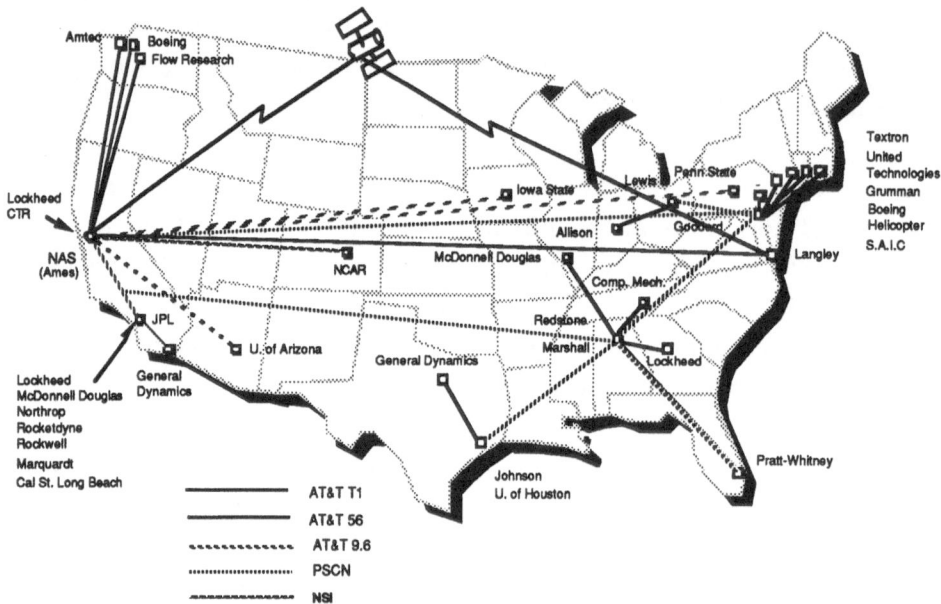

Figure 1. NASnet Configuration

computationally intensive calculation of particle traces through a fluid flow solution is accomplished on a supercomputer, while the results are passed over network links to be graphically displayed on a workstation. The combined use of workstation and supercomputer offers the remote user a 10:1 time savings over the workstation alone in performing complex displays of three-dimensional fluid dynamics solutions. In fact, the NAS system is designed so that all users, whether local or remote, can consider the supercomputer as an extension of their own workstation. Furthermore, new users familiar with UNIX-based workstation environments, such as those offered by Sun Microsystems Inc., Silicon Graphics Inc., Digital Equipment Corp., and others, quickly become productive on the NAS supercomputers.

Current NAS Configuration

The current production configuration of the NPSN is shown in Fig. 2. The Cray Y-MP was installed in 1988 and was the first customer installation for that system. The Cray Y-MP is an eight processor system with a 6.3 nanosecond clock cycle, 32 megawords of main memory and a 256

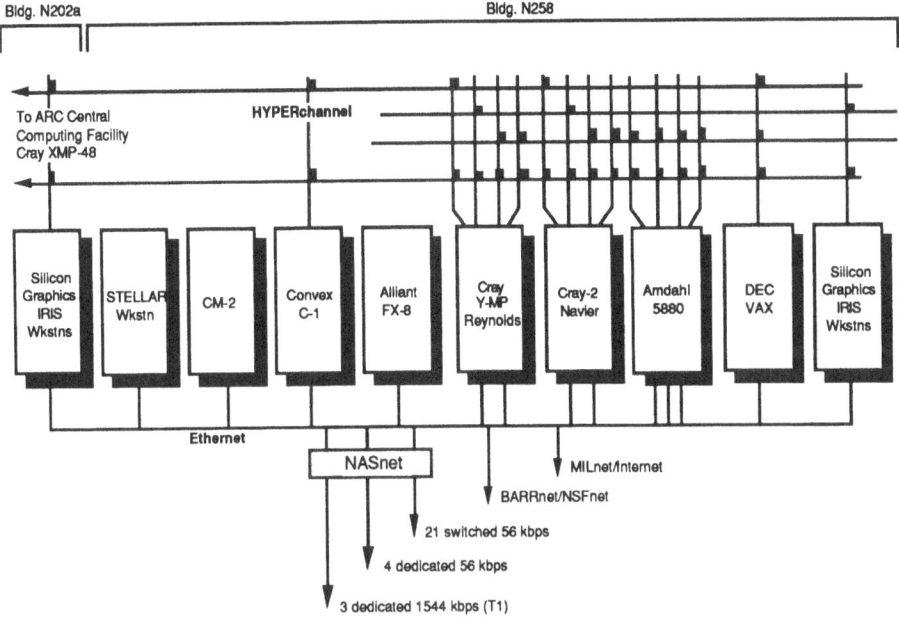

Figure 2. NAS Processing System Network in June 1989

megaword Solid-state Storage Device. A mixed configuration of 40 Cray DD49 and 8 Cray DD40 disk systems give a total local storage capacity of 89.6 gigabytes. During the acceptance period the configuration achieved one billion floating-point operations per second (gigaFLOPS) performance on a standard workload composed of computational fluid dynamics codes. The Cray Y-MP also makes a multi-processing environment available to the general user community. The software provided by Cray Research Inc. supports microtasking, macrotasking and autotasking. Use of microtasking coupled with autotasking has reached a speed 1.68 gigaFLOPS on a general sparse linear equation solver (ref. 6). This achievement earned the authors (Horst Simon, Phuong Vu and Chao Yang) the IEEE sponsored Gordon Bell Award for Parallel Computing. In early 1990, the main memory of the Cray Y-MP will be upgraded from 32 to 128 megawords.

The Amdahl 5880 processor complex provides a mass storage facility as well as general purpose computing capability. Mass Storage devices include 200 gigabytes of 3380 technology DASD and two StorageTek 4400 cartridge tape robots capable of storing 1.2 terabytes each. The mass storage system utilizes both UTS and MVS operating under VM. This dual operating system approach was used to provide a UNIX environment for the user and while maintaining the data handling abilities of the MVS operating system. The Cray-2 to Amdahl disk to disk transfer rate was initially 100 kilobits/sec., but with subsequent revisions, has reached 6 to 8 megabits/sec.

Additional support processing is provided by four Digital Equipment Corporation VAX™ minicomputers, and Convex C-1 and Alliant FX-8 mini-supercomputers. The Convex and Alliant also provide platforms for the high-speed network prototype discussed below. The current configuration also includes 42 Silicon Graphics IRIS workstations and one Stellar workstation. Since three-dimensional, computational fluid dynamics simulations generate an enormous amount of data, it is virtually impossible for the unaided scientist to analyze the raw results and gain any qualitative picture of the physical phenomenon simulated within a reasonable amount of time. Graphics workstations have been employed in solving this problem. The workstations typically consist of a high-resolution, high-throughput color graphics terminals that enable near-realtime manipulation of viewpoint and other display parameters to give three-dimensionality to the results. The software that operates on these workstations (e.g., PLOT3D, RIP and GAS, designed by Ames researchers (ref. 7)) permits scientists to display any facet of their data for easy interpretation and evaluation.

All NPSN computer systems are linked together via Ethernet and Network Systems Corporation HYPERchannel local area networks and are also linked to other computers at Ames Research Center and at remote locations across the United States. The entire NPSN operates under the UNIX operating system with DOD Internet (TCP/IP) network communications provided via the well-known Berkeley UNIX "r" commands. Thus, a user can access any computer, run jobs, and transfer data among computers using a single set of commands on all computers.

Remote users access the NPSN through a wide choice of existing national communication networks including the Department of Defense sponsored MILnet and ARPAnet, and the National Science Foundation sponsored NSFnet. NAS is also a member of the Bay Area Regional Research Network that provides 1544 kilobits/sec. service between Ames Research Center, Stanford University, and the University of California campuses at Davis,

VAX™ is a trademark of Digital Equipment Corporation.

Berkeley, San Francisco and Santa Cruz. MILnet is the main communication path to Department of Defense laboratories while the Internet (principally NSFnet and BARRnet) serves the university community. For NASA and aerospace industry users, NAS has developed NASnet, a unique, NAS-specific network using Ethernet bridging technology to provide fast response and high throughput to remote graphics workstation users. NASnet currently serves 31 remote sites at bandwidths ranging from 56 to 1544 kilobits/sec.

Extended Operating Configuration

The next major NPSN configuration is the Extended Operating Configuration (EOC), scheduled for completion by the end of 1989. EOC is targeted to provide a sustained throughput capacity of one billion floating-point operations per second which represents a factor of four increase in computing performance over IOC. This performance target was achieved with the installation of the Cray Y-MP in 1988. The acquisition of a second-generation scientific workstation system with approximately 10 times the performance level of the Silicon Graphics IRIS 3000 is now in process. A new mini-supercomputer based system for the greatly enhanced data editing and visualization capability required for unsteady, three-dimensional simulations has been defined. Finally, three advanced technology development projects are nearing completion for incorporation into EOC. The three efforts are MSS-II, High-Speed LAN, and dplane

MSS-II

The EOC target for mass storage throughput performance is 85 megabits/sec. In order to achieve this performance goal, a second generation Mass Storage Subsystem (MSS-II) has been designed and is being implemented (ref. 8). To achieve vendor independence and portability, MSS II will be entirely UNIX resident and will appear to users as an ordinary UNIX system with a large number of very fast disks. It is currently being based on Amdahl's implementation of UNIX, UTS. The UTS file system is being enhanced to provide features that are appropriate to a supercomputing environment and which will overcome deficiencies in the vendor-provided software. These include larger i-nodes, a greater number of i-nodes, a larger number of disk blocks, track I/O and striped I/O. In addition, the file systems will be hardened to ensure that no data will be lost after a successful close has occurred. A volume manager will be implemented to provide a generalized tape authentication and queuing capability. The initial configuration will accommodate rapid access to a hierarchy of storage provided by 200 gigabytes of DASD, and two StorageTek Robots with a storage capacity of 1.2 terabytes each. A special link-level driver will be implemented and NSC N220 adapters configured to allow multiple concurrent data transfers. The design will permit future implementations to take advantage of higher performance networks.

High-speed Local Area Network

In anticipation of the increased performance of the Cray Y-MP, second-generation workstations, MSS-II, and even higher performance systems in the future, a high-speed local area network prototype project has been initiated. The project is based on new high-performance network technology developed by Ultra Network Technologies. Ultranet is a tree-structured network that is based upon interconnecting hubs with optical fiber. Devices are attached to the network via interface cards at the hubs. The prototype configuration, which is shown in Fig. 3, connects the Cray-2 , the Cray Y-MP, a graphics frame buffer, special graphics devices, a Convex C-1, an Alliant FX-8, and a small number of VME-based workstations. All connection are through each systems highest speed I/O channel. Network software enhances the existing host-based BSD socket library. When data transfers occur directly over the Ultranet, it bypasses the socket library, preforming data transfers directly with the device driver to hardware protocol processors. Early measurements have shown data transfer rates between the Cray-2 and a frame buffer to be in excess of 715 megabits/sec and between two Cray-2's in excess of 590 megabits/sec. Memory to memory data transfers have exceeded 64 megabits/sec. between the Cray-2 and mini-supercomputers, and 32 megabits/sec. between the Cray-2 and workstations. The limiting factor in these latter cases has been the I/O performance of the systems themselves. The results are very encouraging, and indicate local area networks can be implemented to take full advantage of the supercomputers' high-speed channel capabilities.

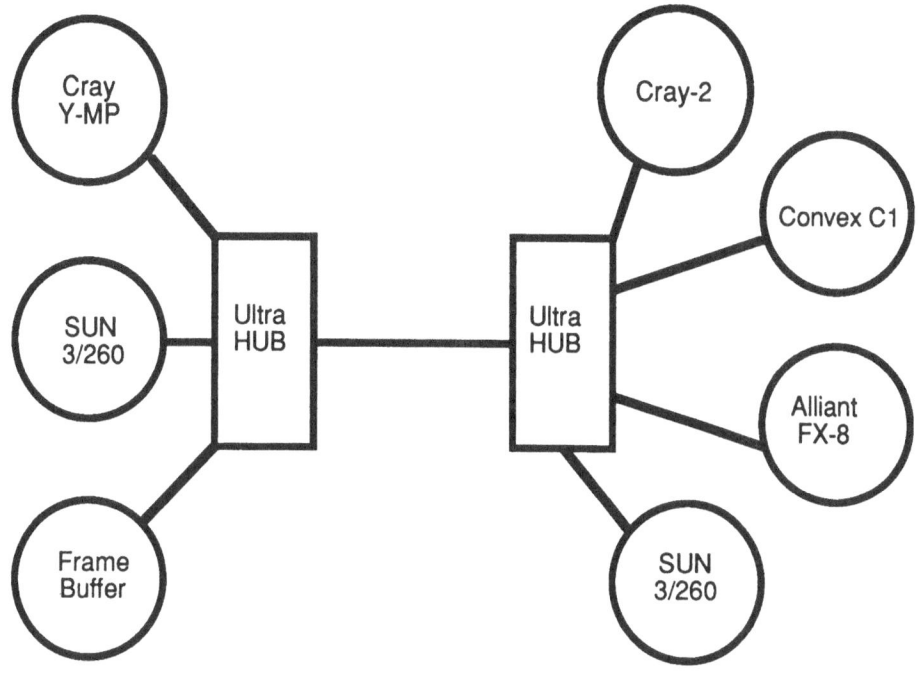

Figure 3. High-speed LAN Prototype.

dplane

The final development project for EOC is "dplane", a distributed graphics application. The dplane application uses the NAS-developed Distributed Library, an application-level network protocol which allows workstation resident programs to invoke remote procedure execution on supercomputers. This allows the computational power of the supercomputer and the graphics capabilities of the workstation to be combined in a single application. dplane greatly facilitates the display of solutions to highly complex computational fluid dynamics problems. A solution and its associated finite difference grid are used to derive and display animated vector and scalar fields in near realtime. A selection of grid "planes" (two dimensional grid surfaces) may be displayed, colored by the normalized value of a physical characteristic of the fluid flow (e.g. pressure, temperature, helicity). Animation is accomplished by looping quickly through a set of planes. The values defining physical flow characteristics on the grid surfaces may be dynamically normalized to enhance visual features of the display. All interaction with the program is through a "Macintosh®-like" user interface consisting of various control panels that contain buttons, sliders, and menus. This highly interactive user interface is key to efficient examination of the behavior of many different quantities over the entire three- dimensional computational domain of fluid flow.

Toward the Future

The inspiration and driving force behind the establishment of the NAS Program was the recognition, during the mid 1970's, that computational simulation of the entire viscous flow field around complex three-dimensional configurations could be accomplished if sufficient computer performance and memory capacity were available. The target performance required to accomplish this goal was set at a sustained computing rate of one billion floating-point operations per second (one gigaFLOPS); and the NAS program set out to provide a production capability to meet this goal. With the completion of the Extended Operating Configuration (EOC) in 1989 this goal will be achieved and so it is timely to formulate a new NAS vision.

As was the case with the original vision for the NAS Program, the new NAS vision is driven by the need to achieve a new plateau in the use of computers for aerospace vehicle design. This goal is no less than the capability to simulate an entire aerospace vehicle system including the coupled interactions of fluid dynamics, chemistry, structures, propulsion, controls and other disciplines that determine the behavior of vehicles in flight. The vision for the NAS Program is to provide the Nation's aerospace research and development community, by the year 2000, with a high-performance, operational computing system capable of simulating an entire aerospace vehicle system, within a computing time of one to several hours. It is estimated that a sustained computing rate of one trillion floating-point operations per second (one teraFLOPS) is required to accomplish this. This new NAS goal is very ambitious, as it represents a 1000-fold increase over today's most powerful computing capability. To accomplish this goal the NAS Program has recently established an advanced technology and applied research program, including parallel processing, mass storage, and remote network research.

Macintosh® is a registered trademark of Apple Computer, Inc.

Parallel Processing Architectures and Algorithms

Research in highly parallel processing architectures and algorithms is an important direction for the NAS Program. At present, highly parallel processing architectures represent the best approach to achieving the goal of a sustained teraFLOPS capability. The objective of the current research is to evaluate parallel processing architectures and to develop efficient algorithms to exploit their use. Beginning in January 1988, a joint NASA/DARPA parallel processing project was undertaken to evaluate the use of massively parallel architectures. A 32 thousand-processor Connection Machine model CM-2 with floating point units has been installed along with other supporting peripheral equipment. The CM-2 is a SIMD (single instruction-stream, multiple data-stream) computer and so each processor is always executing the same instruction as all the other processors, although operating on its own local data. The CM-2 is available to both NASA and DARPA sponsored researchers.

The NASA work to date has been focused on implementing computational fluid dynamics applications. Results for two and three-dimensional, time-accurate Navier-Stokes solvers have been reported (ref. 9 and 10). The approach uses a single virtual processor for each grid point and has been implemented for two different time-stepping methods: explicit third-order Runge-Kutta and implicit approximate factorization. The codes are written in *lisp, a set of parallel extensions to common-lisp, developed for the CM-2 by Thinking Machines, Inc. Results for two and three-dimensional explicit methods show that a 16 thousand-processor CM-2 out performs a Cray X-MP or Cray-2 single processor by a factor of two to three for large grid sizes. Results for implicit time stepping show that the CM-2 is almost as fast as the Cray X-MP.

Research is also being conducted in parallel algorithms for MIMD (multiple instruction-stream, multiple data-stream) hypercubes (ref. 11). Performance on two hypercube computers (a 16-node iPSC/2 and a 512-node NCUBE/10) has been compared with a Cray X-MP for execution of FLO52 (ref. 12), a widely used fluid dynamics code. The 16-node iPSC/2 runs within a factor of 20 for a single Cray X-MP processor and the 512-node NCUBE within a factor of three. The results showed that hypercube computers can provide high-performance, but at the cost of considerable programming effort. Much higher performance is expected with the iPSC/3, to be installed at NAS in 1990.

Network File Server

Another major research area is in high-speed, large capacity, highly reliable storage systems of the scale needed to support teraFLOPS computer systems. NAS systems of the future are envisioned to include network I/O servers that can perform at a level that makes the diskless supercomputer practical. Such servers would be able to support a heterogeneous set of supercomputers as well as workstations in a uniform manner, and at much lower cost, than current schemes. The NAS Program is currently conducting architectural studies of network storage systems and, as described earlier, has made significant progress in increasing network performance. To aid in solving the storage dilemma, NASA and DARPA are jointly funding a grant to the University of California at Berkeley to develop hardware and software I/O architectures for supercomputer networks of the 1990's.

Under the grant, a new I/O architecture will be developed making use of the concept of Redundant Arrays of Inexpensive Disks (ref. 13). By using a combination of arrays of inexpensive disk , file caching, and a log-structured file system, the project will develop a prototype I/O system that can be scaled to achieve a 1000-fold increase in I/O performance over current systems. To demonstrate the technology, a high-performance network file system will be implemented on a commercially available file server, and then interfaced to prototype disk array hardware.

Remote Networks

As a national facility, the NAS serves a nationwide base of users and thus remote access is an essential capability. An important design goal is to provide remote users with the same level and quality of service as is provided to local users. To accomplish this goal, higher performance communications coupled with efficient software tools, including distributed applications, will be developed in order to make the most effective use of the bandwidth available. While current communications technology costs and performance limitations prevent the achievement of complete equality in data communications, it is expected that fiber optic technologies will make communications links up to 45 megabits/sec. practical in the next decade. Perhaps, three gigabits/sec. wide-area communication will be possible by the end of the century.

In an effort to remove one of the technical barriers to higher performance wide area networks, namely gateway performance, NASA, DARPA and DOE have formed a joint project for the development of an advanced communications gateway. The gateway will be capable of connecting to multiple external networks, and will provide more cost-effective utilization of high-bandwidth communications links. This Research Internet Gateway, (RIG), will be able to accommodate:

1. Switched, as well as dedicated, wideband circuits.
2. Terrestrial (low delay) and satellite (high delay) service.
3. Packet switched networks.
4. Interagency internet.
5. Expansion to bandwidths of T2 and greater.
6. Multiple local network interfaces.
7. Support for DoD-internet gateway protocols.
8. Type of service routing.
9. High-speed processing of packets (in the range of 10,000 packets/sec.).

In 1990, NAS will implement a NASA RIG testbed consisting of a prototype gateway unit at each of four NASA centers. The gateways will be connected by both T1 satellite links and 56 kilobits/sec. terrestrial links, as illustrated Fig. 4. The testbed will be used to develop policy as well as type of service

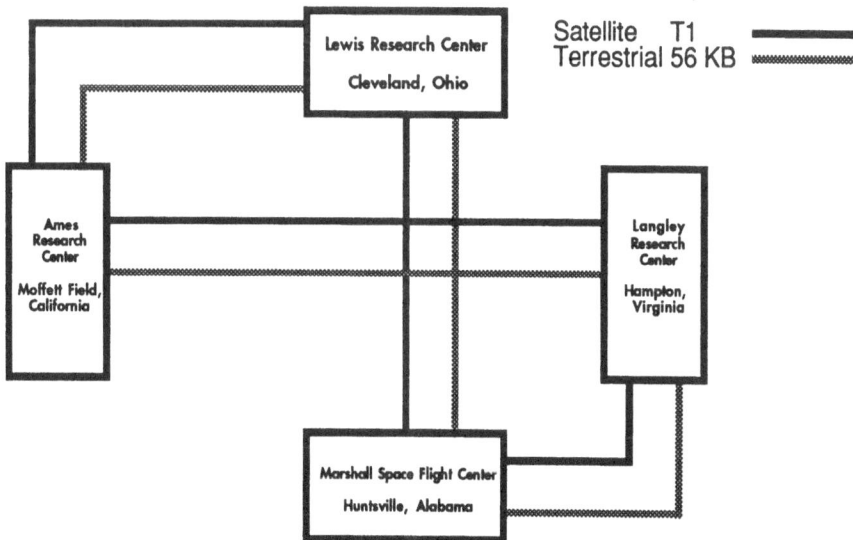

Figure 4. NASA's Research Internet Gateway Testbed Configuration.

routing software, and will serve as the implementation base to meet the additional goals outlined above (ref. 14).

Summary

The NAS Program has achieved the original goals set in 1984. The Program is now providing leading edge computing capability to the American aerospace research community, and is continuing to pioneer new high-performance computing technologies and techniques. A new goal of an operational teraFLOPS capability by the year 2000 has been established and will be applied to the simulation of complete aerospace vehicle systems.

The plan for meeting this aggressive NAS goal requires research and development efforts exceeding those expended to achieve the NAS program's current pathfinding accomplishments. Achievement of the new NAS vision will depend on the exploitation of new massively parallel processing approaches to computing, and the aggressive introduction of this technology into an operational environment. It will mean keeping the NAS production system at the forefront of high-performance computing, to ensure that the aerospace research and development community has the computing tools to achieve the computational physics breakthroughs that will make full aerospace vehicle system simulation a reality.

References

1. Victor L. Peterson; and William F. Ballhaus, Jr.: "History of the Numerical Aerodynamic Simulation Program." Supercomputing in Aerospace, NASA CP 2454, March 1987, pp. 1-11.
2. Blaylock, B. T. and Bailey, F. R., "Status and Future Developments of the NAS Processing System Network." Proceedings of the Third International Conference on Supercomputing, Boston, Mass., May 15-20, 1988.
3. Lasinski, T., Buning, P., Choi, D., Rogers, S., Bancroft, G., and Merritt, F., "Flow Visualization of CFD Using Graphics Workstations." AIAA 87-1180-CP, June 1987.
4. Choi, D. and Levit, C., "An Implementation of a Distributed Interactive Graphics system for a supercomputer Environment." International Journal of Supercomputing Applications, Vol. 1, No.3, winter 1987, pp. 82-95.

5. Rogers, S. E., Buning, P. C. and Merritt, F. J., "Distributed Interactive Graphics Applications in Computational Fluid dynamics." International Journal of Supercomputing Applications, Vol. 1, No.3, winter 1987, pp. 96-105.

6. Simon, Horst D., Vu, Phuong and Yang, Chao, "Performance of A Supernodal General Sparse Solver on the CRAY Y-MP: 1.68 GFLOPS with Autotasking." Submitted to Parallel Computing.

7. Watson, V., Buning, P., Choi, D., Bancroft, G., Merritt, F., and Rogers, S., "Use of Computer Graphics for Visualization of Flow Fields." AIAA Aerospace Engineering Conference and Show, Los Angeles, CA, February 17-19, 1987.

8. Tweten, David E., "Hiding Mass Storage Under UNIX: the NASA Ames MSS-II Project." to be presented at ACM/IEEE Supercomputing '89 Conference, Reno, Nevada, November 13-17, 1989.

9. Levit. C. and Jesperson, Dennis, "Explicit and Implicit Solution of the Navier-Stokes Equations on a Massively Parallel Computer," Computational Structural Mechanics and Fluid Dynamics - Advances and Trends, Special Issue of Computers and Structures, Vol. 30, No. 1/2, Pergamon Press, 1988.

10. Jesperson, Dennis C. and Levit, Creon, "A Computational Fluid Dynamics Algorithm on a Massively Parallel Computer." 9th Computational Fluid Dynamics Conference, AIAA-89-1936, 1989.

11. Barszcz, Eric, Chan, Tony F., Jespersen, Dennis, C., and Tuminaro, Raymond S., "Performance of an Euler Code on Hypercubes." Proceedings of the Fourth Conference on Hypercubes, Concurrent Computers and Applications, March 1989.

12. Jameson, A., "Solution of the Euler Equations for two Dimensional Transonic Flow by the Multigrid Method." Appl. Math. and Comp., 13:327-355, 1983.

13. Patterson, D. A., Gibson, R. H. and Katz, R. H., "A Case for Redundant Arrays of Inexpensive Disks (RAID)," Proc. ACM SIGMOD Conference, Chicago, IL, June 1988.

14. Lekashman, John, "Type of Service Wide Area Networking," to be presented at ACM/IEEE Supercomputing '89 Conference, Reno, Nevada, November 13-17, 1989.

Supercomputing at KFA Jülich:
Experiences and Applications on CRAY X-MP

F. Hossfeld

Zentralinstitut für Angewandte Mathematik (ZAM)
Kernforschungsanlage Jülich GmbH
Postfach 1913, 5170 Jülich
Fed. Rep. Germany

Abstract

KFA Jülich is one of the largest big-science research centers in Europe. At KFA, computational science based on supercomputer techniques has received high priority for many years. Primarily, CRAY supercomputer power has been exploited in the single processor mode by applying vectorization and optimization techniques to numerical kernels and large applications. However, on the multiprocessor vector-supercomputers CRAY X-MP and Y-MP, parallelism - beyond vectorization - can be exploited by multitasking. With increasing number of powerful processors, multitasking strategies like macrotasking, microtasking, and autotasking will become more important and profitable to utilize these multiprocessor systems efficiently. Multitasking results and experiences have been gained by applying these modes to linear-algebra and non-numerical algorithms as well as to large codes. While comparing the results and the concepts of multitasking, the problems, benefits, and perspectives of multitasking programming will be discussed.

Keywords. CRAY X-MP, supercomputing, multitasking, macrotasking, microtasking, parallel programming, linear algebra kernels, graph problems, applications.

Supercomputing at KFA

KFA Jülich was founded in 1956 as a joint nuclear research center for the surrounding universities and other institutions of high education in the Federal State of North Rhine-Westphalia. Today, KFA is one of the largest big-science research centers in Europe. The scope of scientific and technological research and engineering activities ranges from fundamental research to applied sciences focusing with high priorities on three major programmes: basic research in information technology, material research and solid-state physics, and environmental research. Further important fields are biotechnology, plasma and fusion physics, nuclear medicine, and energy technology beyond the original main area of high-temperature reactor development. Last but not least, computer science, mathematics, and electronics are well established fields of R&D at KFA.

KFA currently employs about 4500 people including about 900 scientists. The total annual budget amounts to 500 million Deutsche Mark (which roughly corresponds to almost $300 million) and is covered by the German Federal Government (90 per cent) and the State of North Rhine-Westphalia (10 per cent).

Today, KFA Jülich takes the function and character of a National Research Laboratory with highly interdisciplinary research and manifold interactions and cooperations with universities and industry.

Special emphasis and high priority is dedicated to fundamental research in information technology and solid state physics to investigate and develop new materials and thin-layer systems for information storage, metallizing structures as well as amorphous semiconductors, and new superconductors. This involves major efforts to promote theory and techniques in solid state physics. Hence, computational science has evolved to support and complement theory and experiment as a third strategy and paradigm of scientific investigation based on supercomputing and simulation techniques for complex problems, and has received high-priority as well at KFA for many years. In addition, in 1987 the Supercomputer Center for Science and Research (in German named Hoechstleistungsrechenzentrum: HLRZ) was established at KFA in an initiative similar to NSF's in the US. HLRZ is carried by KFA, GMD (Research Center for Mathematics and Data Processing) and DESY (German Electron Synchrotron Foundation); besides doing supercomputer-oriented research in many-particle and elementary-particle physics at KFA and supercomputer informatics including test-bed activities in new computer architectures at GMD, its mission is to provide, on the highest scale, supercomputer power for large research projects - free of charge - to the German science community. Presently, about 100 approved research projects in computational science are granted by HLRZ on the dedicated supercomputer at KFA, spreading these invaluable resources over universities and research institutions throughout the Federal Republic.

KFA's Central Institute for Applied Mathematics (ZAM) does research and development in computer science and mathematics with a prime focus on parallel algorithms and parallelization tools and techniques and other innovative topics, and runs KFA's computer center including the services for HLRZ. It is also responsible for the development and operation of KFA's computer networks. At present, the two mainframes IBM 3081/K-64 and IBM 3090/E-128 are feeding the two supercomputers CRAY X-MP/416 and CRAY Y-MP/832 (which was installed end of May 1989, and replaced the "old" CRAY X-MP/22 of 1983).

Although multi-CPU systems, the CRAY X-MP supercomputers have been primarily exploited in the single-processor mode by applying vectorization and optimization techniques to numerical kernels and large applications and aiming at the maximum throughput on the multiple CPUs rather than in the parallel processing mode provided by multitasking on the CRAY multi-vectorcomputers. From a strategic point of view on future supercomputing it is considered necessary to explore and investigate the multitasking techniques and their strengths and weaknesses in order to further speedup supercomputing for the effective treatment of large problems.

Multitasking: Parallel Processing on CRAY Multi-Vectorcomputers

In the future, performance enhancements of computer systems will be achieved essentially by parallel architectures. Today, the field of high-speed computing is dominated by vector-supercomputers [11] providing a moderate number of relatively powerful processors which lend themselves to MIMD algorithms [10]. The trends in vectorcomputer architecture will certainly go beyond the four-processor structure of the CRAY X-MP/4x and CRAY-2 and the eight-processor structure of the CRAY Y-MP systems. Thus, vectorization and parallelization will remain issues leading to even more extensive investigations in the design and implementation of parallel algorithms as well as in the development of adequate tools for program parallelization, in addition to vectorization.

The CRAY X-MP and CRAY Y-MP systems are multiprocessor computers with a shared memory. In these vector-supercomputers, parallelism on the programming language level is handled by two modes of multitasking. Macrotasking supports parallelism on the subroutine level. Task creation, synchronization, and communication are specified explicitly by the programmer using subroutine calls. Macrotasking exploits the intrinsic parallelism of a problem by partitioning the computations into N tasks which are simultaneously executed on N processors. Inefficiencies arise when the tasks are not well balanced with respect to granularity and communication [3].
The second strategy which has been implemented recently is called microtasking and works on the statement level. It makes use of compiler directives inserted by the programmer. These directives are given to a preprocessor which generates subroutine calls for the creation of parallel tasks and their synchronization. In contrast with macrotasking, the degree of parallelism in the microtasking mode is bounded only by the maximum number of processors available at run time. This results in an efficient utilization of the computer resources, but accounts less for the intrinsic parallelism of the computational problem. With the microtasking concept, the system overhead is considerably smaller than for macrotasking, although the difficulties of microtasking, as well as macrotasking programming may be considerable. A more detailed description of the multitasking concepts can be found in [12, 21].

In order to explore the respective potential of the macrotasking and microtasking strategies on CRAY multiprocessors, MIMD-type parallel algorithms for linear algebra problems [23] as well as graph problems [6, 15] have been implemented and extensively studied in these multitasking modes with respect to problem size, speedup, and overhead. In addition, a large application program is adapted to macrotasking and microtasking on a four-processor CRAY X-MP supercomputer. These case studies have been useful not only to evaluate the range of performance enhancements which can be achieved by multitasking techniques, but also to explore the limitations of the present realizations on CRAY X-MP systems.

The CRAY X-MP and CRAY Y-MP Multiprocessor Systems

CRAY X-MP machines are vector computers with specialized pipelined functional units which can be utilized in parallel to perform high-speed (floating point) computations; in addition, the X-MP series offers the option of multiple processors with a shared main memory. The supercomputer CRAY X-MP/416 consists of four CPUs (8.5 ns cycle time) and a main memory of 16 megawords. For the main memory bipolar ECL technology is used providing a bank busy time of 34 ns.

The CRAY Y-MP is the successor system to the CRAY X-MP. As far as the logical structure is concerned both CPU types are identical. The major differences between the CRAY X-MP and the CRAY Y-MP series are the faster clock period of 6.0 ns, the enlarged memory (32 MW), and the number of CPUs installed (eight instead of four). The detailed system characteristics of both CRAY multiprocessors installed at KFA are presented in Table 1.

	CRAY X-MP/416	CRAY Y-MP/832
number of CPUs	4	8
clock period	8.5 ns	6 ns
number of functional units	13	13
main memory	16 MW (organized in 64 banks)	32 MW (organized in 256 banks)
memory access time	34 ns	30 ns
SSD	32 MW	128 MW

Table 1. System characteristics of CRAY X-MP/416 and CRAY Y-MP/832

The multiple CPUs of one CRAY X-MP or CRAY Y-MP system are tightly coupled via the shared main memory and five identical groups of registers, called clusters, which may be used in common by all processors. Each cluster contains

- 8 shared address registers (SB),
- 8 shared scalar registers (ST),
- 32 binary semaphore registers (SM).

The attachment of clusters to different CPUs is done by the CRAY operating systems COS or UNICOS [13]. Depending on program requirements either one cluster or multiple clusters, or eventually no cluster will be attached to a CPU.

Today most of these CRAY multiprocessor systems are still used within a multiprogramming environment, where the individual processors execute different jobs which are totally independent of each other. With multitasking, they may also execute different parts of one program in parallel.
In shared-memory systems memory-access conflicts can occur. These section and bank conflicts are widely known; they have been studied for multiprogramming environments already [8, 9, 22]. Using multitasking, these conflicts become even more important because of the similar structure of memory-access within parallel tasks.

The Multitasking Concepts

Macrotasking is the kind of multitasking where parallelism is realized at subprogram level. Within a program subprograms may be executed as different tasks by means of macrotasking. In order to use macrotasking, the application has to be partitioned into a fixed number of tasks; the user has to create the tasks explicitly by TSKSTART calls [12,14]. Task management is done by the library scheduler. The library scheduler gets information about synchronization and communication requirements via the subprogram calls providing event, barrier, and lock primitives. It controls special queues and initiates the necessary activities. The job scheduler attaches the tasks to physically available CPUs [14, 18]. These tasks can be executed in parallel, if there are no prohibitive mutual dependencies concerning correct synchronization and sequencing.

For certain kinds of programs, this concept leads to an efficient use of the multiprocessor machines. But often the user has to deal with three types of problems which may reduce the speedup achievable by the macrotasking strategy.

- For small granularity, the system overhead is too large due to the high synchronization frequency.
- Several tasks cannot be executed in a balanced way on the multiple CPUs.
- The number of processors does not match the fixed number of tasks specified within the macrotasking program.

Several manufacturers of multiple-CPU systems like CRAY and IBM are promoting developments in the field of parallel programming, in particular to reduce synchronization overhead within multitasking [21]. Microtasking is an approach that allows the efficient use of multiple processors even for small granularity [7, 19]. Instead of using a scheduler all communication and synchronization is done by accessing shared variables (registers) directly [18, 21]. Thus the speed of synchronization is increased which allows the use of multitasking features also for small granularity; in addition, load balance can be achieved more easly.

CRAY Microtasking provides parallelization on the statement level. Statements, sequences of statements, and complete subprograms may be executed in parallel on multiple processors of a CRAY X-MP or CRAY Y-MP machine when microtasking is used. Within CRAY microtasking, tasks communicate by means of shared registers which are organized in a cluster. The overhead of the communication via registers is much smaller than the overhead to access corresponding variables in the main memory as realized within the macrotasking library routines.

The microtasking features are provided by preprocessor directives [18]. These directives are coded into the user program; this leads to multitasking programs which are easier to understand than in the macrotasking way. The usage of directives does not affect the portability of programs. Because the directives must be coded with a "C" in the first column, all FORTRAN compilers will interpret such statements as comment statements. The microtasking strategy provides several directives to define microtasking control structures and critical regions. The preprocessor PREMULT [18, 12] translates these directives into additional code and corresponding library calls to assure correct execution of microtasking programs.

The main goal of the microtasking concept is to provide a very efficient method to parallelize programs on CRAY multiprocessors (low overhead, efficient resource usage). Currently there are no standard techniques to verify the correct introduction of microtasking parallelism into programs. Besides specifying the parallel portion of the code, the user must also specify the sequential part of the program explicitly. Otherwise the work is done redundantly by each of the available processors. No automatic analysis support is given for this concept. This problem is partly solved by the CRAY *autotasking* approach [1] available just now under UNICOS.

Experiences with Some Programs Using Multitasking

Using multitasking to take advantage of parallelism within a program gives a chance for appreciable speedups. This is especially true for linear algebra kernels which are heavily used in large application packages. Moreover, multitasking strategies and corresponding problems can be studied in detail with such small programs.

Linear Algebra Problems

Because of the simple structure of matrix multiplication this algorithm is often used to present the different multitasking strategies and to document the potential benefit for such easy-to-use algorithms. There exist several algorithms and implementations for matrix multiplication; the subroutine MXV from SCILIB (CRAY subroutine library) can be used, for example, to perform the matrix-vector operations (see Fig. 1). The usage of microtasking is introduced by marking the DO loop with a DO GLOBAL directive.

```
CMIC$ MICRO
      SUBROUTINE PMXV(A,B,C,N)
      REAL A(N,N),B(N,N),C(N,N)
      INTEGER I
CMIC$ DO GLOBAL
      DO 200 I=1,N
        CALL MXV(B,N,C(1,I),N,A(1,I))
200   CONTINUE
      END
```

Fig. 1. Matrix multiplication using MXV subroutine (SCILIB)

This program is often used in the literature as an example to document the effectiveness of microtasking; there, microtasking is used in a natural and easy way, resulting in a satisfying speedup. Within user programs three nested DO loops (as shown in Fig. 2) are also used as an algorithm for the matrix multiplication. This implementation provides a portable code running on most of the available machines, and microtasking can be specified in an easy way by marking the outer loop with a DO GLOBAL directive.

```
CMIC$ MICRO
      SUBROUTINE NEST3(A,B,C,N)
      INTEGER N,I,J,K
      REAL A(N,N),B(N,N),C(N,N)
CMIC$ DO GLOBAL
      DO 200 J=1,N
        DO 100 K=1,N
          DO 100 I=1,N
            A(I,J) = A(I,J) + B(I,K) * C(K,J)
100     CONTINUE
200   CONTINUE
      END
```

Fig. 2. Matrix multiplication using three nested DO loops

J.J. Dongarra has introduced the unrolling technique [5] for single-task vectorizing programs, in particular to remove memory conflicts. Using 8-way unrolling (see Fig. 3) the memory-section and bank conflicts [22] are significantly reduced.

```
CMIC$ MICRO
      SUBROUTINE UNROLL(A,B,C,N)
      INTEGER N,I,J,K
      REAL A(N,N),B(N,N),C(N,N)
CMIC$ DO GLOBAL
      DO 200 J=1,N
        DO 100 K=1,N,8
          DO 100 I=1,N
            A(I,J) = A(I,J) +
    >           B(I, K ) * C( K ,J) + B(I,K+1) * C(K+1,J) +
    >           B(I,K+2) * C(K+2,J) + B(I,K+3) * C(K+3,J) +
    >           B(I,K+4) * C(K+4,J) + B(I,K+5) * C(K+5,J) +
    >           B(I,K+6) * C(K+6,J) + B(I,K+7) * C(K+7,J) +
100      CONTINUE
200      CONTINUE
      END
```

Fig. 3. Matrix multiplication using DO loops (8-way unrolled)

Figure 4 shows the speedup[1] obtained with microtasking for the library routine MXV as well as for both versions of the DO loop algorithm. Moreover, it documents the benefit using the unrolling technique in combination with microtasking as shown in Fig 3. A detailed discussion of the problem of memory contention which will become more important in conjunction with multitasking is given in [12, 22]. Comparing the timing results for the different matrix multiplication implementations when only one CPU is used the conclusion must be drawn that the SCILIB routines should be used wherever possible.

In addition to matrix multiplication some linear algebra algorithms heavily used in application programs within scientific and technical computing environments are studied. The algorithm for LU decomposition discussed here is based on an implementation published in [16]. Furthermore, the macrotasking implementation (described in more detail in [14]) allows the use of a variable number of CPUs by changing only one FORTRAN statement as usual with microtasking. Figure 5

---

[1] Time measurements were done by calls to IRTC on a dedicated CRAY Y-MP/832 with 6 nsec cycle time and 256 memory banks. The operating system level was COS 1.17 BF 1. The speedup is calculated as the ratio of corresponding times. For kernel measurements always the minimum of three executions is taken as the time result to remove the additional work for the first TSKSTART calls.

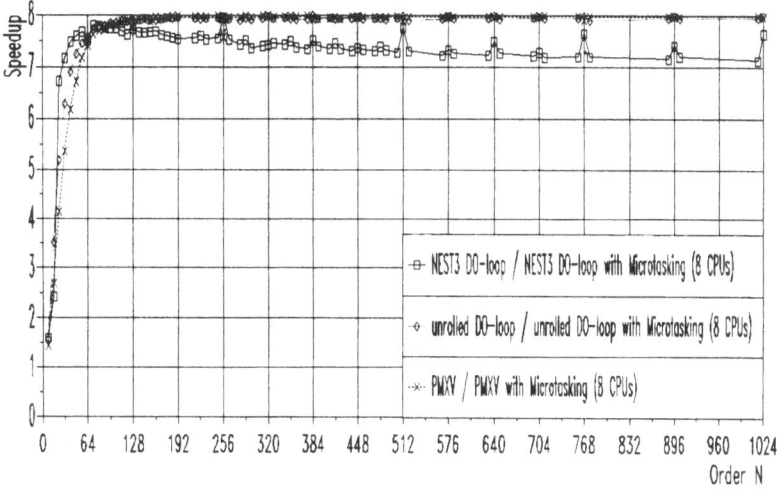

Fig. 4. Matrix multiplication using MXV and DO-loops with microtasking on CRAY Y-MP

shows the speedup results on a CRAY X/MP[2] for the solution of a linear equation system using LU decomposition on four processors with both macrotasking and microtasking. In addition, it compares the sequential version used here with the IMSL routines currently available [8]; the measurements illustrate that our implementation is superior to the library routines. There is a large performance improvement by microtasking in comparison to macrotasking especially for small matrix dimension N due to the smaller overhead of microtasking.

Nonnumerical Problems

To document the benefit of multitasking also for nonnumerical problems some results in the area of graph problems are reported. A more detailed discussion can be found in [6, 7, 15]. Shortest-path problems are the most fundamental and also the most commonly encountered problems in the study of transportation and communication networks. Parallel algorithms have been developed for two types of shortest-path problems:

---

2   Time measurements were done by calls to IRTC on a dedicated CRAY X-MP/416 with 8.5 nsec cycle time and 64 memory banks. The operating system level was COS 1.16 BF 2.

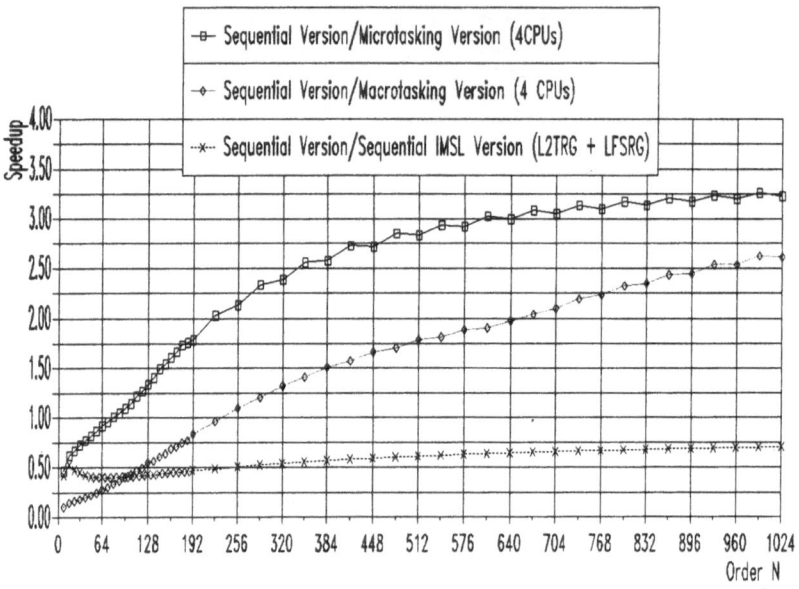

Fig 5. **Solution of a linear equation system using LU decomposition with macrotasking and microtasking on CRAY X-MP**

- Finding the length of the shortest path from a single node to all other nodes in a graph (*single-source shortest-path problem*).
- Finding the length of the shortest path between every pair of nodes in a graph (*all-pairs shortest-path problem*).

For the single-source shortest-path problem, a parallel version of Dijkstra's algorithm has been implemented; results for this algorithm can be found in [12]. One of the best known algorithms for determining the shortest path between all pairs of nodes is due to Floyd, which is based on an earlier algorithm by Warshall. The time complexity of this algorithm is $O(n^3)$ - rather than $O(n^2)$ for Dijkstra's algorithm.

The Warshall-Floyd algorithm works by inserting one or more nodes into paths, whenever it is advantageous to do so. Starting with the $n \times n$ weight matrix W of direct distances between the nodes of the given graph, we construct a sequence of n matrices $W^{(1)}$, $W^{(2)}$, ..., $W^{(n)}$. Matrix $W^{(\ell)}$, $1 \leq \ell \leq n$, may be thought of as the matrix whose entry $w_{ij}^{(\ell)}$ gives the length of the shortest path among all paths from i to j with nodes 1, 2, ..., or ℓ allowed as intermediate nodes. The matrix $W^{(\ell)}$ is constructed as follows:

$$w_{ij}^{(0)} = w_{ij}$$

$$w_{ij}^{(\ell)} = \min \left\{ w_{ij}^{(\ell-1)}, \; w_{i\ell}^{(\ell-1)} + w_{\ell j}^{(\ell-1)} \right\} \text{ for } \ell = 1, 2, ..., n.$$

When implementing Warshall-Floyd's algorithm with macrotasking and microtasking the speedup which can be achieved is quite satisfactory in both modes. Due to the balanced workload which is achieved for all problem sizes and due to the increasing task granularity the influence of the synchronization overhead using macrotasking is negligible.

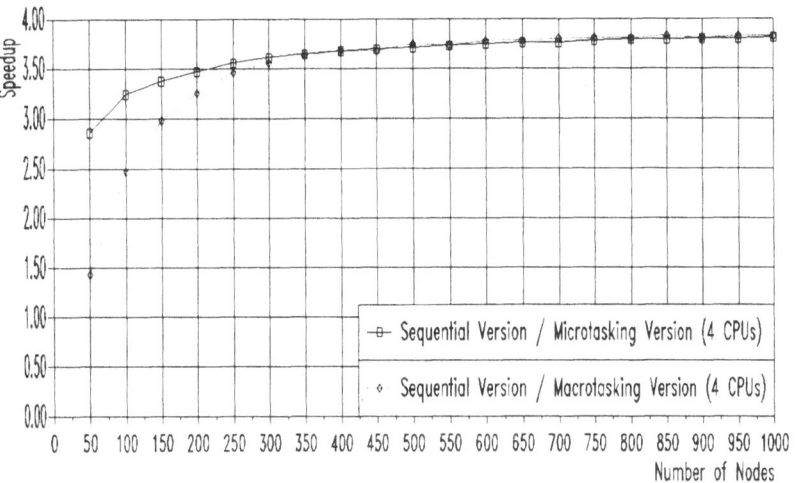

Fig 6. Warshall-Floyd algorithm using macrotasking and microtasking

Experiments with the Czochralski Bulk Flow

For more than ten years numerical simulation of Czochralski bulk flow has been used to get quantitative information about this flow. The program presented here gives a numerical solution of time-dependent Navier-Stokes and convective heat equations governing the forced and free convective flow in a crystal-growth process [17, 20]. It provides information about the qualitative and quantitative difference between the influence of a stationary transverse magnetic field and a vertical magnetic field on the flow and temperature distribution in the (silicon) melt by means of numerical simulations in a three-dimensional mathematical model of Czochralski crystal growth. The geometrical configuration of our model is an idealization of a real Czochralski crystal growth system, and in the model the free surface of the melt and the solid-liquid interface are assumed to be flat at all times. The equations are expressed in finite difference form on the grid according to the well known Marker-and-Cell method [17].

The simulation program is used for parameter studies to get detailed information about Czochralski bulk flow. The CPU time used to solve one of these problems varies from 30 to 100 CPU hours on one processor of a CRAY X-MP, depending on material and geometrical parameters and the underlying time interval. The sequential simulation program for Czochralski bulk flow has a simple structure and is a highly optimized vectorizing program which attains a sustained speed of about 80 MFLOPS.

With macrotasking, for each of the multitasked subroutines four independent tasks are generated with the TSKSTART primitive. These tasks are synchronized with a SYNCH routine. This routine guarantees that all tasks have reached a particular program location, and provides a user interface similar to the new barrier synchronization primitive introduced with the COS 1.17 operating system. The DO loops are partitioned into as many parts as CPUs are available, and each task takes its own work based on a special identifier associated with this task (see Fig 7).

```
      SUBROUTINE VELO(ID)
C     declarations
      COMMON/NOTASKS/NTSKS
      COMMON/EVENTS /EVENT1
C     partition DO loop, starting at 2, ending at LM1
      ISTART=2
      IEND=LM1
      ITER=(IEND-ISTART+1)/NTSKS
      ILAST=(IEND-ISTART+1)-ITER*NTSKS
      IANF=ISTART+(ID-1)*ITER+MIN(ID-1,ILAST)
      IEND=ISTART+(ID  )*ITER+MIN(ID,ILAST)-1
C     . . .
      DO 111 I=IANF,IEND
        DO 111 K=2,NM1
          DO 111 J=2,M
C           . . .
111   CONTINUE
C     . . .
      CALL SYNCH(ID,EVENT1,NTSKS)
C     . . .
      END
```

Fig 7. Macrotasking version of subroutine VELO running with identifier ID

The microtasking parallelism is introduced by several DO GLOBAL directives which specify independent DO loop iterations. Because of the microtasking concept which guarantees explicit synchronization of all active tasks at the bottom of a microtasking control structure (MCS) no additional synchronization is needed [12, 21].

The time[3] is measured to check the results for a time interval of the simulation process which represents a typical situation where the flow is stabilized. This time interval covers a few seconds of the real crystal-growth experiment and corresponds to one CPU hour of one processor of a CRAY X-MP. The results of this time measurement for the macrotasking and the microtasking version are presented in Table 2.

	macrotasking	microtasking
sequential reference (one program version running in dedicated mode)	2747	
4 CPUs — wall clock time	864	822
4 CPUs — Speedup obtained	**3.18**	**3.34**
single-task reference (four identical program versions running on one CPU each)	2941	
4 CPUs — wall clock time	864	822
4 CPUs — Speedup obtained	**3.40**	**3.58**

Table 2.   Speedup results on CRAY X-MP/416 with 64 memory banks (COS 1.16 Bf 3)

The sequential (i.e. single-task) reference represents the wall clock time, i.e. the time the user has to wait for the result on a dedicated machine, needed by the best single task program version simulating 3.4 seconds of the real process. Running the parallel program with only one CPU the overhead introduced by the macrotasking library calls is about 4%, with microtasking the overhead is reduced to 1%. The next rows of Table 2 provide information about the multiple processor version using four CPUs, and the speedup gained with both concepts in comparison to the single-tasking reference version. Because of the partitioning of the loop iterations into four parts the resulting load balance may be a bit unsatisfying, and Amdahl's law limits the speedup to 3.67 using four CPUs. In the experimental results presented here the dominant factor which limits the speedup achieved in this application is memory contention introducing an additional overhead. Using four CPUs, only 91 per cent of the theoretical speedup is obtained by microtasking.

---

[3]   Time measurements were done by calls to IRTC on a dedicated CRAY X-MP/416 with 8.5 nsec cycle time and 64 memory banks. The operating system level was COS 1.16 BF 2.

On the other hand, the sequential reference version is normally executed in a multi-programming batch environment, and the CPU-time used to execute this problem will also increase. For that reason the wall clock time needed by each of four copies of the single-task program version run, each on one CPU, simultaneously on the four-processor system was measured. It is seen from Table 2 that, because of memory conflicts, the average wall clock time for the sequential version is increased by about 200 seconds. Comparing these timing figures with Amdahl's law leads to speedup values of 92% and 97% of the theoretical speedup for the two respective multitasking modes.

Perspective

With the delivery of the new supercomputer CRAY Y-MP/832 running the UNIX-based operating system UNICOS, in addition to macrotasking and micro-tasking, autotasking is becoming available, which basically provides efficient means to exploit fine-grain parallelism in the microtasking sense. Therefore, one can expect that parallel processing on CRAY multi-CPU systems (as well as other manufacturers' multiprocessors like IBM and recently also Fujitsu and NEC - unfortunately no longer CDC/ETA) will penetrate the supercomputing field. On the CRAY Y-MP, new insight will be gained also with respect to more "massive" parallelism by multitasking eight rather than four CPUs.

From an algorithmic point of view, one might expect that research and design of (totally and partially) asynchronous algorithms (see [14, 2]) will be challenged by multitasking also with respect to coarse-grain parallelism, which still suffers from high synchronization overhead in large-scale multiprocessor vector-supercomputers.

Acknowledgement

The author would like to thank W.E. Nagel and R. Knecht for their contributions and K. Wingerath for his continuous support with the Czochralski bulk flow simulation program.

References

1. Autotasking user's guide. CRAY-Research Inc. SN-2088 (1988).
2. Bertsekas, D.P., Tsitsiklis, J.N.: Parallel and distributed computations - numerical methods. Englewood-Cliffs: Prentice-Hall, 425-569, 1989.

3. Calahan, D.A.: Task granularity studies on a many-processor CRAY X-MP. Parallel Computing 2, 109-118 (1985)

4. Dongarra, J.J., Chen, S.S., Hsiung, C.C.: Multiprocessing linear algebra algorithms on the CRAY X-MP-2: experiences with small granularity. J. Parallel and Distributed Computing 1, 22-31 (1984)

5. Dongarra, J.J., Eisenstat, S.C.: Squeezing the most out of an algorithm in CRAY FORTRAN. ACM Transactions on Mathematical Software 10, 219-230 (1984)

6. Gurke, R.: Graphenalgorithmen für MIMD-Rechner. In: Universität Clausthal: Workshop über Parallelverarbeitung: Informatik-Bericht 88/3. 47-62 (1988)

7. Gurke, R.: The approximate solution of the euclidean traveling salesman problem on a CRAY X-MP. Parallel Computing 8, 177-183 (1988)

8. Hake, J.-Fr., Homberg, W.: Linear algebra software on a vector computer. Parallel Computing 10, 65-81 (1989)

9. Hake, J.-Fr., Homberg, W.: The impact of memory organization on the performance of matrix calculations. To appear.

10. Hockney, R.W.: MIMD computing in the USA - 1984. Parallel Computing 2, 119-136 (1985)

11. Hossfeld, F.: Vector-supercomputers. Parallel Computing 7, 373-385 (1988)

12. Hossfeld, F., Knecht, R., Nagel, W.E.: Multitasking: experiences with applications on a CRAY X-MP. To appear in Parallel Computing.

13. Hwang, K., Briggs, F.A.: Computer architecture and parallel processing. New York: McGraw-Hill 1984

14. Knecht, S.: Möglichkeiten des Multitasking zur Beschleunigung von Standardalgorithmen. Kernforschungsanlage Jülich: Jül-Spez-361 (1986)

15. Knecht, R.: Implementation of divide-and-conquer algorithms on multiprocessors. To appear in the proceedings of the WOPPLOT workshop.

16. Lord, R.E., Kowalik, J.S., Kumar, S.P.: Solving linear algebraic equations on a MIMD computer. Journal of the ACM 30, 103-117 (1983)

17. Mihelcic, M., Pirron, Chr., Wingerath, K.: Three-dimensional simulations of the Czochralski bulk flow. Journal of Crystal Growth 69, 473-488 (1984)

18. Multitasking User Guide. Revision D. CRAY-Research Inc. SN-0222 (1987)

19. Nagel, W.E.: Using multiple CPUs for problem solving - experiences in multitasking on CRAY X-MP/48. Parallel Computing 8, 223-230 (1988)

20. Nagel, W.E., Wingerath, K.: Three-dimensional numerical simulations of the Czochralski bulk flow on a CRAY X-MP multiprocessor architecture. In: Proc. 1988 International Conference on Supercomputing, 266-272 (1988)

21. Nagel, W.E., Szelényi, F.: Multitasking on supercomputers: concepts and experiences. IBM Tech. Rep. ICE-VS05. IBM ECSEC 1989.

22. Oed, W., Lange, O.: On the effective bandwidth of interleaved memories in vector processor systems. IEEE Trans. Computers C-34, 949-957 (1985)

23. Ortega, J.M.: Introduction to parallel and vector solution of linear systems. New York: Plenum Press 1988. Also: the ijk forms of factorization methods i. Vector Computers. Parallel Computing 7, 135-147 (1988)

Supercomputing:
Experience With Operational Parallel Processing

Geerd-R. Hoffmann
Head, Computer Division
European Centre for Medium-Range Weather Forecasts
Shinfield Park
Reading, Berks. RG2 9AX
England

Abstract

Operational numerical weather prediction requires that all the
computing power available is harnessed to the one task, otherwise the weather will have changed before it has been predicted. Therefore, in 1983 after the advent of multiprocessor
systems as the most powerful computers, the European Centre for
Medium-Range Weather Forecasts (ECMWF) undertook to modify its
main algorithms and to run its forecast programs in multitasking mode. The decisions for achieving this goal, taken with
regard to program structure and synchronization methods, will
be discussed. The constraints imposed by the operational nature
of the work will be highlighted. A set of requirements will be
developed which will satisfy the needs of ECMWF and similar
centres.

NATO ASI Series, Vol. F 62
Supercomputing
Edited by J. S. Kowalik
© Springer-Verlag Berlin Heidelberg 1990

1. Introduction

Since Richardson in 1922 showed (see Richardson (1922)) that
the weather can be predicted by numerically solving non-linear
equations, computers have been assigned to the task of modelling the atmosphere. As the number of computations needed to
perform a weather forecast is very high - Bengtsson estimates
between 10^{12} to 10^{13} for a forecast of around 10 days (see
Bengtsson (1988)) - the speed of the computers must be
extremely high, otherwise the weather will have happened before
it was forecast. It is therefore no surprise that the main
weather services all over the world are using "supercomputers"
for their work and that parallel processing features were used
by them as soon as they became viable options.

2. The European Centre for Medium-Range Weather Forecasts (ECMWF)

The European Centre for Medium-Range Weather Forecasts (ECMWF)
was established in 1975 as an international organisation, at
present funded by 18 West-European countries. Its headquarters
are situated in Reading, England, and its main functions are
(see Convention 1974):

a) to develop dynamic models of the atmosphere with a view to
   preparing medium-range weather forecasts by means of numerical methods;

b) to prepare, on a regular basis, the data necessary for the
   preparation of medium-range weather forecasts and to make
   available to the meteorological offices of the Member States
   these data;

c) to carry out scientific and technical research directed
   towards improving the quality of these forecasts;

d) to collect and store appropriate meteorological data;

e) to make available to the meteorological offices of the
Member States for their research, priority being given to
the field of numerical weather forecasting, 25% of its computing capacity.

3. Computer Configuration

To carry out its tasks, ECMWF installed as its main computer a
Cray Research, Inc. CRAY-1A in 1978, followed by a CRAY X-MP/22
in 1983 and a CRAY X-MP/48 in 1985. It is planned to replace
the CRAY X-MP/48 by a CRAY Y-MP/864 in 1990.

The CRAY machines are complemented by a number of front-end
services, in particular two Control Data Corporation CYBER 855
for local interactive use, an IBM 3090-150E with Vector Facility as a data handling system for all permanent data, and a
cluster of Digital Equipment Corporation VAX 750, VAX 8350,
and VAX 6210 for supporting the wide area network connecting
the meteorological services of all the Member States of ECMWF.
For internal use, ECMWF has implemented a local area network
based on Ethernet and IBM Token Ring under the supervision of
NOVELL Netware software which supports ECMWF's about 100 IBM PC
based workstations. The configuration and its networks are
depicted in fig. 1. The more detailed description of ECMWF's
functionally distributed computing facilities can be found in
Hoffmann (1983).

Since taking delivery of its first multiprocessing system, i.e.
the CRAY X-MP/22 in 1983, ECMWF has been making use of parallel
features for its operational suite of weather forecasting programs (see Dent (1988)). So far these efforts have resulted in
a speed-up of 3.7 when comparing the execution time of a
weather model on a single processor to that on a four processor
CRAY X-MP.

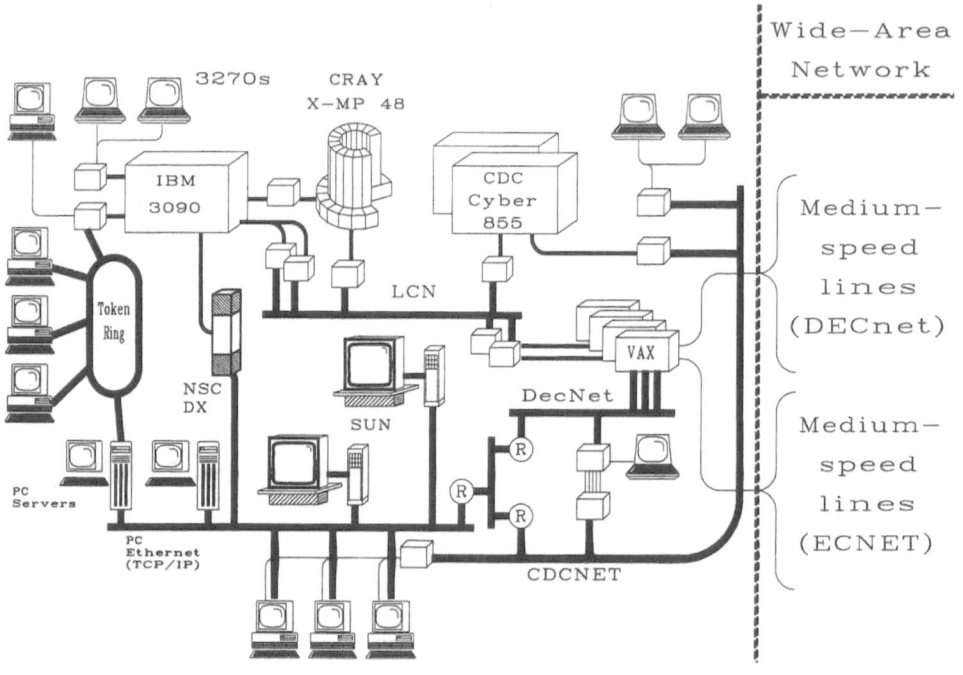

Fig. 1: Computer Configuration of ECMWF

4. Tools for Parallel Processing

When ECMWF started to use the parallel processing features of
the CRAY COS operating system as incorporated by subroutine
calls into the CRAY FORTRAN compiler CFT, it concentrated on
the use of "TSKSTART" as its main synchronization tool (see
Hoffmann (1985)). However, early experience showed that this
use is rather expensive in terms of the elapsed time needed to
perform the necessary functions and that "events" are faster.
Therefore, event handling was introduced, details can be found
in Dent (1988).

In the absence of any other tools provided by Cray Research,
Inc. to ease the use of the parallel features of their multiprocessing systems, ECMWF undertook to write its own set of

subroutines to monitor the use of the different constructs.
These specialized tools allow the print-out of a history file
showing when and from where events are sent or received and the
time spent by each task in its different states. Using these
measurements, ECMWF managed to get its operational programs to
run in parallel mode and to optimize their performance.

5. Operational Aspects of Parallel Processing

The operational use of the parallel features of the CRAY X-MP
systems pose a number of problems which are discussed below.

5.1. Debugging

Since ANSI FORTRAN does not yet support any parallel constructs, no automatic checking of violations in the use of
critical sections is carried out. This may result in
spurious errors when running the programs in the different
environments created by the changing workload of the system.
For example, a memory bank conflict may change the order of
execution of parallel tasks and thus lead to different
results. In the absence of any specialized tools, debugging
such programs is extremely time consuming and is made more
difficult by the difference in rounding errors experienced
when running tasks in a different order.

To overcome this particular problem when debugging, ECMWF
has introduced event handling to guarantee the same order of
execution for the parallel routines. The enforcing of a certain order of tasks entails a penalty in terms of the additional overheads introduced by the event handling routines.
However, this loss of efficiency is more than balanced by
the help provided for the debugging phase of program development.

5.2. Scheduling

In the operating system COS there is no information available about whether a job waiting for initialization is going
to require multiple CPU's to execute efficiently. Therefore,
no automatic scheduling algorithms are available to optimize
the load of the system.

ECMWF has to rely on manual scheduling techniques and the
goodwill of its users to initialize programs using more than
one CPU. During the main operational forecast run overriding
priorities are used to ensure that all resources of the system are made available to the program suite. However, during
the remaining time this solution is not feasible, and,
therefore, wildly varying elapsed times for jobs using parallel features can be seen.

5.3. Efficiency

The efficiency of the use of the CRAY system depends to a
large extent on the scheduling information available. As
discussed above, the use of parallel features can only be
detected once they are invoked by the program, at which time
efficient use of the resources may no longer be possible.
Therefore, there are marked differences in efficiency to be
seen on a normal day, varying between the time the highly
optimized forecast suite runs and the prime shift when users
may submit all sort of jobs indiscriminately. Fig. 2 shows
the CPU utilization of the CRAY X-MP/48 for 24 hours,
depicting this distinction.

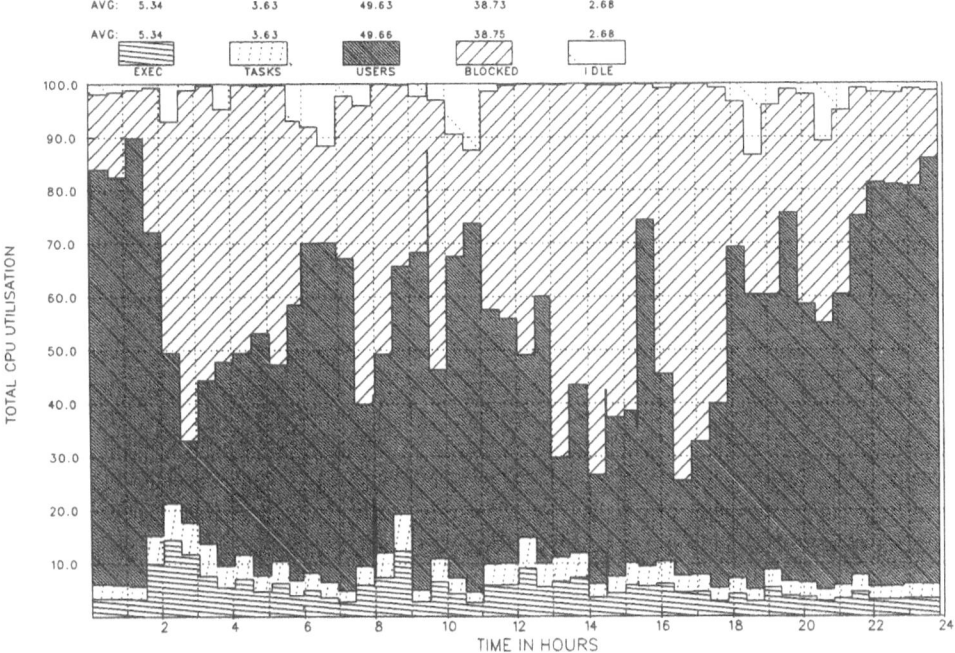

Fig. 2: CPU utilization of CRAY X-MP/48 on 25 September 1989

5.4. Portability

Another problem is introduced by the absence of international or de-facto standards for parallel constructs in FOR-
TRAN and the ensuing uniqueness of CRAY's implementation ,
especially for the international user base of ECMWF.
Programs written and debugged on a CRAY system cannot be
ported easily to other systems. The problem of a portable
environment for programming parallel tasks is discussed in
more detail in Snelling, Hoffmann (1988).

At present, the problem is being alleviated by concentrating all parallel subroutine calls into a small number of programs, thus localizing any necessary changes when porting the code to another system. However, agreed standards have to be implemented soon, otherwise the sharing of programs, which is quite common in the meteorological community, will become very difficult when different multiprocessor systems are installed at the various services.

6. Future Development

From discussions with vendors of "supercomputers" and from the speed of technical development for electronic components it can be deduced that the operational use of massively parallel computers is drawing closer. The problems encountered at present when using only up to 8 processors will become unmanageable when more than 100 processors are involved. Therefore, a concerted effort by both vendors and users is required to standardize parallel FORTRAN constructs and agree on a sensible set of tools for debugging programs using these features.

The recent interest from scientists and industry in the software aspects of using massively parallel architectures is encouraging, and it is hoped that results will be available in time to meet the growing demands of centres like ECMWF which have to rely on parallelism to achieve the performance they require to meet their objectives.

7. Acknowledgements

The author would like to thank his colleagues at ECMWF for their support in preparing this paper and the participants of the NATO Advanced Research Workshop held in Trondheim, Norway, on 19-23 June 1989 for their constructive comments.

8. Literature

L. Bengtsson: Computer Requirements for Atmospheric Modelling. In: Multiprocessing in Meteorological Models. G.-R. Hoffmann, D.F. Snelling (eds.). Berlin Heidelberg New York 1988. pp. 108-116.

Convention establishing the European Centre for Medium-Range Weather Forecasts. Her Majesty's Stationery Office Cmnd. 5632. London June 1974.

D. Dent: The ECMWF Model: Past, Present and Future. In: Multiprocessing in Meteorological Models. G.-R. Hoffmann, D.F. Snelling (eds.). Berlin Heidelberg New York 1988. pp. 369-381.

G.-R. Hoffmann: Die Entwicklung der Rechnerkonfiguration im EZMW. In: Betrieb von DV-Systemen in der Zukunft. 5. GI-Fachgespräch über Rechenzentren. Tübingen, März 1983. M.A. Graef (Hrsg.). Berlin Heidelberg 1983. pp. 176-186.

G.-R. Hoffmann: Supercomputer als Mehrprozessoranlage - Erfahrungen und Ergebnisse. in: Organisation und Betrieb der Informationsverarbeitung. 6. GI-Fachgespräch über Rechenzentren. Kassel, März 1985. W. Dirlewanger (Hrsg.). Berlin Heidelberg New York Tokyo 1985. pp. 245-259.

L.F. Richardson: Weather Prediction By Numerical Process. Cambridge University Press. London 1922.

D.F. Snelling, G.-R. Hoffmann: A comparative study of libraries for parallel processing. Parallel Computing 8(1988). pp. 255-266.

THE EVOLVING SUPERCOMPUTER ENVIRONMENT
AT THE SAN DIEGO SUPERCOMPUTER CENTER

Sidney Karin
San Diego Supercomputer Center
P. O. Box 85608
San Diego, CA 92138
U.S.A.

Establishing the Center

In the early 1980's recognition was growing in the United States that computational simulation using supercomputers was becoming an important aspect of the scientific research process. However, academic researchers in the U.S. had limited access to supercomputers. At that time more European universities had either purchased or had access to American supercomputers than did U.S. universities. A number of studies (the Bardon-Curtis Report[1], the Lax Report[2], the Press Report[3], the FCCSET report[4]) revealed that the United States must take steps to ensure improved supercomputing access for American researchers and scientists.

In late 1983, representatives from General Atomics (a private San Diego firm with extensive supercomputer experience) who were concerned about this almost total lack of access to state-of-the-art computational tools began working with similarly concerned representatives from the University of California, San Diego to draft and send an unsolicited proposal to the National Science Foundation (NSF) to initiate the establishment of national academic supercomputer centers. At about the same time researchers from the University of Illinois drafted a similar proposal to NSF. The NSF did not act directly on these unsolicited proposals, but in fact created a solicitation of their own which then generated an additional twenty proposals. We in San Diego then revised our proposal to match the details of the NSF solicitation.

In late February of 1985, NSF announced awards establishing the original three supercomputing centers: the San Diego Supercomputer Center located at the University of California, San Diego; another center located at the University of Illinois in Champaign; and a third located in Princeton, New Jersey. In addition, the NSF announced an award to Cornell University in Ithaca, New York to develop an advanced prototype computing facility. Eventually the Cornell facility, now known as the Cornell Theory Center, became the fourth supercomputing center. In 1986, a fifth center was added: the Pittsburgh Supercomputing Center (PSC) located in Pittsburgh, Pennsylvania.

NATO ASI Series, Vol. F 62
Supercomputing
Edited by J. S. Kowalik
© Springer-Verlag Berlin Heidelberg 1990

Before establishing these five centers, the National Science Foundation had been purchasing computer time for use by the American academic research community from several institutions including Boeing Computer Services, Purdue University, Colorado State University, AT&T Bell Labs, and the University of Minnesota. NSF referred to this purchase of computer time as Phase 1 of their advanced scientific computing program. The dedicated supercomputer centers described above were considered Phase 2 of the program, and as these centers came on-line, NSF eliminated the Phase 1 program.

We in San Diego received a formal announcement from the National Science Foundation on March 1, 1985 of an award of $58,000,000 over five years to develop and implement the San Diego Supercomputer Center. We broke ground at the site, a field full of weeds on the campus of the University of California at San Diego, in June. Construction began and we took possession of the first phase of the building, over 20,000 square feet composed of the machine room and related facilities, on November 1st -- on schedule. On November 11th., the CRAY X-MP/48 was delivered and the first calculations were made at our facility by Professor Herbert Hamber of the University of California, Irvine Physics Department on December 4, 1985 -- 284 days after receiving the award for the center.

After the building was constructed and the supercomputer delivered, we began providing supercomputer time to the national academic research community. This computing community is truly national in scope: 10% of the users are from the University of California, San Diego, but 90% of the users are off campus. Each year approximately a dozen institutions cumulatively use 1000 or more CPU hours each, and the largest users include the University of Hawaii and the University of Maryland -- 6,000 miles and six time zones apart.

To become a truly national system, we organized a national network to provide remote access to our facilities. Two years in advance of the establishment of the National Science Foundation network (NSFNET), the San Diego Supercomputer Center was the first of the five national centers to establish a network (SDSCnet), both satellite and terrestrial-based, linking remote users all over the country to the resources available at SDSC. The dedicated satellite dishes at Maryland, Michigan, Wisconsin, Seattle, Hawaii, Caltech, etc., allow several thousand widely distributed users at more than 150 institutions in 44 states to communicate with us.

The San Diego Supercomputer Center is administered and operated by General Atomics. Most of the staff, but not all, are employees of General Atomics; some are employees of the University of California. Funding for the Center comes from the National Science Foundation, the State of California, and almost 50 industrial participants. In addition, the University of California provides several million dollars a year and the University of California at San Diego, in particular, also provides substantial financial support. Policy guidance is given by a steering committee composed of representatives from each of the 25 consortium institutions. This steering committee meets quarterly to review progress at SDSC, set policies, and allocate supercomputer time to researchers and scientists. Table 1 lists the 25 SDSC consortium institutions:

TABLE 1
SDSC CONSORTIUM INSTITUTIONS

The Agouron Institute
California Institute of Technology
National Optical Astronomy Observatories
Research Institute of Scripps Clinic
Salk Institute for Biological Studies
San Diego State University
Scripps Institution of Oceanography
Southwest Fisheries Center
Stanford University
University of California at Berkeley
University of California at Davis
University of California at Irvine
University of California at Los Angeles
University of California at Riverside
University of California at San Diego
University of California at San Francisco
University of California at Santa Barbara
University of California at Santa Cruz
University of Hawaii
University of Maryland
University of Michigan
University of Southern California
University of Utah
University of Washington
University of Wisconsin

The Center serves thousands of researchers around the clock. Dozens of users can access SDSC's CRAY at the same time, working from individual workstations in their own offices or laboratories. In each case, the user has real-time interactive access to the supercomputer. Within SDSC itself, experienced staff members of the following five operational departments comprising the operating center help users take maximum advantage of the supercomputer's capabilities:

- Programming and Software Services obtains, installs, and maintains the systems software for all the operating systems on all the computers at the Center, from supercomputers to workstations and PCs.

- Engineering and Operations manages the 24-hour-a-day operations of the Center, supervises the maintenance and installation of equipment (performing much of the maintenance of small pieces of equipment), and performs network engineering and installation.

- Applications and Research supports over 115 applications software packages in the areas of computational chemistry, molecular modeling, mechanics, electrical engineering, and many more. This department includes six senior scientists with Ph.D.'s and research expertise in areas such as Computational Chemistry, Protein Crystallography, Mathematics, Physics, and Biology.

- The Advanced Scientific Visualization Laboratory provides the hardware, software, and support that allows researchers to analyze and convert 3-dimensional scientific data sets into still or animated images. Visualization equipment includes a high-performance graphics engine, a color camera recorder, a video recorder, a film recorder, both color and laser printers, as well as several graphics workstations.

- User Services provides consulting, documentation and training, and installs and maintains applications programs and databases for our local and remote computing community.

In addition to their service roles, members of the staff in all departments are engaged directly in research activities themselves. These activities span the range of the computational sciences and also encompass computer science topics. In particular, a significant portion of the time of each member of the Applications and Research Department is spent in research. Staff research activities include independent projects as well as projects in collaboration with researchers from the SDSC user community and elsewhere.

Setting Our Goals

Many issues came to our attention during the planning stages of the San Diego Supercomputer Center. We discussed these issues and made some decisions that, I believe, have made the center successful.

The first decision was to use "proven systems". With only one supercomputer at our center, we could not afford to use experimental, or unusual, unproven systems. We therefore chose the CRAY X-MP, which was a demonstrated, successful system. We extended the proven-system concept beyond the supercomputer itself to include established production systems for systems software, peripheral devices, communications, output, and storage.

The second decision was to ensure that we were "mission oriented". This means that we employ a staff of professionals who are dedicated to delivering supercomputing resources to the user community.

The third decision we made was critical for an academic supercomputer center: not to charge researchers incrementally for the time they use. This was vital because we wanted to attract academic researchers who would work on important scientific projects and not be limited to those projects that did not require substantial computing resources. Because supercomputers and robust supercomputer support systems can be expensive, it is extremely important that these national supercomputing facilities continue to be fully funded up front, as they have been in the past.

We recognized that high-performance communications was an issue that would have a critical impact on our success. We needed a communications system that would provide our remote users with access to the center. However, "high-performance" is a concept that changes with time. When the highest performance device anyone had access to was a glass terminal limited to 9600-baud, RS-232 communications, then 9600-baud lines were sufficient, and 56-kilobits were capable of multiplexing a number of remote users. When researchers acquired PCs and workstations with Ethernet capabilities, the situation did not immediately change because there were not many Ethernets available to everyone for communications. Users then had devices on their desks that could talk at megabit speeds, but few had access to a megabit local communications network. Consequently, long-haul communications did not need to be any faster either; researchers were continuing to use higher capability devices at 9600-baud. But eventually, when campuses began installing local area networks based on Ethernet, it became necessary to boost inter-campus communications speeds up to T1. In the future, people will probably be using VME buses and fiber optics, and soon T1 will also be inadequate. For the present, however, T1 is adequate because it is not usually possible to distribute higher speeds.

Another issue we discussed was the type of operating system to run on the supercomputer. We thought an interactive operating system was critical because, in our opinion, users needed to interact with their programs running on the machine, particularly during code development. We therefore chose CTSS because it was the only interactive operating system available on a supercomputer in 1985. When our CRAY Y-MP is delivered this fall, we will be running UNICOS which has emerged as the standard for engineering and scientific computing.

Consulting and training were major issues because we planned to provide supercomputing resources to a very diverse user community, both geographically and by discipline. These researchers who would be accessing our facility were not a group of users working in our building or on our local campus where they could take a short walk to our office and sit across the table from us. Consulting and training have proven to be the most challenging and important issues in making the Center successful.

Because we realized that the purpose of a central supercomputing facility is to provide access to more than just a supercomputer that is too expensive for any individual user to purchase, we also paid special attention to procuring appropriate peripherals. Buying a supercomputer was by no means sufficient; we purchased a $15 million supercomputer and spent another $15 million on peripherals. This is roughly the right balance. SDSC maintains several expensive peripheral devices that are used most cost-effectively when they are shared, such as a $500,000 film recorder that aids researchers in the visualization of data. In addition, our large IBM system supports long-term file storage; currently 1 1/2-trillion bytes and 500,000 files are stored in this system. We also have several VAXes that handle communications and hardcopy output, both printed text, graphics, and film.

We also thought it important, in terms of networking, to ask our users how they wanted to communicate with us, and then do the best we could to communicate with them, rather than dictating to them what communications method they should use. For this reason, we are a major node on NSFNET, speaking TCP/IP. We have other TCP/IP connections, including the UCSD campus network, ARPAnet, CERFnet (which SDSC is also responsible for), and we also speak MFEnet protocols. We are a node on SDSCnet (which was originally based on the MFEnet protocols) as well as on BITnet. In addition, we are connected to several of our corporate sponsors' networks as well as the NASA SPAN network and the HEPnet, both of which are DECnet based.

Shaping the Computing Environment

The definition of the supercomputer environment is not precise. It has changed with the evolution of computing and varies depending upon the observer's point of view. Hardware developers, software developers, system administrators, and end users may each provide a different definition of this environment. It is important to consider a very broad definition in order to create the most effective environment for the productive application of supercomputers. At SDSC, we believe that the real supercomputer environment is a synthesis of many considerations: the supercomputer system with its support staff of operators, system programmers, application programmers, trainers, writers, editors, and consultants.

Supercomputers are extremely complex and are generally more demanding than a PC or workstation. Researchers need help to use supercomputers efficiently, so effective use of advanced computing facilities requires strong user support. At SDSC, user services -- a vital link to our user community -- provides consulting, documentation, training, applications programs, and databases to our remote as well as local users.

SDSC's consulting staff ranges from recent college graduates through Ph.D. level senior scientists. A variety of scientific and engineering backgrounds is represented on the staff, including mathematics, chemistry, mechanical engineering, electrical engineering, aerodynamics, geology, physics, biology, and computer science. This breadth is necessary to properly support the wide range of disciplines represented by our users. The consulting hotline is staffed nine hours a day, five days a week and electronic mail is available 24 hours a day, seven days a week. The hotline provides telephone coverage a minimum of six hours a day in six time zones. The number of consulting questions averages 700 per month, and this number does not include the many inquiries directed to staff other than the consultants. The number of consulting contacts has remained high, and, even though the experience of users has increased, many of the questions asked have become more complex and difficult.

SDSC's training staff teaches a series of four introductory training workshops for new users. The entire series takes a user from basic concepts like logging on and features of the operating system through such advanced topics as using the symbolic debugger and accessing the available mathematics and graphics software. Initially, the introductory series was taught once or twice each month, and attendance was high. As the user community became more experienced, the demand for the introductory series decreased. This series is now taught at SDSC at the beginning of each quarter to coincide with the addition of new users.

We also conduct advanced workshops that provide in-depth training on code optimization and vectorization and on using the Common Command Language to write command procedures. In addition, special workshops are taught on topics such as computational chemistry. Future efforts will be devoted to developing more workshops for advanced users in areas such as visualization, distributed computing, engineering codes, and multitasking. SDSC trains users both on-site in our training facility and at remote sites. To date, we have taught 50 workshops attended by approximately 1,000 participants.

In addition to the regular training workshops, SDSC hosts an annual, 2-week Summer Institute on Advanced Computing which is an intensive introduction to advanced computing for researchers and educators throughout the nation. The institute is attended primarily by junior faculty and graduate students who have computational experience but not on supercomputers. In addition to lectures by SDSC staff members, the institutes feature guest speakers who are experts on topics such as parallel processing, visualization, algorithms, and vectorization. During the institutes, participants have an opportunity to work on programs related to their research and can apply the techniques learned during the lectures to their research projects.

Additional training and educational efforts at SDSC include an ongoing student internship program and a regular seminar series. The intern program enables students from local universities to work on research projects with SDSC staff. This gives the students first-hand experience with supercomputing. The seminar series presents talks on many aspects of advanced computing. These seminars are open to and attended by faculty, staff, and students from local universities, research institutes, as well as by industrial representatives.

In addition to training, SDSC staff members also write and edit technical documentation for our user community. This documentation includes introductory user guides and advanced reference manuals. We maintain an extensive set of on-line documents comprising 135 documents totaling over 17,000 pages. We produce and mail a monthly newsletter to 6,000 readers to keep them informed of changes and enhancements at SDSC as well as developments in computational science. We support an on-line HELP package and maintain a total of 35 bulletin boards. Our annual Science Report highlights research projects conducted at SDSC during the year, and, by publicizing the work of researchers at SDSC, we hope to foster greater interdisciplinary collaboration among them.

Producing and maintaining this extensive collection of documentation requires an experienced staff of technical writers and editors. Copies of documents are freely available to university staff and faculty for use in local workshops and classes. The San Diego Supercomputer Center has become a valuable source for technical documentation on advanced computing.

At SDSC we support a rich variety of applications software for our users; over 115 packages are currently available. Access to the SDSC software collection can be just as valuable for many researchers and educators as access to the supercomputer itself because few small groups or universities can afford to maintain comprehensive libraries of math and graphics packages on their local computers. Supporting this wide variety of software requires an experienced, qualified, and dedicated staff. Our support includes providing up-to-date documentation for the packages as well as keeping the software itself up to date through each new release. When we find it necessary to modify packages or to update them, this process is always followed by validation of the software to ensure that it will produce correct results.

We support and provide access to applications programs in Biology, Chemistry, Computational Fluid Dynamics, Electrical Engineering, Mechanical Engineering, Nuclear Engineering, and many other areas. Table 2 is a complete list of the applications software packages supported by SDSC.

TABLE 2
SUPPORTED APPLICATIONS SOFTWARE
(115 Packages + 12 Subpackages)

Biology(5)
DNAMAP
GenBank DB +
IDEAS
NEWAT DB
PIR DB +

Chemistry(32)
AMBER +
BIGSTRN-3
CADPAC
Cambridge DB +
CHARMM +
CHELP
Columbus
DISCOVER +
ECEPP2
FORTICON
GAMESS
Gaussian 82
Gaussian 86
GRADSCF +
GROMOS +
GVB +
MM2
MMP2 +
MMTOOLS
MMX
MODEL +
MOPAC
MPLOT
MXQET
PCK83
POLYRATE
PROLSQ
PSI77
QCFF/PI
RESLSQ
TRIATOM
XPLOR +

Database Mgmt(1)
RIM

Electrical Eng(8)
GREENFIELD +
HSPICE +
PHYLLIS +
PISCES
PRECISE +

SALT +
SPICE
UM-SPICE

Graphics(8)
DI3000
DISSPLA
FPSLIB
GKSGLIB
MOVIE.BYU
NCARGRAPHICS
PLOTXY
TV80LIB

Math(25[+12])
DASSL
ELLPACK
GAMS DB
HARWELL
HOMPACK
IMSL
ITPACKV
LUSP
MATHLIB
MATLAB
MFFT
MINOS +
MINPACK
NAG
ODEPACK
OMNILIB
(EISPACK
LINPACK
SCILIB)
REDUCE
SLAP
SLATEC
(AMOSLIB
BSPLINE
DEPAC
EISPACK
FFTPACK
FISHPACK
FNLIB
FUNPACK
LINPACK
PCHIP
QUADPACK)
SMPAK
SPARSPAK +

TOEPLITZ
TOMS
VECTOR-FFT
VECTOR-FHT

Mechanics(28)
ABAQUS +
ADINA +
ANSYS +
ARC2D
ARC3D
BEASY +
BUCKL
CHARTD
CONTINUSYS +
CSQ
DYNA2D
DYNA3D
FACET
FIDAP +
FLUENT +
INGRID
INS3D
MAZE
MSC/NASTRAN +
NIKE2D
NIKE3D
ORION
PATRAN +
SALE2D
SALE3D
TAURUS
TOPAZ2D
TOPAZ3D

Nuclear Eng(8)
ENDF DB
MCNP
NJOY
ONEDANT
ONETRAN
TIMEX
TRANSX
TWODANT

DB denotes database.
+ requires license or fee for industrial use.

There are many aspects to the vital role the national centers play including computational science research at the many institutions using the centers; research and development at the centers themselves in areas of advanced computing such as distributed processing, visualization, and parallel processing as well as the various branches of computational science and engineering; training and educating scientists and engineers for both academia and industry; providing U.S. researchers, educators, and students access to advanced-computing resources not available elsewhere; and providing an environment for technology transfer to U.S. industry.

At SDSC, the over 3,000 researchers using our computing facilities represent a wide variety of disciplines. We charted the usage-by-discipline from January 1986 to June 1989. Table 3 indicates the cumulative CPU hours used by researchers during this time frame in each of the disciplines:

TABLE 3
CUMULATIVE CPU HOURS USED 1/86 - 6/89

Astronomy 3.4%
Atmospheric Science 10.9%
Biochemistry 11.2%
Chemical Engineering 1.9%
Chemistry 12.2%
Earth Science 3.7%
Electrical Engineering 1.6%
Materials Science 13.3%
Mathematics 1.7%
Mechanics 4.2%
Multidisciplinary 13.3%
Oceanography 2.0%
Other 1.3%
Physics 19.3%

Providing useful software packages, consulting, training, and documentation to such a varied user community places special burdens on an academic center, burdens that a tightly-focused center with a limited number of applications would not normally face.

Supercomputing in the Future

Where is this all going? What does the future hold? What might we find? We might find highly parallel systems such as the Connection machine; we might find supercomputers composed of entirely new architectures such as data flow and neural nets; we might find entirely new technology like optical devices; we might find application-specific hardware like graphics workstations and also the DNA-sequencing chip that is now available in Sun workstations; we might find very intelligent software such as automatic programming.

And what about the problems? As the director of a supercomputer center, I think we can expect inadequate mass storage because growth in memories and CPUs is much faster, and it looks as though it will continue to be much faster, than growth in mass storage. The result will be inadequate mass storage.

We can expect inadequate long-haul communications bandwidth because right now we can connect (and we will be connecting in San Diego shortly) a frame buffer to a 100-megabyte channel into our CRAY and be able to generate real-time, full-color animation on the frame buffer driven through our CRAY, but we will be unable to deliver this performance to a remote site. Thus, the long-haul communications bandwidth will be inadequate.

We can expect inefficiently utilized hardware because we still do not know how to execute our vector machines at full vector performance. In fact, 10-20% on a 24-hour day average seems to be the maximum vector performance achieved to date, and I am distressed that the computer science community is not working to achieve more efficient utilization of vectors.

I see an insatiable demand for all of this advanced technology. Perhaps in a couple of generations we will become saturated with supercomputing. But for now, we will continue to struggle with our greatest problem: trying to simulate more realistic situations that call for more compute power than we are able to provide today. Molecular modeling, weather prediction, and computational fluid dynamics, where scientists attempt to model full airplanes or helicopters with complete physics and geometry in dynamic rather than static situations, still require a great increase in compute power.

What can we look forward to? More users, more supercomputers, new applications, new results, and new science.

1. Bardon, Marcel and Curtis, Kent K., A National Computing Environment for Academic Research, National Science Foundation, July, 1983.

2. Lax, Peter D., Report of the Panel on Large Scale Computing in Science and Engineering, National Science Foundation, December 26, 1982.

3. Press, William H., Prospectus for Computational Physics, Report by the Subcommittee on Computational Facilities for Theoretical Research to the Advisory Committee for Physics, Division of Physics, National Science Foundation, March 15, 1981.

4. Decker, James F., "Report to the Federal Coordinating Council on Science, Engineering and Technology Supercomputer Panel on Recommended Government Actions to Provide Access to Supercomputers". In Supercomputers, Hearings before the Committee on Science and Technology, U.S. House of Representatives, Ninety-eighth Congress, First Session, November 15, 16, 1983, No. 47, pages 403-427.

PART 3

SUPERCOMPUTER ARCHITECTURE

SUPRENUM: Architecture and Applications*

Karl Solchenbach

SUPRENUM GmbH, Hohe Str. 73, D-5300 Bonn

1 Introduction

In the last decade the numerical simulation of physical or chemical processes has gained increasing importance in many fields of science and engineering. Computer experiments, as the numerical simulations are often called, offer the advantage that they are free of principal limitations, whereas experiments are either not always accurate (e.g. wind tunnel) or even not possible (e.g. reentrant space vehicles, astrophysics, elementary particle physics, etc.).

Numerical simulations have become possible with the advent of supercomputers. Their computational power allowed much more complex simulations and their use reduced experimental time and costs considerably.

Computer experiments, however, still require much more computational power for instance in the field of aerodynamical simulations: The present generation of supercomputers allows either simulations of complex flow patterns on simple geometries (e.g. large eddy simulations) or simplified and less accurate models on complex geometrical structures (e.g. Euler equations around airplanes) but not both simultaneously. In order to increase the applicability and accuracy of CFD simulations, faster supercomputers – necessarily based on multiprocessor architectures – are required.

Of course, the present limitations are not only due to lacking computer performance. Also new mathematical models (e.g. for turbulent flow) are necessary as well as new numerical algorithms. What is really needed in the future is the combination of fast numerical methods (like multigrid) with advanced high-performance computer architectures. All attempts to achieve only one of these goals and to neglect the other one have failed. The speed-up of numerical software in the last years has been considerably larger than the speed-up gained by vectorcomputers and any machine which does not optimally support the fastest numerical methods is not really useful.

This central challenge gave rise to the German SUPRENUM project which aimed at the development of a parallel supercomputer for numerical applications. SUPRENUM was developed as a joint project of 14 institutions (universities, research laboratories, industrial companies) and was successfully completed with the SUPRENUM 1 computer. This machine was presented to the public at the Hanover fair in April 1989.

The SUPRENUM architecture has been published many times (see [9], [10], [2], [3]). It can be characterized as a distributed memory multiprocessor system with a medium granularity with

*The work described in this report was funded by the Federal Ministry of Research and Technology of FRG (grant number ITR8601 9) and the Ministry of Economy and Technology of Nordrhein-Westfalen (project number 323-8605200).

NATO ASI Series, Vol. F 62
Supercomputing
Edited by J. S. Kowalik
© Springer-Verlag Berlin Heidelberg 1990

a powerful vector processor in each single node. Hence it combines the advantages of the cost effective SIMD vector processing in each node with the performance capabilities and flexibilty of an MIMD system.

This paper presents the essential features of SUPRENUM. After a short discussion of the basic architectural principals, the hardware is briefly sketched and the software concept as well as some of the most important applications are outlined.

2   Supercomputer architectures

Today many different approaches for a classification of computer architecture, especially with respect to parallel processing, are discussed. A classification may be based on quantitative aspects (the degree of parallelism or the granularity), the structure of the control-flow (SIMD, MIMD, data-driven, demand-driven), the (hardware) technology (VLSI, VHSIC, air cooled, liquid cooled, see [2]), or the topology of the processing elements and the memory units.

We want to emphasize three classification categories:

1. SIMD vs. MIMD

   SIMD operation mode means that parallel functional units execute the same instruction sequence on different data. The two best-known realizations of the SIMD principle are pipelined floating point units and array processors.

   The MIMD principle is the favorite operation mode for parallel computers based on entire and independent processors. Each processor may execute a different instruction stream within the same application.

2. Shared vs. distributed memory

   One of the central problems to be solved in the design of multiprocessor systems is the memory access. Basically, there are two possibilities to organize this:

   - Shared memory
     guarantees fair access to a total memory for each processor.

   - Distributed memory
     means that each processor has direct access only to its own private memory.

   Often both memory organization types are combined to in hierarchical memory systems.

3. Scalar vs. vector floating point units

   Presently, the most cost effective way to achieve floating point rates of up to 100 MFLOPS is vector processing. MIMD multiprocessors which target to the top of supercomputer performance have to employ vector processors as basic floating point units. Therefore, the most powerful architectures today are mixed MIMD/SIMD multiprocessor systems. The efficient use of these architectures requires parallelism on two levels: the coarse grain parallelism related to the global MIMD structure and the fine grain parallelism which ensures efficient vector processing locally.

   The performance relation between scalar and vector processors is currently changing and it has to be expected that in the near future scalar processors achieve the same cost/performance ratio as vector processors.

As will be shown in the next section, SUPRENUM is an MIMD parallel computer with distributed memory and SIMD vector processors.

3 The SUPRENUM system

3.1 The hardware

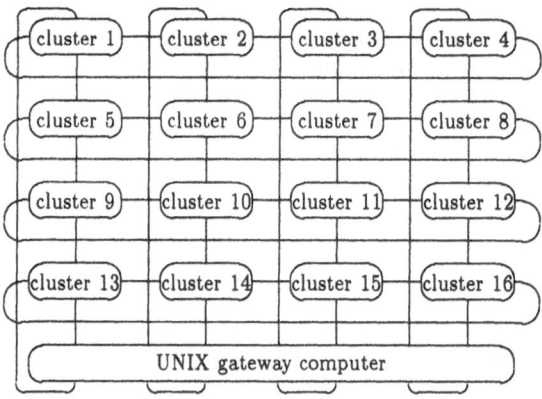

Figure 1: SUPRENUM clusters

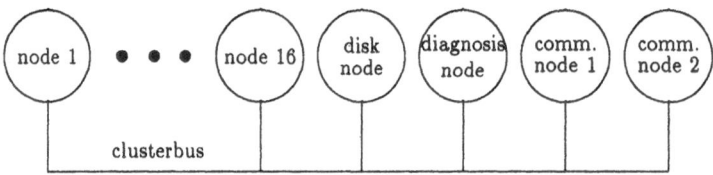

Figure 2: Cluster structure

Figures 1 and 2 show the overall structure of the SUPRENUM system. Up to 256 nodes are connected via a two-level interconnection network of buses.

Each node consists of the MC68020 CPU, 8 Mbyte of private memory, a fast floating-point vector unit (10 Mflop/s peak performance, 20 Mflop/s with chaining) and dedicated communication hardware.

16 computing nodes are combined to a cluster. The nodes in a cluster are connected by the clusterbus (320 Mbyte/s). Additionally to the computing nodes each cluster contains four special nodes: a disk controller node with up to four disks (4.8 GByte), a diagnosis node which supports performance measurements, and two communication nodes which connect the clusterbus to the serial SUPRENUM-bus (see Figure 2).

Any number up to 16 clusters are connected by a matrix of SUPRENUM-buses (200 Mbit/s) and form the high performance kernel (see Figure 1). The SUPRENUM system is completed by a UNIX gateway computer which is linked to standard networks and which is used for operating and maintaining the high performance kernel.

3.2 The system software

The software concept for SUPRENUM is based on a process system and on message-passing communication handling. The process concept (the so-called Abstract SUPRENUM architecture) is a dynamic one which is characterized by the following elements:

- Processes are autonomous program units which run in parallel.

- Processes can terminate themselves and can create, but not terminate other processes.

- Processes communicate only by exchange of messages, and no shared memory is available.

- Applications are started by one initial (or host) process.

- In arithmetic expressions and communication instructions, array constructs are especially supported.

- The user defined process system is homogeneous and independent from the actual hardware configuration. The two-level architecture (cluster structure) is not reflected in the Abstract SUPRENUM architecture and is completely transparent to the user. The processes are mapped to the clusters and nodes at run-time.

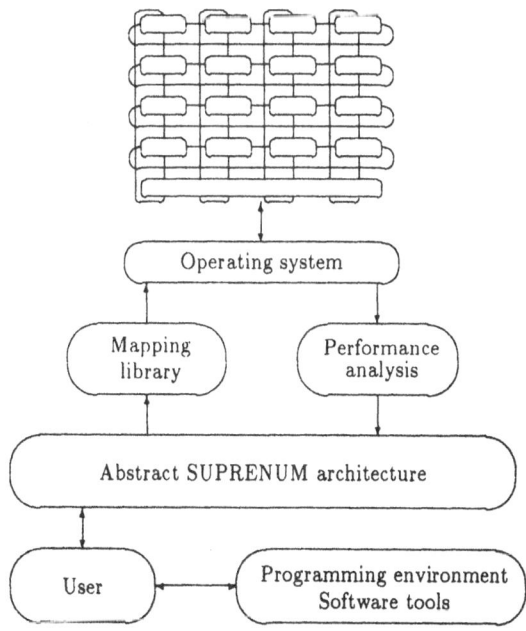

Figure 3: The SUPRENUM system software

Figure 3 shows, that this Abstract SUPRENUM architecture is the central model in the system software. The user should write his or her codes only in terms of processes.

The mapping of processes to nodes is supported by the mapping library. It provides optimal mapping strategies for some standard process systems (like trees, rings, grids) and uses heuristical strategies for other process structures.

The SUPRENUM operating sytem consists of three components residing on the gateway system, the cluster level and on the node level. The gateway system is operated under UNIX V[1]. On the cluster level the operating system supports the local disk, the performance analysis and the connection between the two communication levels. In each node a small operating system (PEACE) is responsible for the process scheduling and the message handling.

The programming language for numerical computations is Fortran ([1], [5]), an extended Fortran 77. The extensions include special process handling and message-passing constructs and an array syntax formulation according to the proposed Fortran-8X standard. Parallel programming is also possible in C where the use of the vector floating point unit is supported by a vector library.

Performance analysis tools collect performance data from each cluster for graphical presentation. This enables the user to analyze the utilization of nodes and buses and to tune the parallel programs.

The SUPRENUM programming environment provides a lot of tools which support the programmer in developing parallel software. The set of tools contains a syntax-directed editor for Fortran which checks the syntactical correctness of a program text during the editing phase.

A state-of-the-art vectorizer automatically generates code for the vector nodes – if the user does not want to use the new Fortran-8X array notation.

A semi-automtatic parallelizer is currently under development which will support the user in constructing host and node processes and which automatically distributes arrays over the nodes.

A communications library for grid applications allows easy, safe and portable programming for regular and block-structured grids.

The SUPRENUM simulator supports program development and first functionality tests (e.g. deadlock detection) on workstations.

The communication between and the synchronization of processes is done via message-passing. SUPRENUM offers an asynchronous message-passing model, i.e. the sending application program does not have to wait until the message has arrived at the receiver and can continue immediately. On the receiving side, the message arrives in the mailbox and can be selected by the application program. The use of the SUPRENUM message passing constructs is demonstrated in [4], [5], [8].

3.3   Applications software

The availability of practical relevant application software is decisive for the scientific and commercial success of a new computer architecture. This software must, of course, make use of the specific advantages of the architecture and translate these advantages into gains of speed. On the SUPRENUM the following applications software is available (see also [6], [7]).

---

[1]registered trademark of AT&T

3.3.1 Basic numerical software

The basic numerical software available on SUPRENUM consists of

- the standard Fortran librarie on each node (including many 8X intrinsics),

- a parallel Linear Algebra package (a subset of LINPACK and EISPACK),

- some solvers for basic partial differential equations (PDEs) using highly efficient parallel multigrid (MG) algorithms,

- a parallel scientific library containing e.g. parallel FFT algorithms.

3.3.2 CFD applications

The emphasis within the application software development lies on computational fluid dynamics (CFD) codes. These cover most of the standard CFD models used today.

- (Nonlinear) Potential equation for subsonic and transonic flow past airfoils. A Finite Volume (FV) discretization is used and a special relaxation is applied at grid points near the shock front (transonic case).

- Euler equations for 3D-computations employing advanced acceleration techniques. The Euler equations are discretized by a FV scheme on a non-staggered grid. The stationary problem is solved by marching in pseudo time-steps (explicit Runge-Kutta method). The time marching can be viewed at as a Jacobi-type relaxation which is accelerated using coarser time-steps in a MG-like manner.

- Euler equations combined with boundary layer methods for the 2D- and 3D- simulation of flow past cars;

- Navier-Stokes equations (incompressible and compressible) for internal flows on general 2D- and 3D-domains. This is the most complex application imposing additional requirements on discretization, MG components and grid refinement strategies.

- Navier-Stokes (compressible) for the full simulation of the flow past cars (3D). This code allows also the simulation of a large wake area which is impossible with combined Euler-boundary layer methods.

- Grid generation codes for the generation of 2D- and 3D-boundary fitted grids. The grid structure is either a single logically rectangular grid or an arbitrary aggregation of blocks each of which is a regular grid. Single boundary fitted grids are generated by the solution of a system of Poisson-like equations (which corresponds to a mapping of the physical domain to a rectangular domain). This step itself may require large computing times and can be solved by (parallel) MG methods.

Besides the CFD applications many different application software packages are currently parallelized for SUPRENUM. Examples are a large commercial Finite Element package, the reactor safety code RELPA5, the GEANT code used in high energy physics, and a quantum chemistry code.

References

[1] **Bolduc, R.:** SUPRENUM Fortran syntax specifications. SUPRENUM Report 7, SUPRE-NUM GmbH, Bonn, 1987.

[2] **Giloi, W.K.:** SUPRENUM – a trendsetter in modern supercomputer development. In [11].

[3] **Giloi, W.K.:** SUPRENUM – the system. Supercomputer 30, Vol. VI,2, pp. 13-19, Amsterdam, 1989.

[4] **Solchenbach, K.:** Grid applications on distributed memory architectures: Implementation and evaluation. In [11].

[5] **Solchenbach, K.:** SUPRENUM Fortran – An MIMD/SIMD language. Supercomputer 30, Vol. VI,2, pp. 25-30, Amsterdam, 1989.

[6] **Solchenbach, K.:** Application software for SUPRENUM. Supercomputer 30, Vol. VI,2, pp. 44-50, Amsterdam, 1989.

[7] **Solchenbach, K.:** Parallel CFD algorithms on SUPRENUM. Proceedings of ISC88 Third International Conference on Supercomputing, Boston, May 15-20, 1988. ISI, St. Petersburg, Florida.

[8] **Thomas, B., Peinze, K.:** SUPRENUM comfort of parallel programming. Supercomputer 30, Vol. VI,2, pp. 31-43, Amsterdam, 1989.

[9] **Trottenberg, U.:** The SUPRENUM project: idea and current state. Proceedings of the SPEEDUP Workshop on Vector and Parallel Computing in Berne, 15 Jan. 88, to appear.

[10] **Trottenberg, U.:** SUPRENUM – the concept. Supercomputer 30, Vol. VI,2, pp. 5-12, Amsterdam, 1989.

[11] **Trottenberg, U. (ed.):** Proceedings of the 2nd International SUPRENUM Colloqium "Supercomputing based on parallel computer architectures". To appear as a special issue of Parallel Computing, North Holland, 1988.

ADVANCED ARCHITECTURE AND TECHNOLOGY

OF THE NEC SX-3 SUPERCOMPUTER

Tadashi Watanabe

EDP Product Planning Division

NEC Corporation

Shiba 5-33-1, Minato-Ku,

Tokyo 180

Japan

1. INTRODUCTION

Based on the expertise of the SX-2 Supercomputer announced in
1983 which was the first supercomputer to break the 1 GFLOPS
performance barrier, the NEC SX-3 Supercomputer has been developed to meet the growing demands for large and high speed computations, offering a maximum performance of 22 GFLOPS. The SX-3
consists of seven models based on the number of processors and
the number of vector pipelines per processor. The models range
from 1.4 GFLOPS uniprocessor entry model to a 22 GFLOPS quad
processor top end model.

The architecture of the SX-3 is an extension from the NEC's first
supercomputer SX-2 to support a powerful multiprocessor system with
up to four processors. To achieve high speed single processor
performance of 5.5 GFLOPS, the SX-3 has a machine cycle time of
2.9nsec and 16 vector pipelines per processor, employing VLSI with
20,000 gates and 70psec switching speed.

An UNIX based operating system, SUPER-UX, with multiprocessor and
vector processing functions is supported in addition to FORTRAN77/
SX with auto-parallelizing and vectorizing functions.

NATO ASI Series, Vol. F 62
Supercomputing
Edited by J. S. Kowalik
© Springer-Verlag Berlin Heidelberg 1990

Table 1 summarizes the major SX-3 characteristics and figure 1 shows the system appearance of the SX-3.

Table 1　SX-3 System Specifications

MODEL	44	42	24	22	14	12	11
Vector Peak Performance	22GFLOPS	11GFLOPS		5.5GFLOPS		2.75GFLOPS	1.37GFLOPS
Number of Artimetic Processors	4		2			1	
Reg — Vector Mask Registers	144KBX4	72KBX4	144KBX2	72KBX2	144KB	72KB	36KB
Reg — Vector Mask Registers	32				16		8
Reg — Scalar Registers	64BitsX128X4		64BitsX128X2			64BitsX128	
Vector Pipelines/Sets	64/16	32/8		16/4		8/2	4/1
Cache Memory	64KBytesX4		64KBytesX2		64KBytes		
Main Memory (Max)	2048MBytes				1024MBytes		512MBytes
Extended Memory (Max)	16GBytes						
I/O Channels (Max)	256						
I/O Transfer Rate	1GBytes/Second						

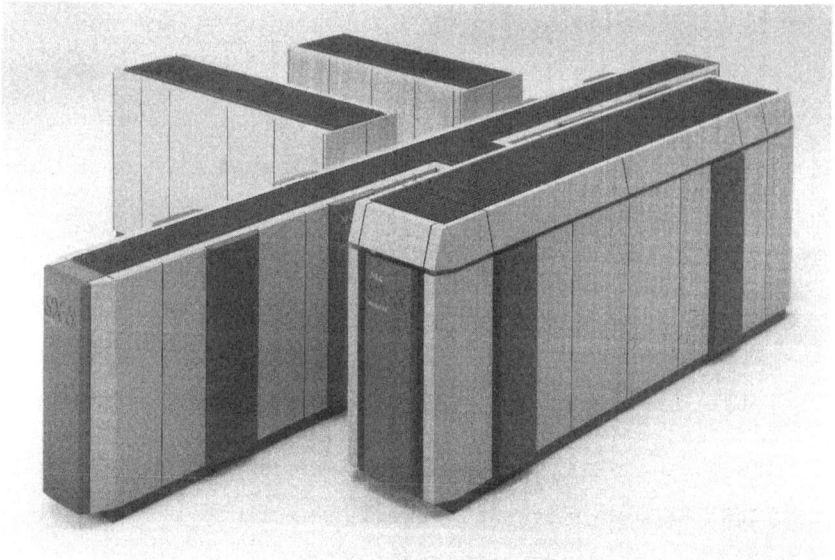

Figure 1　The System Appearance of the SX-3/44

2. HARDWARE TECHNOLOGIES FOR THE SX-3 SUPERCOMPUTER

One of the key factors to realize high speed processing is to shorten a machine cycle time as much as possible. To shorten a machine cycle time, crucial points are to get fast switching device and to shorten wire lengths to connect among devices. In other words, we need high speed VLSI technology and high density packaging system.

In the SX-3 Supercomputer, the proven but highly advanced silicon CML (Current Mode Logic) technology is used for the basic VLSI chips. Figure 2 shows an enlarged photo of a VLSI chip employed which has 20,000 gates and 70psec switching speed. This gate switching speed is comparable to that of GaAs, but the gate density of this silicon chip is far higher than that of the currently available GaAs chip.

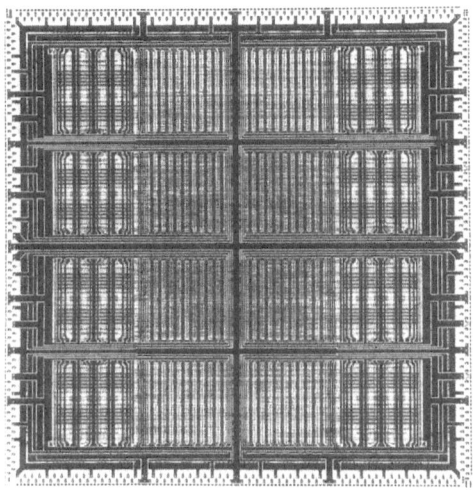

Figure 2   VLSIChip with 20,000 Gates

Figure 3   Multi-chip Package with 100 VLSIs

To mount LSI chips, we have traditionally employed a multilayer ceramic substrate with fine wiring patterns. In the SX-3 Supercomputer, the size of the ceramic substrate is enlarged to 22.5cm x 22.5cm from 10cm x 10cm employed in the former SX-2 Supercomputer. This ceramic substrate can mount up to 100 VLSI's as shown in figure 3. The total number of gates on the package amounts to 2 million which is comparable to that of a processor of large scale mainframe computer. The substrate consists of 15 ground and power layers. On the surface of the substrate, there are four signal layers insulated by poly-imide organic material with low di-electric constant which results in lower signal propagation delay on the wiring compared to inorganic materials. Four signal layers are sufficient to contain all wirings for connections among 100 VLSI's because of fine wiring pattern technology with 25 micron width. The power dissipation of a VLSI chip with 20,000 gates is about 33 watts which results in a total power of 3.3kwatts in a multi-chip package. The water cooling system is employed to cool this package.

3. SYSTEM CONFIGURATION AND PROCESSOR ARCHITECTURE

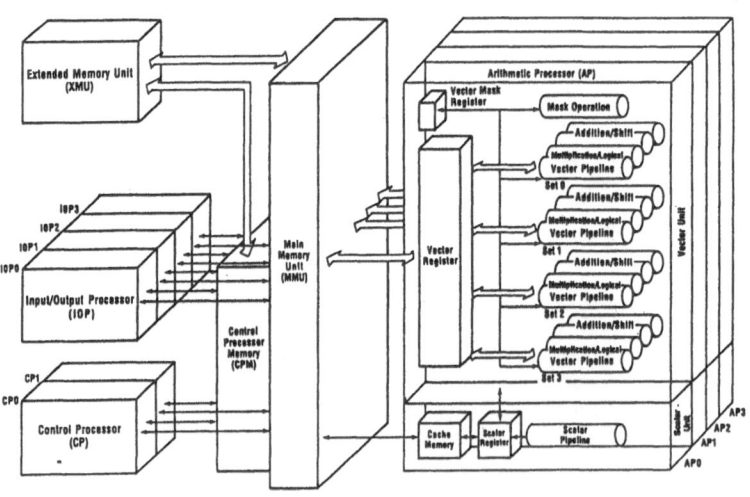

Figure 4   System Configuration of the SX-3/44

Figure 4 shows a maximum configuration of the SX-3/44 top end model. The Arithmetic Processor (AP) is a high speed computational engine which has a scalar and vector unit internally, and has a vector peak speed of 5.5 GFLOPS. The AP executes all codes compiled by Fortran. In a maximum configuration, the SX-3 can configure four Arithmetic Processors.

The Control Processor (CP) performs supervisory functions such as operating system functions and I/O buffer control. The CP has a dedicated memory, the Control Processor Memory (CPM), with up to 256Mbytes which works as a large I/O buffer and disk cache memory.

The Extended Memory Unit (XMU) with a capacity of up to 16GBytes works as a fast access disk device. The XMU is used for various file memory requiring high speed access such as staging/destaging files, job swapping files and disk cache memory.

The SX-3 can configure up to four Input/Output Processors (IOP's) each of which supports up to 64 channels including 20MBytes/sec channels for high-speed disks, and has a total I/O throughput of 1GBytes/sec. Four IOP's with a maximum of 256 channels and various peripherals as well as large capacity disk units allow us to configure large-scale supercomputer system.

As shown in figure 4, the Arithmetic Processor (AP) is functionally divided into the Scalar and Vector Unit. The Scalar Unit fetches and decodes all instructions to be executed on the AP, and dispatches them to the appropriate execution unit. The execution unit in the Scalar Unit consists of one set of pipelined arithmetic units which include integer add, logical, shift, floating point add, multiply and divide. For fast access to scalar data and instructions, the Scalar Unit has 64kbytes of cache memory, 4kbytes of instruction buffer and 128 scalar registers. Together with the instruction buffer, is a branch prediction capability incorporated where the branch direction, go or no-go, is quickly decided and the targeted instruction is pre-processed in advance, based on the previously branched target.

From the architectural point of view, the Scalar Unit employs the so-called RISC architecture, in which the high-speed scalar processing is realized by simplified architecture and hardware design. The performance of the Scalar Unit is approximately as twice as that of the former SX-2. The Fortran compiler makes scheduling of instruction sequences to fully exert hardware capabilities by analyzing and resolving data dependencies.

In the maximum configuration of the Vector unit, the Arithmetic Processor has four identical sets of vector pipeline units as shown in figure 4. Each pipeline set consists of two adds and two multiplies which results in 16 vector pipelines in total. All sixteen pipelines can execute in parallel clocked in 2.9nsec, resulting 5.5 GFLOPS peak performance per processor.

The Vector Unit supports various vector functions such as masked vector operations, compress/expand functions, and list vector capability for handling scattered vector data to achieve higher vectorizing percentage.

Like a cache memory and scalar registers for scalar data, the Vector Unit has a maximum of 144kbytes vector register for storing and buffering the intermediate results of vector operations.

4. SOFTWARE FOR THE SX-3 SYSTEM

The SX-3 supports two kinds of the operating system, SXOS and Super-UX. The SXOS is NEC's proprietary operating system which has extensions of the general purpose mainframe operating system, ACOS. The extensions include high-speed I/O functions such as parallel/ asynchronous I/O, support of the XMU and the connectivity with UNIX systems. The SXOS is enhancing the connectivity with UNIX systems by having functions of TCP/IP and ftp/telnet.

In the Super-UX operating system based on AT&T UNIX System V, various extensions are also made to support supercomputing environments, as shown in figure 5.

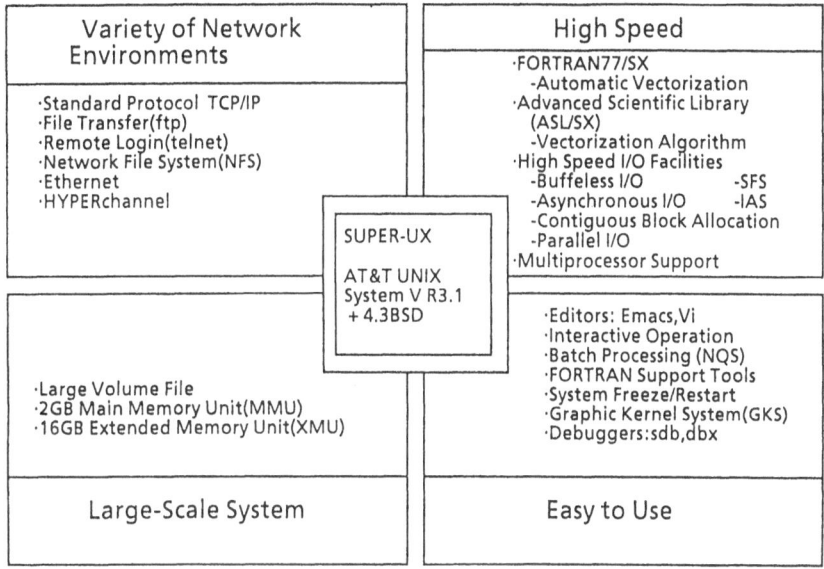

Figure 5   Major Characteristics of Super-UX

Major extensions are:

- Multiprocessor control function,

- Support of large and high-speed I/O devices such as the Extended Memory Unit and large disk units with a transfer speed of 20Mbytes/sec.

- High-speed I/O functions including bufferless I/O, parallel I/O and asynchronous I/O.

- Large-scale file system of virtual volume function and file allocation function to continuous area,

- System operation management such as accounting, file space management and file back-up function.

FORTRAN77/SX is a vectorizing and parallelizing Fortran for the SX-3 system. This compiler has parallelizing functions such as multi-tasking and auto-parallelizing as well as auto-vectorizing and optimizing functions. Vectorizing functions include vectorization for DO loops with conditional IF statements and list or indirect vectors in addition to simple arithemetic operations.

Table 2   Major Macro-tasking Libraries

Libraries	Meanings
CALL PTFORK	Create a subtask and invoke subtask
CALL PTJOIN	Wait for completion of subtasks
CALL PLLOCK	Set the lock or wait for the lock cleared
CALL PLUNLOCK	CLEAR the lock
CALL PEPOST	Post the event
CALL PEWAIT	Wait for the event posted
CALL PBASGN	Allocate a barrier and set the number of tasks necessary to break it.
CALL PBSYNC	Wait until all other tasks join.

Two kinds of multi-tasking functions, macro-and micro-tasking, are supported. The macro-tasking is a kind of top-down approach for parallel processing in a subroutine level. The macro-tasking is supplied by various library routine calls as shown in table 2. Syncronization for parallel processing as well as subroutines executable in parallel can be controlled and specified by using these Fortran callable routines.

The micro-tasking is a parallel processing in DO loop and statement level where the compiler analyzes the data dependencies in DO loops and decomposes those DO loops into multiple tasks. When the compiler does not detect the parallelizable parts, users can specify those parts by directives for parallelization. In multiple nested DO loops, the outer loop is, in principle, micro-tasked, and the inner loop is vectorized.

Various performance tuning tools are also supported to fully utilize hardware functions and to get higher performance. They include a tool for analyzing program profile and finding a focal point to be tuned, an interactive tool for vectorization and parallelization, and an interactive debugging tool.

5. CONCLUSION

An overview of the NEC SX-3 Supercomputer was presented. Our basic approach is that we pursue a fast single processor performance to the extreme, and then constitute a multiprocessor system by combining those fast processors. It should be noted that a multiprocessor system with a small number of fast processors and shared memory system brings higher performance over a broad range of vectorization and parallelization degree, and easy programming environments.

References

1) T. Furukatsu, T. Watanabe and R. Kondo, "Supercomputer SX System with a Peak Performance of 1.3 GFLOPS and 6 Nanosecond Cycle Time, "Nikkei Electronics, 356, pp. 237-272, Nov. 9, 1984. (In Japanese)

2) T. Watanabe, "Design Concept for High-speed Vector and Scalar Processing Architecture of the NEC Supercomputer SX System, "Proc. of I CCD, pp.38-41, 1986.

3) T. Watanabe, H. katayama and A. Iwaya, "Introduction of NEC Supercomputer SX System, "Supercomputers, Class VI Systems, Hardware and Software, pp. 153-167, S. Fembach (Editor), Elsevier Science Publishers B.V., Amsterdam, 1986.

4) T. Watanabe, "Architecture and Performance of NEC Supercomputer SX System, "Parallel Computing, 5, pp. 247-255, 1987.

The UNIX operating system was developed and licensed by AT&T. NQS is the Network Queueing System, originally developed by Sterling Software for the NASA Ames Reserach Center. NFS is a trademark for Sun Microsystems, Inc. The Emacs editor program was developed by Dr. R.M. Stallman. Ethernet is a trademark by Xerox Corporation. HYPER channel is a trademark of Network Systems Corporation.

PART 4

PROGRAMMING TOOLS AND

SCHEDULING

PROCESS SCHEDULING:
Parallel Languages Versus Operating Systems

Harry F. Jordan
University of Colorado
Campus Box 425
Boulder, CO 80309-0425
USA

ABSTRACT

This paper deals with multiple processes performing a parallel program, how they are scheduled on physical processors, and their effectiveness in completing the total work to which the individual processes contribute. The current ideas in process scheduling have, to a large extent, grown out of operating systems research on scheduling relatively independent processes on one or a few processors, with the goal being maximal utilization of available resources, especially the processors. Multiprocessor parallel programs may trade processor utilization efficiency for decreased completion time and may employ very fine grained interaction among processes. This yields a different set of requirements on process scheduling than those arising from optimal use of shared resources among weakly dependent or independent tasks. It is argued that coscheduling specifications are a good way to coordinate the possibly conflicting requirements of parallel program completion and multiple job resource utilization. Coscheduling requirements specify the simultaneous availability of a set of resources, processors being the foremost example. Coscheduling requirements may be met by simultaneous assignment of the required number of units or by fine grained time multiplexing of fewer units.

Introduction

The concept of a process evolved from the idea of a virtualized processor. Different programs shared the resources of a single computer system by time multiplexing the use of processor, memory and other hardware resources. Each different program operated in an environment which simulated that of a stand-alone virtual machine. As the environment associated with a program evolved into being more distinct from the real hardware, it became useful to use the term process for the virtual processor which runs a given program. A good process model can quite accurately mimic the operation of multiple independent computers, except for the inescapable fact of time multiplexing.

Actual relative timings between multiple processes are not pertinent to running multiple, independent, sequential programs. They are important when multiple processes cooperate in running a parallel program for a multiprocessor. Time multiplexed operation appears parallel on a time scale which is long compared to the average length of a quantum of the multiplexing. In independent sequential programs sharing I/O resources, this tends to be the length of time for an elementary I/O operation. In multiprocessor parallel programs, the assumed model is one of multiple physical processors on which the cooperating processes of a parallel program execute concurrently. The concurrency of execution is important to the parallel program on the time scale of interaction between the processes. This time scale is usually referred to as the granularity of a parallel program.

Coscheduling is a term often used to express the fact that time multiplexed processes are behaving as if they were making simultaneous progress on different physical hardware. The

NATO ASI Series, Vol. F 62
Supercomputing
Edited by J. S. Kowalik
© Springer-Verlag Berlin Heidelberg 1990

appearance of simultaneous progress depends on the time scales involved. Processes may be coscheduled with respect to the sharing of I/O resources without being coscheduled with respect to the fine grained sharing of computed results. Coscheduling is thus not precisely defined, but it is a useful way of expressing the idea that multiple processes all make some progress between time points separated by a given sized quantum. Processes assigned in a static manner to separate physical processors are coscheduled at any time scale of importance in a computation, and it is usually this model which is assumed in tightly coupled parallel programming. Processes managed by a scheduler derived from a uniprocessor style operating system are usually coscheduled at the time scale of I/O operations, often driven by terminal interaction rates.

The two different process domains, sequential processes sharing resources and parallel processes sharing data, make different demands on process scheduling. As already pointed out, the granularity of one is that of an I/O transfer and of the other that of data exchange. Resources to be managed in the multiple sequential case include:

physical devices,
I/O channels,
buffer space, and
throughput.

In the parallel processing case the resources to be managed are:

shared data items,
primary memory,
synchronization delay time, and
time to completion of the parallel program.

It is useful to call attention to the distinction between managing throughput and managing time to completion of a parallel program. The purpose of writing a parallel rather than a sequential program is to reduce the program's running time by using multiple processors. If the processors are subjected to management by a process scheduler, it is important that the scheduler treat parallel program completion time as a resource to be managed. The completion time of a parallel program is tied up with the efficiency of utilization of processors and memory. In a multiprogramming environment, efficiency of processor and memory use implies using the processor in each time quantum, because the swap overhead per quantum is small, and requiring the working set of a process to be in memory before scheduling it. The processor resources demanded are on behalf of one process for one quantum. In a multiprocessing environment, efficiency implies infrequent swapping of processes, on the time scale of data exchange, and a multiple process working set in memory, which includes the information needed for a set of processes to make simultaneous progress. The processor resources are thus demanded on behalf of many processes and can often involve the requirement to coschedule a specific set of interacting processes over a period of many time quanta.

Coscheduling of Parallel Processes

The need for coscheduling of parallel processes at the level of a small time quantum is ultimately based on the fine grained sharing of data among processes. An intermediate result produced by one process is required by another to make further progress. In shared memory multiprocessors, this requirement is usually implemented in a program by means of synchronization operations. Synchronizations require the cooperation of multiple processes, sometimes many of them. The barrier synchronization, for example, requires all processes to reach a synchronization point before any of them may proceed past it. If processes participating in a barrier are not coscheduled, then those which are active and reach the barrier cannot make useful progress until swapped out processes are run up to the point of reaching the barrier. If the parallel processing barrier synchronization is independent of the operating system process scheduler, the barrier will not be satisfied in a timely manner. Even if the barrier interacts with the operating system to give up processors until the synchronization has been satisfied, context switching might occur for every process before the operation is complete. This would impose a large penalty for using the barrier in the common case where the amounts of work per process and the process speeds are approximately equal and the synchronization merely insures against erroneous results in the low probability case of early completion by one or a few processes.

This problem with barrier synchronization is clearly shown in the measurements of Benten [1] on the Encore Multimax [2] running a multiprogrammed operating system and with barriers implemented using spin locks. The curve in Fig. 1, taken from his measurements, shows that on a 20 processor machine, the operating system does not coschedule more than 19 user processes, the 20th processor being occupied with system functions. The data shown was obtained on an otherwise empty system. Measurements on the same system with a multiprogramming load show the same sharp increase in execution time starting at fewer processes and subject to more statistical variation.

While the barrier affords an extreme example of the need for coscheduling, any synchronization between two processes requires both of them to be active before it is complete.

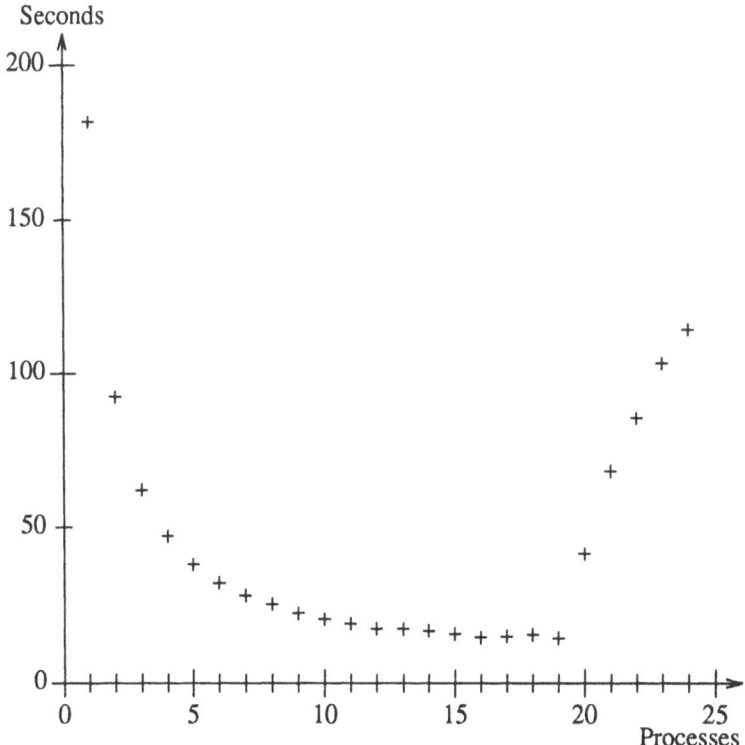

Figure 1: Execution Time for Barrier Synchronized Program

The synchronization most closely related to data sharing, which is the root source of the requirement, is that of producer/consumer. The producer may store a value and set the full indication without the consumer being active, but the consumer is blocked if the producer is not active. More significantly, since a given process does not always serve as producer or consumer with respect to the same second process, the collection of all two process produce/consume synchronizations should be viewed as a more global, complex synchronization. Figure 2 shows two patterns of pairwise synchronizations whose overall effect is that of a barrier. Thus patterns of process synchronization, and more fundamentally, patterns of cooperative data sharing among processes, give rise to coscheduling requirements at the granularity of data interchange.

There are various ways to ensure that the coscheduling requirements of a program are met by the multiprocessor system. The one which is effective at any level of granularity, and is most in tune with the philosophy of parallel processing, is to use enough physical processors. The maximum number of parallel instruction streams should be specified or determined by the compiler and that number of processes assigned to the program for the duration of the execution. The assignment of processes to processors should be static and non-interruptible, otherwise the coscheduling property will not hold at the granularity of interrupt service time.

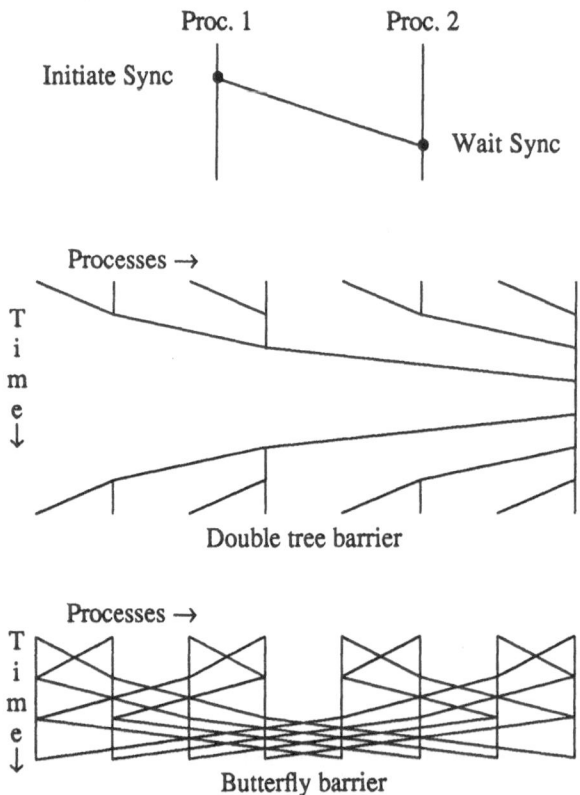

Figure 2: Barriers Built from Pairwise Synchronizations

Another way to ensure coscheduling at any level above that of single instruction execution is to use hardware supported, fine grained scheduling. The CDC6600 peripheral processors[3] used this mechanism, but most of the operating system code written for them made little use of fine granularity interaction. A machine which used this mechanism for fine grained parallel scientific programming was the Denelcor HEP[4]. This machine also supported low overhead synchronization waiting in hardware in which a waiting process occupied some hardware registers and used some memory bandwidth but did not use execution cycles.

The third possibility for satisfying the coscheduling requirements of a multiprocessor program is to communicate these requirements to the operating system and to let its scheduler take them into account in managing the CPU resources of the multiprocessor. Typically, the design specifications for an operating system scheduler are oriented towards coscheduling at the granularity of I/O operations and do not always take into account the needs of finer grained data sharing. Within the realm of operating system involvement in parallel process synchronization, there are two ways to communicate the program's needs to the system. Requirements may be determined by a combination of user specification and compiler analysis and communicated statically to the operating system at the beginning of execution. The simplest example of such communication would be the specification of one or a few sets of processes to be coscheduled, i.e., no process of the set should be assigned execution resources unless all of them are. A second method is to involve the operating system dynamically in every synchronization. The system can keep records of synchronization activity and attempt to schedule accordingly. Problems with this approach include the often high overhead of operating system invocation and the fact that the assumptions of statistical independence and time invariance which are so successful in managing resource sharing in the multiprogramming environment are questionable when applied to multiprocessing.

Operating System - Parallel Program Interaction

The idea of granularity has been mentioned several times. This concept appears in the process scheduler in the form of "weight" of a process. A heavy process takes many cycles to activate on a given processor. A light weight process can be made active quickly. Any uncoordinated scheduling activity on processes which interact on a time scale finer than the time needed to assign a processor to a process of that weight will lead to unacceptable performance for the parallel program.

Early multiprocessor systems adapted a multiprogramming operating system such as Unix† for use with multiprocessing. The memory sharing required by parallel processes was provided as an operating system extension. Synchronizations were already supported in the form of channel events and timed delays, but these had a time granularity on the order of a disk transaction time, or in some cases that of a terminal response. With the shared memory, other synchronizations using a busy wait could be constructed, but they were independent of the scheduler. Scheduling policy was normally a strong function of the I/O to compute time ratio. Although parameters of scheduling could be adjusted, a major change in policy was difficult. Coscheduling of a fixed set of processes was not normally available in these minimal extensions of a multiprogramming operating system.

One way to try to meet the needs of both multiprogramming and multiprocessing is to support processes of two different weights. Several different groups have used the idea of lightweight "threads" running on the considerably heavier Unix processes. The thread is the

---

†   Unix is a trademark of Bell Laboratories

schedulable unit of parallel computation. The scheduler for threads has a low overhead for assigning a process to a thread. Although a thread has its own private state, scheduling a thread usually does not affect memory occupancy. Only volatile processor registers are saved and restored. Parallel processing synchronizations interact with the thread scheduler while processes are managed by Unix in the usual manner with little concession to multiprocessing. Coscheduling is usually not available in these systems, and the behavior shown in Fig. 1 should not be expected to change.

A few multiprocessors were delivered with distinct multiprocessing operating systems. The Flexible Computer Systems Flex/32, for example, used the MMOS[5] operating system for multiprocessing and could dedicate one processor to running a stand alone version of Unix for multiprogramming. The MMOS operating system supported shared memory and synchronization. Processes were assigned to fixed processors, and although limited capability for time multiplexing more than one process on a processor was provided, processes did not move between them. Since there was a fixed assignment of processes to processors, coscheduling was the norm.

A common characteristic of the three styles of parallel program interaction with the operating system mentioned above is that there is little information explicitly supplied to the operating system about the coscheduling needs of the program. The information needed is ultimately determined by the data sharing among processes of the parallel program. In control flow based parallel languages, the sharing is coordinated by synchronization operations which directly determine the coscheduling needs. In a data flow architecture, the instruction scheduling mechanism must implicitly meet the coscheduling needs in order to realize a good schedule. It is possible for compilers to do an analysis of either the data dependences, the synchronization structure or both to help determine coscheduling needs. Explicit user information to the operating system should not be ignored. Part of a programmer's task in explicit parallel programming is to develop a synchronization scheme. Coscheduling requests could be based on it. It should be borne in mind that only the performance of a parallel program is affected by coscheduling, provided it is correct. The important problem of correctness of explicitly synchronized programs is not considered here.

Along with the question of supplying information on parallel program behavior to the operating system, there is the question of whether the system can gather such information dynamically. The difficulty is that many multiprocessor programs undergo radical changes in their synchronization behavior, so the past is a poor predictor. Since the key problem is that of interdependent operation of processes, the assumptions of statistical independence, which are so useful in multiprogramming, do not hold. Another problem with monitoring the program's behavior to gather information is that the monitoring affects coscheduling if it is done using the same processors on which the program is scheduled. A good example of this was observed on the Flex/32 by Arenstorf[6]. In this experiment, 18 processes were statically allocated to 18 physical processors, each of which responded to a separate, asynchronous clock interrupt. The ratio of interrupt service time to interrupt interval was 1.5%. By changing the clock period, it was possible to determine that barrier completion was delayed by $27\% = 18 \times 1.5\%$ with the normal clock period. This corresponds to a simple probabilistic model for the availability of processes during barrier synchronization.

One approach to the operating system problem is to use an explicitly parallel language whose constructs do not require coscheduling. Instead of the "process barriers" described above, it is possible to use "work barriers," which do not wait for processes to arrive, but only for previous work to be done. Work barriers were used in the IBM EPEX language[7] in the limited context of DOALLs. Benten[8] extended the concept to a more general setting by supplying work reporting mechanisms for a range of parallel constructs. Completion reports replace the arrival of processes at a barrier as the synchronization. Progress is then insensitive to process suspension, except as it affects reporting completion of work explicitly allocated to that process. When run using work barriers, the program of Fig. 1 performed as in Fig. 3. It can be seen that the extreme effects of the failure of coscheduling for more than 19 processes have been considerably reduced, and only moderate swapping overheads appear for 20 or more processes.

Drawbacks to trying to solve the coscheduling problem at the parallel language level are twofold. First, although reporting of work completion may be simple, as in the case of the barrier separated DOALLs in Figs. 1 and 3, it can be costly for more complex work scheduling mechanisms. Benten defined the concept of a computational phase containing various parallel constructs and terminated by a work barrier separating it from the next phase. The need to report completion constrains the parallel constructs which can be included in a phase. The second problem with this approach is that all synchronization must be done using the mechanism which has been designed to be operating system insensitive. As pointed out in connection with Fig. 2, global complexes of simple synchronizations restore the sensitive dependence on the operating system scheduler. Eliminating simple, low cost synchronizations entirely from the language is probably worse than the original coscheduling problem.

It is tempting for a parallel programmer to argue for a minimal operating system which statically allocates processes to sufficient processors and then gets out of the way, but there are positive aspects to involving the operating system in process scheduling. Parallel

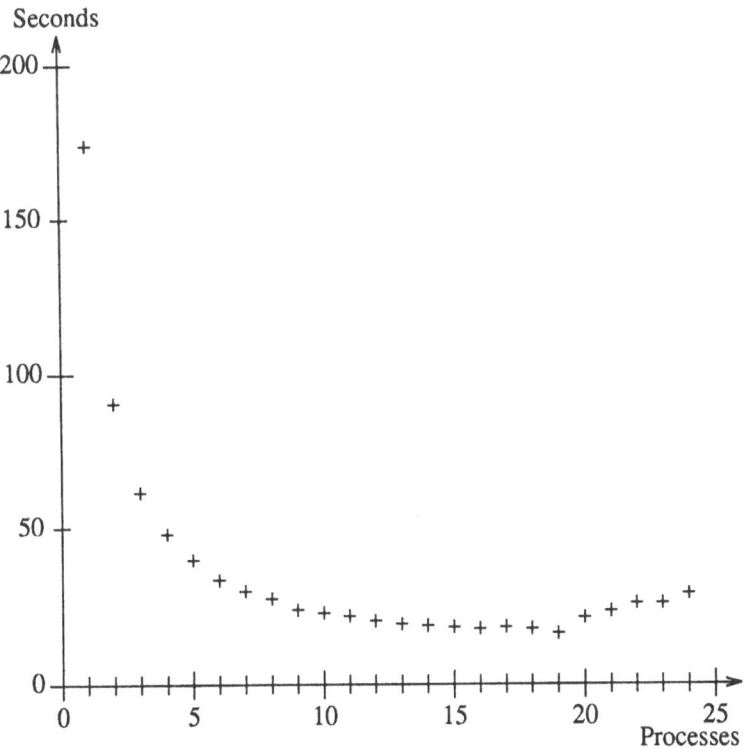

Figure 3: Execution Time for Work Barrier Synchronized Program

programs also do I/O, which is often a sequential part of the program and amenable to standard scheduling techniques. Parallel I/O extensions, such as disk striping, can benefit from the methods known from the multiprogramming environment. The operating system can manage processors as groups instead of treating one processor as central to all activity. Furthermore, it is not really necessary that the operating system run on the same processors used for parallel execution. It can run non-intrusively on separate hardware, as in the CDC6600[3].

Performance Issues

The crux of the need for coscheduling is performance. Poor coscheduling does not affect the correct completion of an otherwise correct parallel program. The reason to write a parallel program is to reduce the time to completion of the computation. This is not the same as reducing the number of CPU cycles used or to reducing the space-time product of memory utilization. It is quite possible that the real time to completion of a parallel program might trade off against overall throughput in a multiprogrammed parallel system. It is important to keep in mind that the normal environment for running a parallel program is one which has sufficient processors. This sufficiency is not based on the maximum parallelism which might be extracted from the algorithm, but on the degree of coscheduling required for efficient execution.

The question then becomes how to efficiently involve the operating system in the scheduling of the processes of a parallel program. A technique to be avoided is invoking the operating system at every potentially blocking synchronization, as this has a strong negative impact on simple, tightly coupled parallel constructs. An example is the parallel construction of a histogram shown as a program in the Force[9] language in Fig. 4. Processes increment elements of the histogram array in parallel, using producer/consumer synchronization to prevent conflict when two processes attempt to update the same element simultaneously. The thing to bear in mind when considering the performance of this program is that, for a synchronization wait to occur, two processes must first obtain the same value of K and then must reach the pair of statements (Consume, Produce) simultaneously. The example is typical of well written parallel programs in that there is a low probability of blocking at most synchronizations. Synchronizations are usually only insurance against wrong behavior in rare cases. Exceptions to this rule are usually major features of program structure, such as barriers separating large computational phases.

A more reasonable approach, used in some parallel systems, i.e. [5], is to invoke the operating system at a blocking synchronization only after spin waiting for a time. The time limit can be tuned statically, or even dynamically, to achieve optimal performance. More precise control over performance can be obtained by supplying static information to the

```
        Asynch INTEGER HIST(100)
        Presched DO  10  I = 1,N
              REAL = FUN(I)
              K = INT(SCALE*REAL + OFFSET)
              Consume HIST(K) into ITMP
              Produce HIST(K) = ITMP + 1
10      End Presched DO
```

Figure 4: Parallel Construction of a Histogram

operating system at the beginning of a run of a parallel program. Coscheduling can satisfy the needs of most synchronizations used in tightly coupled portions of the program, while major global synchronizations can explicitly invoke the system.

The technique for unifying the process scheduling requirements of both parallel language and operating system which is advocated in this paper is a combination of user and compiler supplied information to communicate the resource requirements of a parallel program to the operating system. Information from the user is appropriate when coscheduling needs are based on the overall strategic plan for the parallel computation. More local needs can be derived by a compiler analysis of data dependency and/or synchronization structure. Major synchronizations, such as barriers, which involve all or many of the processes, should have internal implementations which are tailored to the architecture and explicitly involve the operating system's scheduler where necessary for improved performance.

The other position advocated in this paper is that the information to be supplied to the operating system often takes the form of coscheduling requirements. While these are usually requirements for processors, they extend to simultaneous availability of any resources needed to allow tightly coupled processes to make simultaneous progress. A coscheduling specification might involve a request for:

K physical processors,
M words of shared memory,
L locks,
A produce/consume variables,
D disk stripes, and
a time slice of at least T.

Unavailability of resources to satisfy the coscheduling requirements leads to poor performance, not deadlock. Deadlock is a separate issue, except in so far as deadlock can be classified as an extreme case of poor performance.

Conclusions

We have seen that minimization of the use of the real time resource by a parallel program may compete with optimization of the use of other resources. The structure of a parallel program interacts with process scheduling to have a distinct impact on performance. This feature is not present in sequential programs and is not well predicted from the past behavior of the parallel program. Without some knowledge of the structure of a parallel program, the operating system running it may make scheduling decisions that result in extremely poor performance.

Parallel language constructs and restrictions can be developed to limit the impact of operating system scheduling policies on parallel performance. Such techniques lead to minimized interference between parallel work allocation to processes in the program and scheduling of processes by the operating system. The non-interference is difficult to maintain with flexible, general purpose parallel languages.

In preference to non-interference, we advocate active cooperation between the program and the operating system. A combination of user directives, compiler derived program structure information, and calls to the operating system at key synchronizations can support the cooperation. Much of the information needed to optimize performance can be expressed as coscheduling requirements. It is perhaps not surprising that a program written for parallel execution requires some degree of actual parallelism in its hardware resources to execute efficiently.

REFERENCES

[1] M. S. Benten and H. F. Jordan, "Multiprogramming and the Performance of Parallel Programs," *Proc. 3rd SIAM Conf. on Parallel Processing for Scientific Computing*, Los Angeles, CA (Dec. 1987).

[2] *Multimax Technical Summary*, Encore Computer Corporation, Marlboro, MA (1986).

[3] J. E. Thornton, *Design of a Computer: The Control Data 6600*, Scott, Foresman and Co. (1970).

[4] J. S. Kowalik, Ed., *Parallel MIMD Computation: The HEP Supercomputer and its Applications*, MIT Press (1985).

[5] *Multicomputing multitasking operating system (MMOS) reference manual*, Flexible Computer Corporation, Dallas, TX (1986).

[6] Norbert S. Arenstorf and H. F. Jordan, "Comparing Barrier Algorithms," to appear in, *Parallel Computing*, (1989).

[7] F. Darema-Rogers, D. A. George, V. A. Norton and G. F. Pfister, "VM/EPEX - A VM environment for parallel execution," *IBM Res. Rept. RC11225 (#49161)*, T. J. Watson Res. Ctr. (Jan. 1985).

[8] M. S. Benten, *Multiprogramming and the Performance of Parallel Programs*, Ph.D. Thesis, University of Colorado, Boulder (May 1989).

[9] H. F. Jordan, "The Force," Chap. 16 in *The Characteristics of Parallel Algorithms*, L. Jamieson, D. Gannon and R. Douglass, Eds., MIT Press, Cambridge, MA (1987).

SARA: A CRAY ASSEMBLY LANGUAGE SPEEDUP TOOL

Robert G. Babb II

Department of Computer Science and Engineering

Oregon Graduate Center

19600 NW Von Neumann Dr

Beaverton, Oregon 97006

USA

1. Introduction

SARA (Single Assignment Register Assembler) is an extended form of CAL (Cray Assembly Language) meant for obtaining near optimal performance from relatively short (100's of instruction) Cray X-MP basic block code sequences. The SARA Optimizing Preprocessor (informally referred to also as "SARA") converts SARA source files into a form that is acceptable as input to standard Cray Research Inc. CAL Assemblers. The SARA Optimizing Preprocessor can greatly speed up the job of CAL coding by automating the difficult, tedious, and error-prone tasks of assigning registers and ordering instruction sequences to take maximum advantage of the Cray X-MP architecture.

2. History

Several years ago, while working as a consultant for one of the first commercial customers for the Cray X-MP, I was asked to attempt to "speed up" their computing throughput by coding lower precision (16 bit) scalar and vector "VFUNCTION" versions of x**y and 1/sqrt(x) in CAL, for specified ranges of values of x and y. I accomplished this task manually (achieving speedups ranging from 1.7x to 2.3x for scalar and vector versions over the standard Fortran library functions), but found the job of scheduling instruction sequences for optimal performance on the X-MP to be very difficult. The complex timing and conflict characteristics for Cray-style architectures makes it difficult to remember them all, and people are not very good at producing optimum schedules for instruction streams of this type. The job of scheduling instructions and assigning registers is handled fairly well by the latest generation of Fortran compilers for Crays, but when coding in Cray Assembly Language, programmers still can not benefit from this capability. SARA was created to help with this problem.

NATO ASI Series, Vol. F 62
Supercomputing
Edited by J. S. Kowalik
© Springer-Verlag Berlin Heidelberg 1990

3. Language Syntax

3.1 SARA ON/OFF Blocks

SARA source files consist of alternating "SARA ON" and "SARA OFF" sections. CAL statements in SARA OFF sections are copied unchanged. However, as discussed later, SARA-assigned register values can be referenced symbolically in subsequent SARA OFF code sections. The currently implemented versions of SARA are basic block schedulers only. Branch instructions must occur only in SARA OFF code sections. Within SARA ON blocks, CAL macros or pseudo-ops for storage definition are also diagnosed as unrecognized.

3.2 Pseudoregisters and Single Assignment

Since SARA is an extended form of CAL, SARA input source lines generally follow standard CAL formats, with the addition of "*:" control lines and a notation for "pseudoregisters". In SARA, data dependencies between machine instructions are specified explicitly via the use of pseudoregister identifiers. For example, the SARA form of the CAL scalar floating point add instruction

```
S1    S2+FS3
```

could be represented in SARA as

```
S.X   S.Y+FS.Z
```

A pseudoregister is like a variable that can be assigned a value only once (hence "single assignment"), but whose value (once assigned) can be referenced as many times as necessary. Each SARA instruction therefore typically puts its result into a new pseudoregister. SARA allows a CAL programmer to pretend that the machine has an infinite supply of scalar (S), vector (V), and address (A) registers available. Obviously, many different pseudoregisters will be used in a typical basic block.

One result of the operation of the SARA Optimizing Preprocessor is a mapping from pseudoregister identifiers (X, Y, and Z in the example above) to actual registers. Pseudoregister identifiers can consist of any combination of upper case letters and numeric digits. Identifiers beginning with a letter can be up to 8 characters long, identifiers beginning with a digit can be up to seven characters long, and are prefixed automatically with a "%" character.[*1]

---

[*] There are a few other minor restrictions on pseudoregister identifiers. They can not duplicate a symbol (such as a location label) in the same program. SARA does

Register assignments in the SARA output file are specified via CAL "SET" statements. For the example above, if SARA assigned the registers S1, S2, and S3 to pseudoregisters S.X, S.Y and S.Z, the SARA output would look like:

```
        .
        .
        .
*:SARA OFF
X           SET         O'1
Y           SET         O'2
Z           SET         O'3
        .
        .
        .
    S.X  S.Y+FS.Z
```

SARA programmers can either let the SARA preprocessor assign specific registers, or they can "pre-assign" specific register numbers. In the example above, suppose that S2 and S3 had been assigned values corresponding to Y and Z outside this SARA ON block. Then the example could be modified to look like:

```
    S.X  S2.Y+FS3.Z
```

to force the correct register assignments.

SARA assigns registers automatically only for the S, V, and A registers. All other register designators (such as T, B, SB, SM, and ST) must have a number pre-assigned by the programmer (T70.XYZ for example). A tag on a T register corresponds to a T register load "lifetime". For most registers the pseudoregister to actual register binding need not be explicitly expressed. The exceptions to this rule are the special registers A0, S0, VL and VM. These registers must always be specified as 'sreg.preg' where sreg is A0, S0, VL or VM, for example: S0.100, A0.FINAL, VL.00, and VM.LEFT.

3.3   Input Declarations

All pseudoregister operand identifiers used in a SARA ON block must be defined previously in the file. This can occur in one of three ways:

1) defined as a result in a previous instruction in the same block

---

not check for this. If the identifier corresponds to one of the reserved Cray register names:

A0-7, B00-B77, S0-7, T00-T77, V0-V7, SB0-7, SM00-37, ST0-7, CA, CE, CI, CL, MC, RT, SM, VL, VM, or XA

SARA will prefix the name with a "%" character.

2) declared explicitly as a "*:IN" input value

3) declared automatically as a *:IN value passed in from a previous SARA ON block.

In case 1), a previously unencountered pseudoregister is automatically defined by its appearance on the left (result) side of a CAL operation instruction. In case 2), a *:IN declaration is used to specify pseudoregister identifiers for previously computed register values that exist at the point that a SARA ON block is entered. In the example above, to specify that Y and Z have been previously computed, the SARA input would look like:

```
*:SARA ON
*:IN     S2.Y
*:IN     S3.Z
   S.X   S2.Y+FS3.Z
   .
   .
   .
```

Note that this floating point add statement could also have been written:

```
   S.X   S.Y+FS.Z
```

since specification of pre-assigned register numbers, once defined in either an assignment statement or a declaration statement, is optional.

The third possibility will be discussed below as part of the explanation of the "*:PASS" and "*:SAVE" SARA directives. Note that it is an error for a pseudoregister identifier to make its first appearance in a SARA source file on the right (operand) side of a CAL instruction, unless it has been declared via a *:IN statement.

3.4 Unites

Although SARA uses the single-assignment paradigm as a basis for its syntax, the "unite" operation allows specifying that two different pseudoregisters be assigned to the same actual register. The unite operation has the form:

```
   r.preg1~preg2     <right-hand-side>
```

where:

'r'	is S, V, or A
'~'	is the unite operator,
'preg1'	is a "new" pseudoregister identifier, and
'preg2'	is a previously defined pseudoregister.

The unite operator forces the SARA Optimizing Preprocessor to assign the same actual register to different pseudoregisters. A unite is necessary, for example, when multiple execution paths exist. The single assignment rule prohibits one pseudoregister from being used for both result value names. However, subsequent statements need to be able to refer to the result of either computation. Uniting the two pseudoregister identifiers allows either name to be used to refer to the result. This mechanism must also be used whenever the effect of an "update in place" is desired, such as within the body of a loop. The simplest example where a unite is required is the scalar shift operation, which *must* be specified using a unite, because of a restriction of the Cray instruction set, as in:

```
S.19        O'40060
S.FXR~19    S.19<O'60
```

3.5 Output Declarations

In order to schedule instructions within SARA ON blocks, the SARA optimizer needs to know which instructions compute results that are "outputs" of the SARA block. The "*:OUT" declaration is used to accomplish this. The *:OUT statement must be placed immediately before the statement that computes the result that is to be an output of the block. If the output is a memory store operation (which does not have a register as a result), the convention is to specify "MEM" as the output. For example:

```
    .
    .
    .
*:OUT   V.XYZ
        V.XYZ       S.ONE!V.24&VM.23
*:OUT   MEM
        ,A0.35,1    V.39
*:OUT   A1.00
        A1.00       A2.ANM1X+A.KK
```

Every instruction in a SARA ON block must either be used to compute a register value that is an output of the block, or a value that is stored as a MEM output. Otherwise, the SARA Optimizing Preprocessor report that one or more instructions are "unconnected".

3.6 PASS and SAVE Declarations

A programmer who desires to pass a pseudoregister identifier/register number association across a SARA OFF section, can substitute "*:PASS" for "*:OUT". An appropriate *:IN statement will be automatically generated when the next *:SARA ON statement is encountered. The "*:SAVE" declaration is like the *:OUT declaration except that the former is used when to continue a previously PASSed pseudoregister definition to the next SARA ON block. The SAVE declaration can occur anywhere in the block and can be continued across more than one block.

An example of a (non-useful) sequence of SARA blocks illustrating use of *:OUT, *:PASS, and *:SAVE follows:

```
================
INPUT TO SARA:
================
      .
      .
      .
*:SARA ON
*:IN   A1.00
*:IN   A2.01
*:OUT  A3.02
       A3.02      A1.00+A2.01
*:PASS A.03
       A.03       A.02
*:PASS A4.XYZ
       A4.XYZ     A.02*A.00
*:SARA OFF
      .
      .
      .
*:SARA ON
*:IN   A5.ST
       A0.XX      A.03+A.XYZ
*:OUT MEM
       ,A0.XX     A.03
*:SAVE A.XYZ
*:SARA OFF
      .
      .
      .
*:SARA ON
       A0.YY      A.XYZ
*:OUT MEM
       ,A0.YY     A.XYZ
*:SARA OFF

============
SARA OUTPUT
============
      .
      .
      .
*:SARA ON
* p00a: SARA source:
*
*   1. *:IN   A1.00
*   2. *:IN   A2.01
*   3. *:OUT  A3.02
*   4.        A3.02      A1.00+A2.01
*   5. *:PASS A.03
*   6.        A.03       A.02
*   7. *:PASS A4.XYZ
```

```
*   8.          A4.XYZ       A.02*A.00
*:SARA OFF
%00      SET        O'1
%01      SET        O'2
%02      SET        O'3
XYZ      SET        O'4
%03      SET        O'1
* p26a: SARA CAL - CRAY-XMP   1.6bf5  FBCON=1
*                                                    SN  FRT  FIT   PLAIN CAL
   A.%02 A.%00+A.%01                                 4.   2    6    A3 A1+A2
   A.XYZ A.%02*A.%00                                 8.   4    4    A4 A3*A1
   A.%03 A.%02                                       6.   2    2    A1 A3
   .
   .
   .
*:SARA ON
* p00a: SARA source:
*
*   1. *:IN  A1.%03       ** PASSED IN **
*   2. *:IN  A4.XYZ       ** PASSED IN **
*   3. *:IN  A5.ST
*   4.       A0.XX     A.03+A.XYZ
*   5. *:OUT MEM
*   6.       ,A0.XX    A.03
*   7. *:SAVE A.XYZ
*:SARA OFF
%03      SET        O'1
XYZ      SET        O'4
XX       SET        O'0
ST       SET        O'5
* p26a: SARA CAL - CRAY-XMP   1.6bf5  FBCON=1
*                                                    SN  FRT  FIT   PLAIN CAL
   A.XX A.%03+A.XYZ                                  4.   2    2    A0 A1+A4
   ,A.XX A.%03                                       6.  -1    0    ,A0 A1
   .
   .
   .
*:SARA ON
* p00a: SARA source:
*
*   1. *:IN  A4.XYZ       ** SAVED IN **
*   2.       A0.YY     A.XYZ
*   3. *:OUT MEM
*   4.       ,A0.YY    A.XYZ
*:SARA OFF
XYZ      SET        O'4
YY       SET        O'0
* p26a: SARA CAL - CRAY-XMP   1.6bf5  FBCON=1
*                                                    SN  FRT  FIT   PLAIN CAL
   A.YY A.XYZ                                        2.   2    2    A0 A4
   ,A.YY A.XYZ                                       4.  -1    0    ,A0 A4
```

In SARA output, such as that shown above, the SARA ON blocks are listed twice--first with comment asterisks (reflecting the original input order), then, following the "SETs", re-ordered and with diagnostic information appended. Any original comment fields are not shown in the re-ordered output CAL code.

Following the transformed and re-ordered CAL instruction are a series of comment fields:

-- The "SN" column gives the SARA input statement number (before reordering).

-- "FRT" stands for First Result Time. This corresponds to the number of clocks after issue that a computational result is available for scalar instructions, or the time the first in a stream of vector result values should be available.

-- "FIT" stands for Forward Issue Time. An estimate of the latest time that the corresponding instruction could issue (in clocks before the end, or T=0) and still have all computational work completed. (This estimate is not always exact, since SARA views time as running backwards!) While some instructions will issue earlier than SARA predicts, the largest (first) FIT gives a good estimate of the running time of overall sequences from beginning to end since instructions on the critical path will issue at the intervals predicted.

-- The column headed "PLAIN CAL" shows the equivalent simple CAL instruction with assigned register numbers filled in.

4. Outline of the SARA Optimization Strategy

Scheduling and register assignment by the SARA Optimizing Preprocessor are done in reverse time order (backward in time), in the following steps:

1. Determine the "latest forward issue times" for all instruction trees based only on "binary conflicts". (It is possible to determine for any two Cray X-MP instructions the closest they could issue based on the resources they require).

2. Working backwards in time, "bubble" instructions (and their predecessors) in order to account for:

   --function unit reservations/conflicts

   --operand register reservations/conflicts

   --issue time conflicts

   --register assignment conflicts

   The last item is treated as co-equal to the other conflicts, and corresponds to exhausting available register "bandwidth".

3. Choose an instruction to schedule based on a combination of global and local criteria. "Local" means that instructions are chosen as candidates for scheduling from among the set of all mutually conflicting instructions whose issue could tie up a resource at a particular issue time slot. The global criteria is that we bubble the instructions which would have the least worst effect on the resulting set of changes to earliest forward issue times. The remaining instruction is "frozen" (scheduled).

SARA does not guarantee that the resulting schedules are optimal (this is an NP-complete problem [1]) but has performed very well against good hand-coded CAL, and does better than most production compilers. For other studies of various aspects of the instruction scheduling problem for Cray architectures, see [2], [3], and [4].

5. Examples

5.1 Scalar Code Example

SARA preprocessing for scalar code sequences tends to result in an instruction stream that is "issue-time limited". This means that a new instruction will issue at (almost) every available clock tick. Some Cray instructions take more than one clock to issue. An example, taken from an experiment to see whether SARA could be used to automatically improve the output of a compiler, is shown below:

```
*:SARA ON
* p00a: SARA source:
*
*   1.          S.W0       W,
*   2.          A.J0       J,
*   3.          S.W1~W0    S.W0>A.J0
*   4.          S.M07      <07
*   5.          S.K0       S.W1&S.M07
*   6. *:OUT MEM
*   7.          K,         S.K0
*   8.          A.AK       S.K0
*   9.          S.C0       C,A.AK
*  10.          S.M01      <01
*  11.          S.C1       S.M01+S.C0
*  12.          A.AK2      S.K0
*  13. *:OUT MEM
*  14.          C,A.AK2    S.C1
*  15.          S.J1       J,
*  16.          S.MM01     <01
*  17.          S.J2       S.MM01+S.J1
*  18. *:OUT MEM  AFTER: S.J1
*  19.          J,         S.J2
*  20.          S.C57      57
```

```
*   21. *:OUT    S0.TEST
*   22.          S0.TEST    S.C57-S.J2
*:SARA OFF
W0       SET       O'3
J0       SET       O'3
J1       SET       O'1
W1       SET       O'3
M07      SET       O'5
K0       SET       O'2
AK       SET       O'2
C0       SET       O'4
MM01     SET       O'7
C57      SET       O'6
J2       SET       O'5
M01      SET       O'3
TEST     SET       O'0
C1       SET       O'1
AK2      SET       O'1
* p26a: SARA CAL - CRAY-XMP   1.6bf5   FBCON=1
*                                          SN   FRT  FIT    PLAIN CAL
*
   S.W0 W,                                 1.  14    39     S3 W,
   A.J0 J,                                 2.  14    37     A3 J,
   S.J1 J,                                15.  14    24     S1 J,
   S.W1 S.W0>A.J0                          3.   3    23     S3 S3>A3
   S.M07 <07                               4.   1    21     S5 <07
   S.K0 S.W1&S.M07                         5.   1    20     S2 S3&S5
   A.AK S.K0                               8.   1    19     A2 S2
   S.C0 C,A.AK                             9.  14    17     S4 C,A2
   S.MM01 <01                             16.   1    14     S7 <01
   S.C57 57                               20.   1    12     S6 57
   K, S.K0                                 7.  -1    10     K, S2
   S.J2 S.MM01+S.J1                       17.   3     9     S5 S7+S1
   S.M01 <01                              10.   1     7     S3 <01
   S.TEST S.C57-S.J2                      22.   3     6     S0 S6-S5
   J, S.J2                                19.  -1     4     J, S5
   S.C1 S.M01+S.C0                        11.   3     3     S1 S3+S4
   A.AK2 S.K0                             12.   1     2     A1 S2
   C,A.AK2 S.C1                           14.  -1     0     C,A1 S1
```

In this case, the speedup (measured with Cray X-MP performance monitor [5]) over the original compiler code was 2.3x.

5.2 Vector Code Example

Optimum instruction schedules for the Cray X-MP typically depend upon vector length. Unless directed otherwise (through a *:VL directive), SARA schedules for VL 32. (The directive was put in for this example, even though unnecessary, to allow direct comparison of statement numbers with the next example after this . one).

```
*  SARA version  1.6bf5   CRAY-XMP. FBCON=1
* Mandelbrot Set Test Case
        IDENT    BENOIT
```

```
BENOIT    PROGRAM
          A2        32
          VL        A2
*:SARA    ON
* p00a: SARA source:
*
*    1. *:VL      32
*    2. *:IN      A7.00       CURRENT VECTOR START ADDRESS FOR Z
*    3. *:IN      A6.01         "         "         "         "         "    LAMBDA
*    4. *:IN      A5.02         "         "         "         "         "    ITER
*    5. *:IN      T70.04      - BOXSIZE PARAMETER
*
*    6.           A0.11       A7.00             BASE FOR Z
*    7.           A2.12       2                 STRIDE FOR COMPLEX
*    8.           V.XP        ,A0.11,A2.12   XP IS REAL(Z)
*    9.           A0.13       A7.00+1
*   10.           V.YP        ,A0.13,A2.12   YP IS AIMAG(Z)
*   11.           A0.15       A6.01
*   12.           V.RL        ,A0.15,A2.12   RL IS REAL(LAMBDA)
*   13.           A0.17       A6.01+1
*   14.           V.RM        ,A0.17,A2.12   RM IS AIMAG(LAMBDA)
*   15.           V.20        V.RL*RV.XP
*   16.           V.21        V.RM*RV.YP
*   17.           V.SUBX1     V.20-FV.21
*   18.           V.22        V.RL*RV.YP
*   19.           V.23        V.RM*RV.XP
*   20.           V.SUBX2     V.22+FV.23
*   21.           S.30        1.
*   22.           V.31        S.30-FV.XP
*   23.           V.32        V.31*RV.SUBX1
*   24.           V.33        V.YP*RV.SUBX2
*   25.           V.XN        V.32!FV.33
*   26.           V.34        V.31*RV.SUBX2
*   27.           V.35        V.YP*RV.SUBX1
*   28.           V.YN        V.34-FV.35
*   29.           A0.40       A5.02             BASE FOR ITER
*   30.           V.ITERI     ,A0.40,1
*   31.           S.41        1
*   32.           V.ITERN     S.41+V.ITERI
*   33.           V.42        -FV.XP
*   34.           VM.43       V.XP,M
*   35.           V.44        V.42!V.XP&VM.43
*   36.           S.45        T70.04
*   37.           V.46        S.45+FV.44
*   38.           VM.47       V.46,M
*   39.           S.48        VM.47
*   40.           V.52        -FV.YP
*   41.           VM.53       V.YP,M
*   42.           V.54        V.52!V.YP&VM.53
*   43.           V.56        S.45+FV.54
*   44.           VM.57       V.56,M
*   45.           S.58        VM.57
*   46.           S.59        S.48&S.58
*   47.           VM.60       S.59
*.  48.           V.XF        V.XN!V.XP&VM.60
*   49.           V.YF        V.YN!V.YP&VM.60
```

```
*   50.              V.70       V.ITERN!V.ITERI&VM.60
*   51. *:OUT     MEM
*   52.              ,A0.40,1   V.70
*   53.              A0.71      A7.00
*   54. *:OUT     MEM
*   55.              ,A0.71,A2.12   V.XF
*   56.              A0.72      A7.00+1
*   57. *:OUT     MEM
*   58.              ,A0.72,A2.12   V.YF
*:SARA    OFF
%00        SET        O'7
%11        SET        O'0
%12        SET        O'2

   .
   .
   .

%34        SET        O'3
YN         SET        O'1
YF         SET        O'0
%72        SET        O'0
* p26a: SARA CAL - CRAY-XMP   1.6bf5   FBCON=1
*
```

	SN	FRT	FIT	PLAIN CAL
A.%11 A.%00	6.	2	661	A0 A7
A.%12 2	7.	1	660	A2 2
V.XP ,A.%11,A.%12	8.	17	659	V1 ,A0,A2
VM V.XP,M	34.	36	642	VM V1,M
V.%42 -FV.XP	33.	11	607	V6 -FV1
V.%44 V.%42!V.XP&VM	35.	7	572	V4 V6!V1&VM
A.%13 A.%00+1	9.	2	555	A0 A7+1
V.YP ,A.%13,A.%12	10.	17	553	V2 ,A0,A2
VM V.YP,M	41.	36	536	VM V2,M
V.%52 -FV.YP	40.	11	501	V5 -FV2
V.%54 V.%52!V.YP&VM	42.	7	466	V0 V5!V2&VM
S.%45 T.%04	36.	1	442	S6 T70
V.%46 S.%45+FV.%44	37.	11	441	V3 S6+FV4
VM V.%46,M	38.	36	430	VM V3,M
A.%40 A.%02	29.	2	408	A0 A5
V.ITERI ,A.%40,1	30.	17	406	V7 ,A0,1
V.%56 S.%45+FV.%54	43.	11	404	V6 S6+FV0
S.%48 VM	39.	1	394	S3 VM
VM V.%56,M	44.	36	393	VM V6,M
S.%41 1	31.	1	390	S5 1
V.ITERN S.%41+V.ITERI	32.	8	389	V5 S5+V7
S.%58 VM	45.	1	357	S4 VM
S.%59 S.%48&S.%58	46.	1	356	S2 S3&S4
VM S.%59	47.	1	355	VM S2
V.%70 V.ITERN!V.ITERI&VM	50.	7	354	V4 V5!V7&VM
,A.%40,1 V.%70	52.	-1	347	,A0,1 V4
A.%15 A.%01	11.	2	346	A0 A6
V.RL ,A.%15,A.%12	12.	17	344	V3 ,A0,A2
V.%22 V.RL*RV.YP	18.	12	327	V0 V3*RV2
A.%17 A.%01+1	13.	2	310	A0 A6+1
V.RM ,A.%17,A.%12	14.	17	308	V6 ,A0,A2
S.%30 1.	21.	2	292	S1 1.
V.%21 V.RM*RV.YP	16.	12	291	V7 V6*RV2

	SN	FRT	FIT	
V.%31 S.%30-FV.XP	22.	11	290	V5 S1-FV1
V.%23 V.RM*RV.XP	19.	12	255	V4 V6*RV1
V.%20 V.RL*RV.XP	15.	12	219	V6 V3*RV1
V.SUBX1 V.%20-FV.%21	17.	11	184	V3 V6-FV7
V.%32 V.%31*RV.SUBX1	23.	12	149	V7 V5*RV3
V.SUBX2 V.%22+FV.%23	20.	11	148	V6 V0+FV4
V.%33 V.YP*RV.SUBX2	24.	12	113	V4 V2*RV6
V.XN V.%32+FV.%33	25.	11	101	V0 V7+FV4
V.%35 V.YP*RV.SUBX1	27.	12	66	V4 V2*RV3
V.XF V.XN!V.XP&VM	48.	7	53	V7 V0!V1&VM
A.%71 A.%00	53.	2	40	A0 A7
,A.%71,A.%12 V.XF	55.	-1	38	,A0,A2 V7
V.%34 V.%31*RV.SUBX2	26.	12	30	V3 V5*RV6
V.YN V.%34-FV.%35	28.	11	18	V1 V3-FV4
V.YF V.YN!V.YP&VM	49.	7	7	V0 V1!V2&VM
A.%72 A.%00+1	56.	2	2	A0 A7+1
,A.%72,A.%12 V.YF	58.	-1	0	,A0,A2 V0

For comparison, shown below is the result schedule for the same test case but with VL set to 2:

```
* p26a: SARA CAL - CRAY-XMP    1.6bf5   FBCON=1
*
```

	SN	FRT	FIT	PLAIN CAL
A.%15 A.%01	11.	2	163	A0 A6
A.%12 2	7.	1	162	A2 2
V.RL ,A.%15,A.%12	12.	17	161	V6 ,A0,A2
A.%11 A.%00	6.	2	160	A0 A7
V.XP ,A.%11,A.%12	8.	17	158	V5 ,A0,A2
A.%13 A.%00+1	9.	2	156	A0 A7+1
V.YP ,A.%13,A.%12	10.	17	154	V2 ,A0,A2
A.%17 A.%01+1	13.	2	153	A0 A6+1
V.RM ,A.%17,A.%12	14.	17	151	V4 ,A0,A2
V.%20 V.RL*RV.XP	15.	12	140	V3 V6*RV5
V.%21 V.RM*RV.YP	16.	12	134	V1 V4*RV2
V.%23 V.RM*RV.XP	19.	12	128	V0 V4*RV5
S.%30 1.	21.	2	123	S6 1.
V.%22 V.RL*RV.YP	18.	12	122	V7 V6*RV2
V.%31 S.%30-FV.XP	22.	11	121	V4 S6-FV5
V.SUBX1 V.%20-FV.%21	17.	11	115	V6 V3-FV1
V.SUBX2 V.%22+FV.%23	20.	11	109	V1 V7+FV0
V.%32 V.%31*RV.SUBX1	23.	12	104	V3 V4*RV6
V.%33 V.YP*RV.SUBX2	24.	12	98	V0 V2*RV1
V.%34 V.%31*RV.SUBX2	26.	12	92	V7 V4*RV1
V.%42 -FV.XP	33.	11	87	V1 -FV5
VM V.XP,M	34.	6	82	VM V5,M
V.XN V.%32+FV.%33	25.	11	81	V4 V3+FV0
V.%44 V.%42!V.XP&VM	35.	7	76	V0 V1!V5&VM
V.%52 -FV.YP	40.	11	75	V3 -FV2
VM V.YP,M	41.	6	70	VM V2,M
V.%54 V.%52!V.YP&VM	42.	7	64	V1 V3!V2&VM
S.%45 T.%04	36.	1	60	S5 T70 .
V.%46 S.%45+FV.%44	37.	11	59	V3 S5+FV0
V.%56 S.%45+FV.%54	43.	11	53	V0 S5+FV1
A.%40 A.%02	29.	2	50	A0 A5
V.ITERI ,A.%40,1	30.	17	48	V1 ,A0,1
VM V.%46,M	38.	6	42	VM V3,M

```
S.%48 VM                              39.  1   36   S2 VM
VM V.%56,M                            44.  6   35   VM V0,M
S.%41 1                               31.  1   32   S4 1
V.ITERN S.%41+V.ITERI                 32.  8   31   V3 S4+V1
V.%35 V.YP*RV.SUBX1                   27. 12   30   V0 V2*RV6
S.%58 VM                              45.  1   26   S3 VM
S.%59 S.%48&S.%58                     46.  1   25   S1 S2&S3
VM S.%59                              47.  1   24   VM S1
V.%70 V.ITERN!V.ITERI&VM              50.  7   23   V6 V3!V1&VM
V.YN V.%34-FV.%35                     28. 11   18   V1 V7-FV0
,A.%40,1 V.%70                        52. -1   16   ,A0,1 V6
V.XF V.XN!V.XP&VM                     48.  7   15   V3 V4!V5&VM
A.%71 A.%00                           53.  2   10   A0 A7
,A.%71,A.%12 V.XF                     55. -1    8   ,A0,A2 V3
V.YF V.YN!V.YP&VM                     49.  7    7   V0 V1!V2&VM
A.%72 A.%00+1                         56.  2    2   A0 A7+1
,A.%72,A.%12 V.YF                     58. -1    0   ,A0,A2 V0
```

It is interesting to note that the Mandelbrot set example can be speeded up further by *adding* read statements. Many vector code sequences are limited (as far as SARA optimization is concerned) by the number of vector registers available. Since there are two vectors of values that are read once at the beginning of the Mandelbrot example (V.VY and V.VX) and then used several times, adding re-reads of these values can have the effect of freeing up vector registers for intermediate results. The speedup in this case from adding the re-reads is about 1.08x for vector length 32. This is an example of the kind of experimental code tuning that is possible with the aid of SARA, but which would be very tedious otherwise.

5.3 The SARA Coloring Game

One way of checking whether the (latest possible) issue times predicted by SARA make sense is to find the conflict (or other timing rule) that explains why each instruction is scheduled at the clock specified. For example, if two vector instructions both use the floating point adder, they must be scheduled at least VL+4 apart. If the instruction sequence shows them further apart, then there must be another explanation for their separation, such as a vector register reservation conflict. The SARA coloring game consists of using a variety of colored pens to code the various explanations, and diagramming these effects. The resulting patterns can form not only interesting artistic designs, but can also give a global idea of what resources are in short supply at which points in the execution sequence. Such analysis frequently leads to ideas for speeding up the sequence still further.

6. Conclusions

6.1 Limitations

SARA currently does not recognize data dependences through memory. Hence, results may be incorrect if a value is stored and then read from memory within the same SARA block, since SARA may schedule the read *before* the write!

Another limitation of the current version of SARA (September 1989) is the problem of "runaways". Runaways occur when SARA attempts to evaluate too many overlapping expression trees in parallel. When a runaway occurs, a gap arises in the predicted instruction schedule that is larger than the largest possible interval between instructions for a Cray X-MP. SARA recognizes runaways, but currently does not have the ability to fix the problem automatically because SARA uses a "persistent" scheduling strategy. The SARA programmer must manually force the completion time of one or more trees backward in time from the end (T=0) by assigning a minimum separation in clocks as a MINSEP parameter on the appropriate *:OUT declarations.

6.2 Results

SARA is written in Fortran for portability. It runs on a wide variety of UNIX and non-UNIX machines, including Crays under CTSS and COS, as well as DEC VAX computers under VMS. The speed at which scheduling and register assignment can be accomplished is somewhat problem dependent, but ranges from 20-30 statements a second on a SUN 3/80 to hundreds of statements a second on a Cray X-MP. The performance of the SARA Optimizing Preprocessor could be greatly improved without affecting its functionality.

6.3 Future Work

Although SARA was originally designed to function as a CAL programmer's assistant, we are also investigating the possibility of using the technology to improve the performance of existing production Cray higher level language compilers. In its original mode, in addition to its use for squeezing that last bit of performance out of computationally intensive kernels, it could also serve as a library maintenance and portability aid for Cray-like architectures, since optimum schedules for mathematical library routines are often invalidated by different instruction timings in newer machines, even if the instruction set is upward-compatible.

We plan to enhance SARA in the near future to include the ability to accept raw CAL as an input language (by generating appropriate internal pseudoregister identifiers), to add the ability to breakup runaways automatically, and to support other Cray X-MP-like architectures. SARA currently supports both the X-MP and the Scientific Computer Systems SCS-40. We are currently adding support for the Cray Y-MP and may also add support for the Cray 2.

7.   References

[1]   S. Arya, "An Optimal Instruction-Scheduling Model for a Class of Vector Processors", IEEE Trans. Computers, Vol. C-34, No. 11, Nov. 1985, pp. 981-995.

[2]   D. Bernstein, H. Boral, and R. Y. Pinter, "Optimal Chaining in Expression Trees (Preliminary Version)", SIGPLAN Notices, Vol. 21, No. 7, July 1986, pp. 1-10.

[3]   J. Tang and E. S. Davidson, "An Evaluation of Cray-1 and Cray X-MP Performance on Vectorizable Livermore Fortran Kernels," ACM Proc., 1988, pp. 510-518.

[4]   S. Weiss and J. E. Smith, "A Study of Scalar Compilation Techniques for Pipelined Supercomputers," SIGPLAN Notices, Vol. 22, No. 10, Oct. 1987, pp. 105-109.

[5]   E. A. Williams, "Measurement of Two Scientific Workloads Using the CRAY X-MP Performance Monitor", Technical Report SRC-TR-88-020, Supercomputing Research Center, Lanham, MD., Nov. 1988.

MODELING THE MEMORY OF THE CRAY2 FOR COMPILE TIME OPTIMIZATION

C. Eisenbeis, W. Jalby, A. Lichnewsky
INRIA
Domaine de Voluceau
78153 LE CHESNAY CEDEX
FRANCE

Abstract

In a previous work [3], a cyclic scheduling method was shown efficient to generate vector code for the Cray-2 architecture, and compared to existing compilers. This method was using the framework of microcode compaction through a simplified model of the Cray-2 vector instruction stream. In this paper, we further elaborate on how to model the machine architecture within the underlying cyclic scheduling method. The impact of the choice of the model on the code generated is analyzed and performance results are presented.

1 Introduction

The Cray-2 architecture provides both vector registers and a local memory, which must be used to hide the main memory latency, and reduce the demand on main memory throughput, to obtain very efficient programs. Similar characteristics occur for several modern vector processors, but this is much more critical on the Cray-2 (because of the use of dynamic memory and of a very short CPU cycle) than on other processors, like the Cray X-MP or Y-MP. To exploit this memory hierarchy, the code generator faces a significant resource allocation and scheduling problem, which is made even more critical by the absence of chaining in the Cray-2 architecture.

Our approach to these problems relies on several ideas:

- the approximation of the NP-complete original optimization problem [5] [9] by a sequential procedure involving three natural subgoals: the functional unit scheduling, data spilling, and register allocation,

- usage of a simplified model of the architecture, that permits to concentrate on the key architectural features, and reduce the complexity of the involved algorithm.

- a careful scheduling of the functional units performing vector operations in parallel,

- a register allocation procedure that attempts to make efficient use of the limited size of the register set,

This overall approach is described in our previous paper [3], and we will focus here on the choice of the model used to present an abstract and simplified description of the machine.

Although our presentation is restricted to the optimization of vector loops for the Cray-2 architecture, our techniques have much wider applicability, and are of interest for RISC and VLIW architectures. The fact that the common memory latency of the Cray-2 is much larger than on most current vector machines permits us to explore a direction of growing importance. Indeed, the relative importance of both memory delays and transmission latency is increasing because of the evolution of semiconductor technology and the decrease in cycle time. The architectural evolution toward larger multiprocessor configurations will also result in the usage

NATO ASI Series, Vol. F 62
Supercomputing
Edited by J. S. Kowalik
© Springer-Verlag Berlin Heidelberg 1990

of interconnection networks that will have longer latencies, and possibly very complex dynamic behavior. Some of these characteristics are already present on the Cray-2, and our experimental results show how they can be tackled.

Our presentation will adopt the following outline. In section 2, a model for the execution of the vector instructions on the Cray-2 is introduced. It captures the RISC-like nature of the Cray-2 vector architecture, and allows the use of a framework and techniques developed for code optimization on horizontally microcoded processors. Then in section 3 we present the principles of operation of our code optimizer (VASCO). In section 4, we analyze the impact of the simplifying assumptions of our model on the performance and we show how this model can take into account parameters describing the common memory and its practical behavior.

Although the experimental results and the code optimizer used are specific to the Cray-2, most of the ideas may be carried to other machines with similar memory hierarchies.

2 A model of the Cray-2 architecture

In this section, we present a model of the instruction execution of the Cray-2. The key point is to develop a model which has the ability to handle simultaneous execution of several instructions provided that they do not use the same hardware resource. Such a model is provided within the classical framework of microcode optimization, and enables us to use the techniques developed in that context [14] [5] [11] [9].

For the convenience of the reader, the main characteristics of the Cray-2 are summarized hereafter. The main memory is shared by the 4 CPU's, organized in four quadrants linked to the processors by a X-bar switch and involves adequate access arbitration and buffering mechanisms. Each CPU, privately owns a local memory of 16 Kwords, 8 vector registers of 64 elements each, 8 floating point registers, and 8 address registers. The transfers between all these storage elements are entirely managed by explicit transfer instructions, emitted by the the compilers. The peak data rate from common memory is one word per cycle per processor, with a high latency (depending on the exact memory options). However, two independent vector memory accesses can be partially overlapped resulting in an "ideal" rate of 72 cycles per block of 64 elements (210 Mwords per second). The local memory provides the same peak access rate but has a much smaller latency and the absence of conflicts guarantees a practical bandwidth equal to its peak. However, the aggregate data rate from these two memory levels is not sufficient to saturate the two floating point units, making use of the vector registers a crucial part of code optimization for the Cray-2. The vector register file can support up to 8 accesses simultaneously. Moreover, due to the lack of instruction chaining, the vector registers have to be managed very carefully as buffers in order to exploit the parallelism between the functional units.

A classical notion in the framework of microcode optimization is the reservation table which allows description of the use of resources for a series of discrete time steps. These tables specify, for each time step, the occupancy of the hardware resources subject to reservation. In addition, all the instructions (or microoperations) are associated with elementary reservation tables (called in the sequel templates) which describe the impact of the instruction execution on the reservation tables. Each template may specify reservations of several resources spanning several time steps. Scheduling the basic operations and checking for the effect of interlocks now amounts to placing the templates onto the reservation table in a conflict-free manner. The role of the code optimizer is to determine the templates scheduling without violating dependence and resource constraints while minimizing the execution time.

The key advantage of microcode compaction framework is the use of reservation tables which can take precisely into account simultaneous activities of the various components of the machine. A first solution to embed the Cray-2 architecture in such a model is to consider all the instructions (scalar and vector) and to associate with each of them a template describing it at the processor

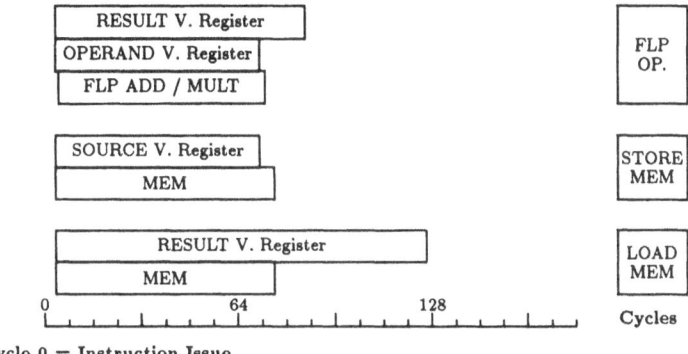

Cycle 0 = Instruction Issue

Figure 1: Instructions and Resource Reservations

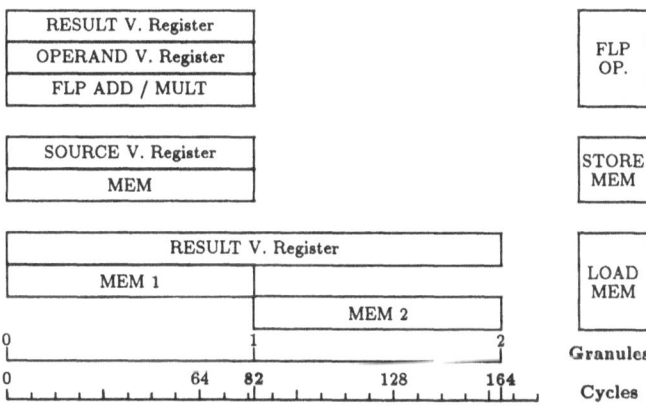

Figure 2: Templates with a granule of 82 cycles

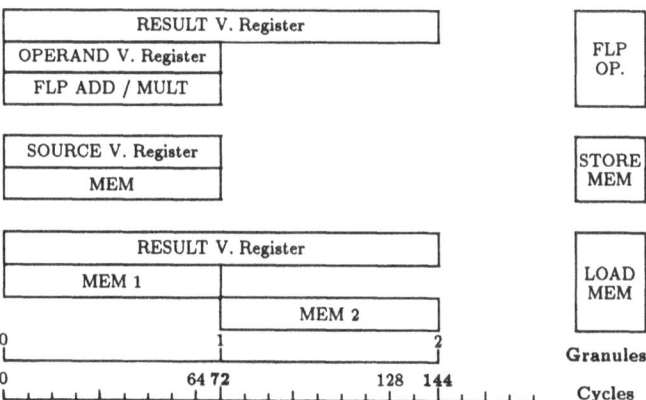

Figure 3: Templates with a granule of 72 cycles

cycle level. Figure 1 shows such templates for the most common Cray-2 vector instructions. Such a fine grain model was successfully used by Arya [1] for optimizing code for the Cray-1. The main drawback of such a solution is the cost of the resulting optimization phase which leads very quickly to optimization problems over the integers of high computational complexity.

We chose to go a step further in the modeling process by simplifying it to reduce the complexity of the optimization phase. We are also able to take advantage of this modeling strategy to describe the practical characteristics of main memory access. The rationale and key model characteristics are summarized hereafter:

- First, the model takes only into account vector operations, with full vector length. As a result, we proceed in two phases. The first one only considers vector instructions, schedules them, and allocates vector registers and vector temporaries. In a final pass, we emit and schedule scalar code. This is justified by the fact that we are primarily interested in vector loop optimization for which most of the time is spent in the vector instructions. Neglecting to optimize vector length less than 64 is not considered too severe since it occurs only once per loop after the classical strip mining loop transformation.

- Second, for the scheduling of the vector instructions, we first define a time unit called macrocycle. All the timings during the scheduling of the vector instructions are expressed as integer multiples of this time unit. In Figure 1, we give the exact timings of the vector instructions together with their representations as templates using a macrocycle of 82 cycles (Cf. Figure 1) and a macrocycle of 72 cycles (Cf. Figure 2). The introduction of the macrocycle greatly simplifies the reservation tables, and therefore their optimization. However, for the scheduling of scalar instructions, we work with a time granularity of a single cycle, which is almost two orders of magnitude smaller. It should be noted that for instructions which do not involve common memory, the timings can easily be obtained from the manufacturers specifications. The case of instructions involving common memory transactions is more complicated: due to conflicts, the real time spent in a vector load can vary largely. This implies that mapping loads or stores into templates requires approximations within the modeling procedure.

- Third, for scheduling, we will assume that the vector instruction issue is exactly described by the program's code according to our breakdown into macrocycles, and therefore entirely under our control. Several vector instructions can be executed at the same macrocycle provided there is no resource conflict. This differs from the actual hardware issue policy, which is greedy, and therefore will start an instruction as soon as it will not cause any resource conflict. This difference is studied in detail in [4].

The templates are derived directly from the description of the instructions, by marking the resources occupied during each time macrocycle. The busy times are simply rounded up to the next granule. For example, Figure 1 (respectively Figure 2) gives the templates with a macrocycle of 82 (respectively 72) clock periods. The number 82 corresponds to the longest reservation time, in processor cycles, which occurs for the result register in a floating point add. (Cf. Figure 1). The number 72 corresponds to the minimal latency between successive vector memory access.

Due to the choice of the macrocycle (72 or 82), all the resources in the templates are considered as single stage pipeline, except the common memory access, which is considered as a 2-stage pipeline. Thus, in the case of a main memory store or load, the access is split into 2 macrocycles and main memory accesses might be partially overlapped. The local memory access is single stage. We clearly see the advantage of the macrocycle which results in extremely simple templates (busy times of either one or two macrocycles), greatly simplifying the complexity of the vector scheduling phase, at the price of an approximation of the processor timings.

```
      DO 1 i=1,1200,1
        z(i) = (a+x(i)*y(i))*b
1       CONTINUE
```

Figure 4: Fortran code of MV1

BODY

	0	1	2	3	4	5	6	7	8	9	10	11	12	13	14	15	16	17	18	19	20
MEM1	X1	Y1		X2	Y2		X3	Y3		X4	Y4	Z1			Z2			Z3			Z4
MEM2		X1	Y1		X2	Y2		X3	Y3		X4	Y4									
ADD						+1			+2			+3			+4						
MUL				*1			*2	*1		*3	*2		*4	*3			*4				

Figure 5: Reservation table for MV1 (VAS72)

BODY

	0	1	2	3	4	5	6	7	8	9	10	11	12	13	14	15
MEM1	X1	Y1		X2	Y2		X3	Y3	Z1			Z2			Z3	
MEM2		X1	Y1		X2	Y2		X3	Y3							
ADD					+1			+2			+3					
MUL				*1		*1	*2		*2	*3		*3				

Figure 6: Reservation table for MV1 (VAS82)

For the vector scheduling phase (which only deals with macrocycles), the only difference between a macrocycle of 72 and 82 is the occupancy of the result register: two macrocycles in the first case, and one in the second case. However, this minor difference at the level of the resulting models hides deeper effects which will appear both at compile-time (register usage) and at run-time. The implications of the choice of the macrocycle will be discussed in detail in the next section.

3 VASCO: a code optimizer for the Cray-2

This section outlines the organization and the principles of VASCO, our code optimizer for the Cray-2. As explained in the previous section, our code optimization strategy distinguishes between vector operations and scalar operations: in a first coarse granularity pass, vector operations are scheduled, then, in a subsequent pass, scalar operations are scheduled, working at the machine cycle granularity. The reader should by now be aware of the approximate nature of our "optimization" procedure, and should acquire a deeper insight into the rationale of our approach by considering the related set of heuristics that support the problem decomposition steps exposed below.

The optimization of vector operations itself is decomposed in two consecutive phases:

- scheduling of the functional units assuming an infinite number of pseudo-registers.

- management of the storage units: spilling in local memory, physical register allocation

During the first phase, the activities of the virtual registers are very precisely recorded, first, to preserve the dependencies within the program, and second, to gather all the information necessary for the allocation and spilling phases. The allocation of physical registers is performed after the functional unit scheduling for two major reasons: first, it avoids the premature introduction of unnecessary dependencies due to register reservations, and second, it simplifies the scheduling pass since we do not have to take into account reservation conflicts of the registers.

3.1 Preprocessing

We are working on a vector loop which has already been detected and processed by a vectorizer (in our case VATIL, [10]). The dependence graph [8] is available at this stage and several optimizations, including strip mining, have been performed. The dependence information (intra- and inter-iterations) is used during the scheduling process to preserve the semantics of the program. The strip mining operation consists in breaking (blocking) all the vectorized loops by blocks of 64 iterations. For a loop of N iterations, the result is an outermost loop of $\left\lceil \frac{N}{64} \right\rceil$ and an innermost loop of 64 iterations.

Since vector operations are considered as atomic blocks, the term iteration (respectively loop) will exclusively denote outermost iteration (respectively outermost loop).

3.2 Code generation of the loop body

In this phase, we generate an intermediate vector code for the loop body, which is used as a basic pattern for the cyclic scheduling of the iterations.

The registers are handled in a virtual manner according to a single assignment rule: the number of registers is assumed to be unbounded, and each time a register is needed to hold a value resulting from a load or an operation, a new register is allocated. In the cyclic scheduling, each iteration of the loop body will be represented by a similar copy of the generated intermediate code. The virtual registers defined in the generic loop body are named R_1 through R_k. The copy corresponding to iteration i will create the registers numbered $R_1(i)$ through $R_k(i)$. The virtual registers which either live on loop entry or are invariant are named $R_l(0)$. Globally, as a consequence of the single assignment rule, each virtual register is written once, but may be read several times. The main advantage of such a technique is to avoid introduction of artificial dependencies due to bad allocation [2].

3.3 Scheduling

The scheduling of the whole loop involves two intimately related subproblems: scheduling of the generic loop body code, and then scheduling of the successive iterations. We use a cyclic scheduling technique which was developed for array processor microcode compaction [11] [9], and which tackles the two issues simultaneously.

In this method, all the iterations are exactly scheduled following the same pattern derived from the generic loop body. Therefore, each iteration contributes to the reservation table by a translated copy of the generic loop body reservation table RT. Iterations are started with a constant unknown period of d macrocycles. The global scheduling problem is to build a reservation table, RT, for the loop body and find a period, d, such that:

- the sequence $RT + kd$ does not conflict(at each macrocycle, a given functional unit is not used more than once).
- the semantic dependence constraints are satisfied.
- the total execution time is minimized.

Note that the execution of one iteration may span more than d macrocycles and therefore, at a given cycle, several successive iterations might be concurrently executing. Consequently, the local scheduling has to take *cyclicity constraints* into account in order to avoid resource conflicts between iterations. This simply means that dependencies are satisfied under the periodic initiation mode, and cyclicity constraints can readily be computed from the dependence graph as a function of d. The payoff of that additional complexity is the perfect chaining between iterations.

For determining d, we first compute for each functional unit, t, the number of macrocycles, n_t, where t is used during one iteration. The initial value for d is chosen to be the maximum value of n_t. This corresponds to the saturation of one of the functional units. Then for this period, d, we try to schedule the loop body according to a list scheduling strategy modified to take into account both semantic and cyclicity constraints. If it appears that no cyclic scheduling is possible with period d, we increment it by one and try again. However, if the loop is vectorizable, it is always possible to find a cyclic scheduling with the initial value of d (for more details see [11], [9]). The technique used is the exact equivalent of the introduction of delays in a pipeline [13].

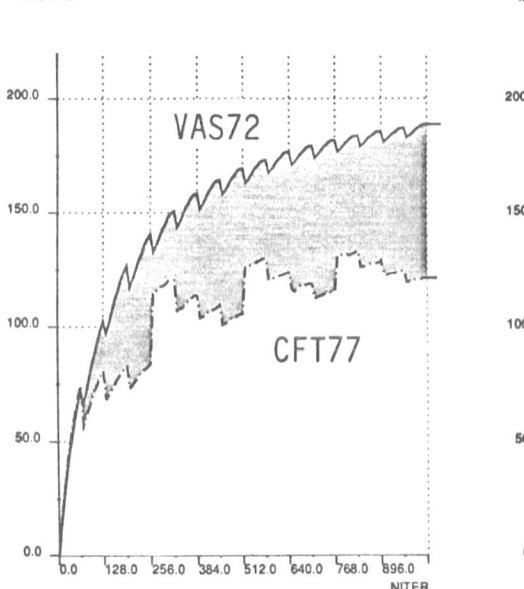

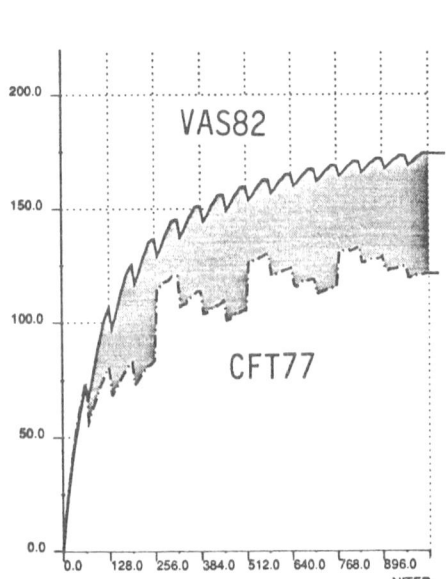

Figure 7: Performance of VAS72 compared with CFT77: MV1

Figure 8: Performance of VAS82 compared with CFT77: MV1

It should be noted that since we are dealing with virtual registers, the computation of d does not take into account the occupation of the registers. Moreover, the compaction process as described above is architecture independent in the sense that all the architectural characteristics (which are just used as parameters by the compaction procedure) are embedded into the template description. So, for example, our two different models of the Cray-2, which differ by the choice of the value of the macrocycle, will use the same scheduling procedure but with a different set of templates: VAS72 (respectively VAS82) corresponds to the version of the code optimizer using the templates as described in Figure 2 (respectively in Figure 3). An example of the result of the compaction procedure using the two models is given in Figures 5 and 6.

The success of such a strategy of scheduling is mainly due to the fact that the templates are extremely simple: most functional units are not used more than one macrocycle.

Figures 9, 10 and 11 give examples of the performance of the codes generated by VAS72 and VAS82 in comparison with codes obtained through CFT77 compiler. Our code code generation results in a better usage of the memory bandwidth, more systematic benchmarking results are given in [3] and [4].

M1	$y(i)$	$=$	$(a * x(i)) + b$
MV1	$z(i)$	$=$	$((x(i) * y(i)) + a) * b$
MVF1	$t(i)$	$=$	$((x(i) * a) + y(i)) * z(i)$
MCOM	$zr(i)$	$=$	$xr(i) * yr(i) - xi(i) * yi(i)$
	$zi(i)$	$=$	$xr(i) * yi(i) + xi(i) * yr(i)$
DROT	$x(i)$	$=$	$c * z(i) - s * t(i)$
	$y(i)$	$=$	$s * z(i) + c * t(i)$
MVC1	$a(i)$	$=$	$a(i) + b(i,1) * c1$
MVC2	$a(i)$	$=$	$a(i) + b(i,1) * c1 + b(i,2) * c2$
MVC3	$a(i)$	$=$	$a(i) + b(i,1) * c1 + b(i,2) * c2 + b(i,3) * c3$
MVC4	$a(i)$	$=$	$a(i) + b(i,1) * c1 + b(i,2) * c2 + b(i,3) * c3 + b(i,4) * c4$
MVC5	$a(i)$	$=$	$a(i) + b(i,1) * c1 + b(i,2) * c2 + b(i,3) * c3 + b(i,4) * c4 + b(i,5) * c5$
MVC6	$a(i)$	$=$	$a(i) + b(i,1) * c1 + b(i,2) * c2 + b(i,3) * c3 + b(i,4) * c4 + b(i,5) * c5$ $+ b(i,6) * c6$
MVC7	$a(i)$	$=$	$a(i) + b(i,1) * c1 + b(i,2) * c2 + b(i,3) * c3 + b(i,4) * c4 + b(i,5) * c5$ $+ b(i,6) * c6 + b(i,7) * c7$
MVC8	$a(i)$	$=$	$a(i) + b(i,1) * c1 + b(i,2) * c2 + b(i,3) * c3 + b(i,4) * c4 + b(i,5) * c5$ $+ b(i,6) * c6 + b(i,7) * c7 + b(i,8) * c8$
MVC9	$a(i)$	$=$	$a(i) + b(i,1) * c1 + b(i,2) * c2 + b(i,3) * c3 + b(i,4) * c4 + b(i,5) * c5$ $+ b(i,6) * c6 + b(i,7) * c7 + b(i,8) * c8 + b(i,9) * c9$
MVC10	$a(i)$	$=$	$a(i) + b(i,1) * c1 + b(i,2) * c2 + b(i,3) * c3 + b(i,4) * c4 + b(i,5) * c5$ $+ b(i,6) * c6 + b(i,7) * c7 + b(i,8) * c8 + b(i,9) * c9 + b(i,10) * c10$
LL1	Lawrence Livermore Kernel 1		
LL7	Lawrence Livermore Kernel 7		
LL12	Lawrence Livermore Kernel 12		
LL21	Lawrence Livermore Kernel 21		
LL183	Third loop of Lawrence Livermore Kernel 18		

Table 1: Program kernels used in the experiments

3.4   Spilling and register allocation

During this step, we consider the schedule, ($\mathcal{FU}$-sched) obtained in the previous step as given, and will endeavor to find a physical register allocation scheme which permits to execute it as planned. Clearly, it may become necessary to insert delays in the schedule, or transfer data to and from local memory because of a shortage in registers, thus obtaining a schedule ($\mathcal{FU}$-sched*). We are in effect decoupling the scheduling problem by imposing the constraint that ($\mathcal{FU}$-sched*) is derived from ($\mathcal{FU}$-sched) by delay insertion and register spilling.

The key characteristic of our register allocation procedure is that, for vector loops, it requires a number physical registers equal to the number of simultaneously alive virtual registers implied by the functional unit schedule [1] (this number is called CRQ and is entirely dtermined by

---

[1] In the absence of spilling, given a functional unit scheduling, it thus generates the minimal number of registers needed.

($\mathcal{FU}$-**sched***). This allocation is obtained by unrolling the loop and is described in detail in [4].

Therefore after having generated the functional unit schedule ($\mathcal{FU}$-**sched**), the quantity CRQ is evaluated. If CRQ is less or equal to 8, the register allocation procedure is directly applied. Otherwise, data spilling and delay insertion is performed (generating a new functional unit scheduling ($\mathcal{FU}$-**sched***)), in order to reduce CRQ down to 8 (cf [3]). Then the register allocation is applied.

4 Discussion of the model

The quality of the code produced by VASCO not only depends upon the algorithms and heuristics used for the optimization, but also upon our model of the Cray-2 architecture. In this section we will analyze how the choice of the model influences the code generated and its performance; this last point brings the problem of evaluating how much the model differs from the real behavior.

Four major simplifications can be found in our model, and need to be validated:

1. We used a model which is explicitly synchronized on macro-cycles intervals although the execution appears asynchronous when analyzed at this level of granularity. This hides a more complex issue mechanism which works at a much lower time scale and involves a much more detailed description of the machine. In reality, we do not have an entire control over the scheduling of the instructions issued as soon as there is no resource conflict.

2. The loop optimization procedure takes into account only vector instructions; the scalar and address instructions are scheduled in a subsequent pass.

3. The choice of the macrocycle duration introduces some errors in that the timing of every instruction is rounded up to the next multiple of a macrocycle length (discretization error)

4. While the timing of non-memory instructions (instructions whose all operands are in registers or local memory) can be accurately described, the timing of the common memory instructions has to be modeled: it is impossible to statically predict all memory conflicts statically, in addition to its being too complex.

In this paper the impact on performance of the last three points mentioned above is analyzed in more details. An analysis of the first aassumption can be found elsewhere [4]. The experimental results were obtained using the codes described in Table 1. The sequence of kernels MVC1 through MVC10 was chosen because these kernels correspond to a gradual increase in the complexity of the loop body and allow a precise determination of the models' impact on resource usage.

Codes	VAS72		VAS82	
	Sc Ov	Chime	Sc Ov	Chime
M1	0%	72	0%	74.5
MV1	2.5%	74.7	0.8%	81.6
MVF1	0.7%	73.25	0%	72
MCOM	0%	73.3	0.8%	81.3
DROT †	0.5%	71.2	1.2%	84.25
MVC1	0%	72	0.9%	73.6
MVC2	0%	72	0.6%	79
MVC3	0.4%	72.8	1%	82
MVC4 †	0.7%	62.8	1.2%	82.5
MVC5 †	1.2%	59.1	0.7%	81
MVC6 †	6.6%	71.4	1.8%	83
MVC7 †	4.9%	69.9	0.8%	82.3
MVC8 †	6.9%	64.2	0.2%	82.3
MVC9 †	9.4%	66.9	0.7%	82.3
MVC10 †	10.7%	70.2	0.5%	81.4
LL1 †	1.3%	74.7	0%	82.5
LL7 †‡	7.8%	73.9	1.3%	78.8
LL12	0.5%	72.8	0.5%	72.8
LL21	0%	72	0.7%	78.7
LL183	0%	72	0.4%	75.3

Codes	MWORDS		
	VAS72	VAS82	CFT77
M1	66.8	68.7	59.5
MV1	67.3	68.5	55
MVF1	69.6	70.0	65.0
MCOM	68.1	68.1	59.4
DROT	68.9	68.7	55.3
MVC1	69.8	70.0	58.4
MVC2	68.9	69.7	58.7
MVC3	69.1	69.3	60.1
MVC4	67.9	68.7	59.7
MVC5	67.8	68.8	60.1
MVC6	63.0	68.2	61.1
MVC7	65.5	67.6	62.3
MVC8	60.7	67.8	63.2
MVC9	63.4	68.1	63.7
MVC10	54.2	67.8	64.2
LL1	66.6	68.2	63.6
LL7	63.3	68.4	59.3
LL12	69.6	69.6	64.5
LL21	67.3	67.7	57.9
LL183	68.7	69.4	69.4

Table 2: Latencies (in number of cycles) of two different models and impact of scalar instructions (stride 1); the presence of a † (resp. a ‡) following code name means that VAS72 (resp. VAS82) used spilling.

Table 3: Memory bandwidth usage (stride 2, vector length 1024 and timings performed in dedicated mode)

4.1 Impact of scalar instructions

Ignoring the scalar instructions in the optimization of the loop body greatly simplified the optimization procedure. This was justified by the fact that most scalar instructions take a number of cycles an order of magnitude smaller than the full vector length vector instructions. However, a legitimate question arises as to how much the post pass scheduling of the scalar instructions costs, and whether it excessively perturbates the execution of vector instructions. To obtain an experimental estimation of the cost, we first measured the latency (denoted $Lat1$ and expressed in cycles) between two consecutive blocks of 64 iterations for the codes generated by VASCO; theses codes were then trimmed down by suppressing all scalar instructions but the jumps and vector length decrementations (i.e., this corresponds to discarding almost all the scalar instructions), and the latency for these codes was measured (denoted $Lat2$). By comparing these two latencies and computing the scalar overhead $ScOv = (Lat1 - Lat2)/Lat1$, we obtained an estimate of the impact of the scalar instructions on the performance (cf. Table 2). It turns out that the cost in most cases is less than 5%, the only exceptions worth mentioning being codes involving spilling (these codes are marked with a single or a double dag in Table 2) for which the cost might be as high as 11 %. The reason for this is that, in these cases, the pointers to the local memory locations used for spill have to be spilled themselves due to the lack of address registers, which drastically increases the number of scalar instructions.

4.2 Choice of the macrocycle

The choice of both macrocycles (72 and 82 cycles) was first motivated by the resulting simplification of the templates, and therefore, the table reservation optimization process. Consequently all the elementary reservation tables for both models were extremely simple. Although the difference between the two models may seem minor, the impact on the generated code and performance is far from being negligible. In the sequel, we shall designate by VAS72 (respectively VAS82) the result of applying VASCO with the 72 (respectively 82) cycles macrocycle templates. CFT77 will stand for the output of the Cray CFT77 3.0 compiler.

Codes	CRQ		MWORDS			MFLOPS		
	VAS72	VAS82	VAS72	VAS82	CFT77	VAS72	VAS82	CFT77
M1	6	4	185	176	148	185	176	148
MV1	6	5	188	174	118	188	174	118
MVF1	7	5	196	203	165	147	152	124
MCOM	8	8	193	177	143	193	177	143
DROT †	9	8	163	170	108	244	255	162
MVC1	6	4	201	197	143	100	98	71
MVC2	8	5	199	184	135	133	123	90
MVC3	8	6	197	180	143	148	135	107
MVC4 †	10	7	153	180	143	122	144	114
MVC5 †	10	8	154	180	143	128	150	120
MVC6 †	11	8	102	171	148	87	147	127
MVC7 †	11	8	115	172	151	101	150	132
MVC8 †	12	8	92	174	154	82	155	137
MVC9 †	12	8	97	174	156	87	157	140
MVC10 †	13	8	76	175	160	69	159	145
LL1 †	9	6	176	170	133	220	212	166
LL7 †‡	10	9	128	170	137	205	272	219
LL12	4	4	198	198	167	66	66	56
LL21	6	4	188	177	135	99	88	67
LL183	7	6	201	197	189	134	131	126

Table 4: Register requirements and memory bandwidth usage (stride 1, vector length = 1024 and timings performed in dedicated mode): the presence of a †(resp. a ‡) following a code name indicates that VAS72 (resp. VAS82) used spilling)

First, for the code optimizer, the only difference between the two models is that VAS72 considers the result register of the vector floating point add or multiply reserved for 2 macrocycles while VAS82 considers it reserved just for one macrocycle. As we expected, VAS72 turns out to be consuming more registers (see Table 3, the columns CRQ of which show the number of virtual registers simultaneously alive). As a result, codes produced by VAS72 will more often require spilling than codes produced by VAS82.

The situation with performance results is more subtle. Table 3 gives the memory bandwidths obtained when running codes obtained by VAS72, VAS82, and CFT77. The memory bandwidth was computed by dividing the total number of memory accesses actually performed by the time measured. The timings were done in a dedicated environment and on a Cray-2 with a dynamic memory. Since all the codes in this table are memory bound, the Megaflop rate can be obtained by straightforward scaling. Depending upon the code, we distinguish two cases; either VAS72 was able to produce a code requiring no spilling (i.e., $CRQ \leq 8$), or not. In the first case, VAS72 performs better than VAS82, while in the second, the situation is reversed. The second case is relatively easy to explain: the cost of the spilling code required by VAS72 reduces the performance so severely that the advantage of VAS72 is lost.

The situation in the first case (no spilling for VAS72) is directly related to the choice of the macrocycle and the induced approximation. For VAS72, the reservation time for both memory and functional unit is modeled accurately. The only major error introduced is in the reservation of the result registers which are considered two macrocycles (i.e., 144 cycles) although they are, in reality, reserved for only 82 cycles. As a result, the memory and the functional units are well modeled, and their usage very well optimized via VAS72.

On the other hand, the choice of 82 as a macrocycle implies that memory and functional unit timing will be overestimated, resulting in an under utilization of these resources. In the contrast the register reservations are more accurately modeled.

However, the impact of such an overestimation is not uniform, as evidenced by consideration of the following two extreme cases:

1. all the instructions scheduled at macrocycle $i + 1$ do not use a value produced by one of the floating point units at macrocycle i

2. one of the instructions scheduled at macrocycle $i + 1$ uses a result produced by a floating point unit at macrocycle i

To get a better appreciation of this effect, we compute the apparent chime which is defined as the number of cycles between the starting of two consecutive blocks of 64 iterations of the loop body divided by the latency expressed in macrocycles, as predicted by the model (cf. Clumns *Chime* in Table 2).

In the case 1 mentioned above, although the VAS82 model assumes a length of 82 cycles, all the instructions will be executed in around 72 cycles (cf MVF1 in Table 2 where the apparent *Chime* is 72 cycles for code generated by VAS82). In the second case, the presence of dependence and reservations involving the result register will enforce the value of 82 as the time spent at execution for macrocycle i. An example of such a case is MV1 (cf Figure 6) where the addition (+2) of the second macrocycle depends on the multiplication (+2) executed during the previous cycle. As a result, the apparent *Chime* for MV1 code generated by VAS82 will be 81.6 cycles (cf Table 2). This stems from the fact that the critical path of the code produced by VAS82 is constituted by the concatenation of triples of dependent floating point operations (*2, +2, *2). Conversely VAS72 has scheduled the operations corresponding to the same triple in non consecutive macrocycles, resulting in a better utilization of the memory bandwidth (cf Tables 2 and 4)

For VAS72, when no spilling is required, the model is very close to the apparent chime because the critical path of code produced by VAS72 is mostly comprised of operations all of which last 72 cycles. As described elsewhere [3], the spilling procedure we used introduces macrocycles during which no operations are scheduled: they correspond to waiting for the liberation of result registers. In such cases, the model is considerably inaccurate because such macrocycles are accounted for a full macrocycle, while in reality, they last only about 10 cycles (82 -72).

For VAS82, the apparent chime rate varies between 72 and 82 cycles (cf Table 2). This variation is due to the phenomenon described above, i.e, the presence of dependencies between instructions scheduled in consecutive macrocycles. An extreme case is MVF1 where the apparent chime rate is 72, which turns out to be a result of the scheduling process, that systematically inserted a macrocycle between the production of a result by one of the floating point units and its usage. On the other hand, for the loop body of MV1, any result produced by a functional unit was used in the following macrocycle, resulting in an apparent chime rate of 81.6 cycles.

4.3 Modeling the memory behavior

The two models we used were constructed assuming an almost ideal memory behavior in terms of bandwidth and assuming the processing of a vector memory request every 72 or 82 cycles. Although code optimized for such models may seem inadequate because of over optimistic memory behavior modeling, in practice it turned out to generate high performance code. The reason is that most of the codes are memory bound on the Cray-2. Therefore, it is important to be able to issue memory transactions as soon as possible to saturate the memory bandwidth. Our models fulfill that requirement by trying to schedule a memory transaction every macrocycle, achieving a perfect overlap between the memory operations and the floating point computations. This also means that the constraints imposed by the model are more severe than in reality. If we assume that all vector memory access are more than 82 cycles long, it is easy to show that the critical path in the codes generated by VASCO will mostly be constituted by the memory access. To validate that assumption, we ran the generated codes, with all the vector accesses performed with stride 2 (this is equivalent to a slowdown of the memory access by a factor of 3). As shown in Table 4, the codes produced by VAS72 and VAS82 are still saturating the available memory bandwidth.

4.4 Strict limitations of the common memory model

There are two issues of common memory access that are clearly overlooked by our simple model:

- non-unit strides memory accesses have been modeled, a priori, identically to unit-stride, (or to even stride) accesses: the same load template is used for all cases. Although such a choice does not waste memory cycles as shown in the previous subsection for the stride 2 case (i.e. the code generated still saturates the memory bandwidth available), it may result in wasting register space. This could avoided in many cases where stride is known at compile-time (either explicitly or after global interprocedural analysis) by using different templates for loads with strides resulting in a degradation of the memory performance (i.e., stride multiples of 2, 4, 8 etc...).

- Our pipelined model ignores many of the fine grain details of processor and memory system architecture: bank level reservation, data path buffering, hardware resources multiplexing (like pseudo-banking). This can be justified in light of the satisfactory use we can make of this model in our search for faster codes. However, we can also show examples where our generated code clearly shows that the limit of the simplified model has been reached.

It is interesting to note that a very simple experiment suffices to illustrate this point. It can be performed by using the following piece of code:

```
c       0 < IGAP < 512
c       N < 4096   therefore no dependency
        DO I = 1, N
1           A(I+IGAP+4096) = (B + A(I)) * C
```

where IGAP is a parameter influencing the memory offset of memory writes with respect to memory reads (This also offsets quadrant controllers and pseudo banking common hardware (modulo 4 and modulo 256)).

The results (Figure 10), obtained in a dedicated mode with dynamic memory configuration, show that our optimized code is sensitive to changes in IGAP whereas CFT77 code does not stress memory access enough to exhibit the phenomenon.

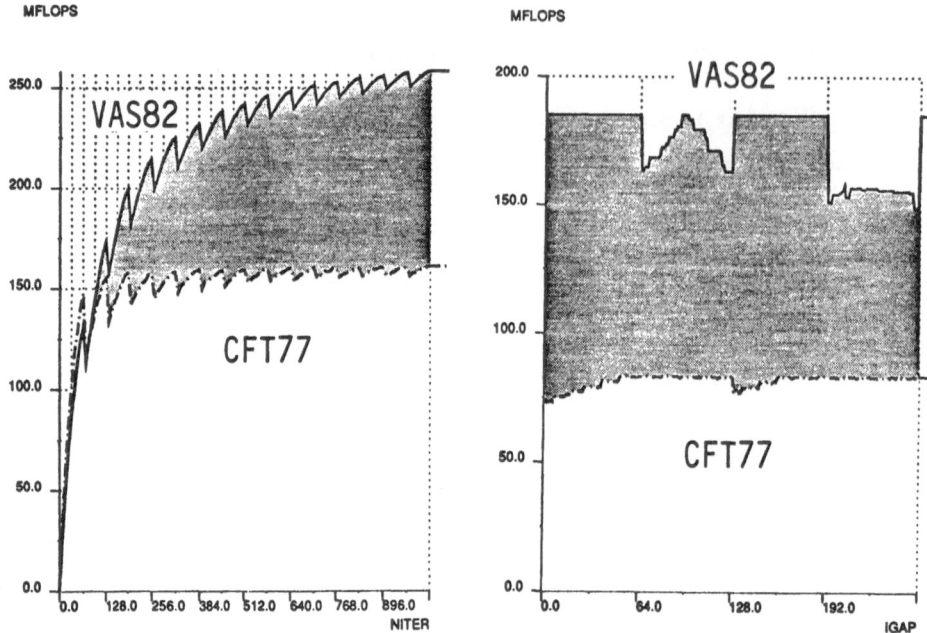

Figure 9: Performance of VAS82 compared with CFT77: DROT

Figure 10: Influence of IGAP on MV1 performance

Such an effect is relatively difficult to take into account at compile-time because in most cases, it depends upon the data layout. The only effective solution is to unroll the loop and try to schedule successive loads on the same vector into consecutive macrocycles, although the price in terms of registers space is prohibitive when compared with the benefits.

5 Conclusion

We have shown the applicability of a technique based on microcompaction framework for generating efficient vector code on the CRAY2. The experimental results presented demonstrate the interest of a sophisticated register allocation procedure which must be efficient enough to make good use of the vector registers which appear to be a very scarce resource on the Cray architectures.

The modelling procedure we used gives some insights on the relation of key architectural features to practical performance of general vector code. This can be used to improve the tradeoff that has to be made between purely architectural considerations, and compiler related issues, following a methodology much popularized by the RISC movement.

More generally, supercomputers are more and more using complex hierarchical memory systems to achieve a data rate matching the arithmetic rate. Although such memory organizations result in a substantial improvement at least in peak performance, the efficiency may become highly dependant upon the usage of the different storage levels of the hierarchy. This implies a major change in the algorithm design, the larger applicability of high-level restructuring transformations, and an evolution of compiler technology. At that level, several problems have to be solved:

- code optimization when memory may have a very long, impredictable and highly fluctuating latency,

- optimization of the usage of all the various levels of storage,

- minimization of the transfers between the different levels, which includes generation and scheduling of the corresponding code.

References

[1] Arya, S., *Optimal Instruction Scheduling for a Class of Vector Processors: an Integer Programming Approach*, Report CRL-TR-19-83, University of Michigan, 1983

[2] Cytron, R., Ferrante, J., *What's in a name? The value of renaming for parallelism detection and storage allocation*, Proc. ICPP, 1987

[3] Eisenbeis, C., Jalby, W., Lichnewsky, A., *Squeezing more CPU performance out of a Cray-2 by vector block scheduling*, Proc. Supercomputing 88, Kissimee, Florida 1988.

[4] Eisenbeis, C., Jalby, W., Lichnewsky, A., *Compile-time optimization of memory and register usage on the CRAY2*, INRIA rep. to appear 1989.

[5] Fisher, J.A., *The optimization of horizontal microcode within and beyond basic blocks: an application of processor scheduling with resources*, Phd thesis, New York Univ, 1979.

[6] Fisher, J.A., Ellis, J.R., Ruttenberg, J.C., Nicolau, A., *Parallel processing: a smart compiler and a dumb machine*, Proc. SIGPLAN Symp. on Compiler Construction, 1984.

[7] Kennedy, K., *Automatic Vectorization of Fortran Programs to Vector Form*, Technical Report, Rice University, Houston, TX, October 1980.

[8] Kuck, D.J., Kuhn, R., Padua, D., Leasure, B., Wolfe, M., *Dependence Graphs and Compiler Optimizations*, Proc. 8th ACM Symp. POPL, Williamsburgh, VA, 1981.

[9] Lam, M.S.L., *A systolic array optimizing compiler*, Phd Thesis, Carnegie Mellon University, 1987.

[10] Lichnewsky, A., Thomasset, F., *Techniques de base pour l'exploitation automatique du parallelisme dans les programmes*, Rapport de Recherche INRIA, N 460, 1985.

[11] Lichnewsky, A., Thomasset, F., Eisenbeis, C., *Automatic Detection of Parallelism in Scientific Programs with Application to Array-Processors*, Proc.of IBM Institute, North Holland, 1986.

[12] Nicolau, A., *Uniform Parallelism Exploitation in Ordinary Programs*, Proc. of ICPP, 1985.

[13] Patel, J.H., Davidson, E.S., *Improving the throughput of a pipeline by insertion of delays*, Proc., 3rd Ann. Symp. Comp. Arch., 1976.

[14] Touzeau, R.F., *A Fortran Compiler for the FPS-164 Scientific Computer*, SIGPLAN Notices, Vol. 19, N 6, 1984.

ON THE PARALLELIZATION OF SEQUENTIAL PROGRAMS

Swarn P. Kumar
and
Janusz S. Kowalik

Boeing Computer Services
Scientific Computing and Analysis
P.O. Box 24346
Seattle, Washington 98124

ABSTRACT

Parallelization of sequential programs for MIMD computers is considered. In general, the major steps included in the parallelization process are: program partitioning into a task system, derivation of a parallel task system, scheduling and execution of this task system on a multiprocessor system. We present a general framework for an automatic maximally parallel task system generator which may be useful as a component of the code parallelization process. The framework is based on the concept of maximally parallel task systems, a concept which has been mainly used for designing operating systems.

1.0   INTRODUCTION

Automatic parallelization of serial off-the-shelf programs is an important practical problem. A natural way for deriving local parallelism is by considering Fortran DO-loops. Recently, commercial software packages able to handle loop-based concurrency have started to emerge. For example, the Alliant FX/series preprocessor for automatic parallelization is capable of concurrent-outer-vector-inner analysis. That is, for nested loops, the Alliant system will vectorize the inner loop and execute the outer loop in parallel. Other vendors offer parallelizing compilers with analogous capabilities.

NATO ASI Series, Vol. F 62
Supercomputing
Edited by J. S. Kowalik
© Springer-Verlag Berlin Heidelberg 1990

The problem of global parallelization is more difficult. By global parallelism we mean techniques based on considering task systems defined on the complete code or a large part of it. The main difficulty lies in the sequential code decomposition and derivation of a suitable precedence graph required for executing tasks by multiple processors. We suggest some concepts that may prove useful for deriving automatically a precedence graph that can be synchronized, scheduled, and executed on a multiprocessor computer system.

We assume here that our parallel machine is a general purpose Multiple Instruction Multiple Data (MIMD) multiprocessor with shared memory. Furthermore, for simplicity, we assume that each processor is capable of fetching data from the memory, performing any binary arithmetic operation and storing the result in a unit of time called step.

The following notation and definitions will be used:

p : Number of processors

t_1 : Number of time steps for an algorithm executed on a uniprocessor machine

t_p : Number of time steps for a parallel algorithm executed on a parallel computer with $p > 1$ processors

S_p : The speed-up t_1/t_p

E_p : The efficiency of p processors S_p/p

2.0 COMPUTATIONAL TASK SYSTEM

We follow Coffman and Denning [1973] and introduce definitions of terms used in the subsequent sections.

A *task T* is an indivisible unit of computational activity specified only in terms of its inputs, outputs, and execution time. We assume that a task is *uninterpreted*; i.e., the internal operation of the task is not specified.

A *Task System C* is the pair $C = (J, <)$ where $J = \{T_1, T_2, \ldots, T_n\}$ is a set of tasks and $<$ is a precedence relation (an irreflexive, antisymmetric, transitive binary relation) on the set J. The precedence relation $<$ defines the operational precedence among the tasks of J, i.e., $T_i < T_j$ means that the task T_i is to be completed before the execution of the task T_j can start. The ordered pair (T_i, T_j) means $T_i < T_j$.

A task system can pictorially be represented by a task graph (also called precedence graph). A *task graph* G of a task system $C = (J, <)$ is a directed graph $G = (J, E)$ with the elements of the task set J as its vertices and the edges in E defined by the precedence relation between the vertices. The edge (T_i, T_j) from T_i to T_j is in E, if and only if, $T_i < T_j$ and there exists no T such that $T_i < T < T_j$.

A sequential program can be viewed as a task system (usually not unique). The program is segmented into a set of computational tasks along with the given precedence of their execution. An example of one such segmentation of a Fortran program is shown in figure 1. If the tasks T_j, T_{ji}, i, j = 1, 2, . . . , n are defined by the computations as shown in figure 1, the task system of figure 1 is given by

$C_1 = (J, <)$ where $\hspace{6cm}$ (2.1)
$J = \{T_{start}, T_{ji}, T_j, i, j \le n; T_{end}\}$ and
$< = \{ T_{start}, T_{11}), (T_{ji}, T_{j, i+1}), (T_{jn}, T_j), (T_n, T_{end}), 1 \le i \le n-1, 1 \le j \le n\}$

The task graph G_1 of C_1 is shown in figure 2.

3.0   MAXIMALLY PARALLEL TASK SYSTEMS (MPTS)

With each task T of a task system are associated two ordered subsets of memory cells. One is called the *domain* D_T and the other the *range* R_T. Executing a task T can be regarded as reading the values stored in the domain cells and writing its value into its range cells. The range and domain cells can overlap.

An *execution sequence* of an n-task system is any sequence $(a = a_1 a_2 \ldots a_{2n})$ of task events satisfying the precedence relation and consisting of exactly one initiation and one termination event for each task. Let $M = \{M_1, M_2, M_3, \ldots M_m\}$ be the physical system of ordered memory cells on which a task system executes. Let $a = (a_1 a_2 \ldots a_{2n})$ be an execution sequence of C. The *value sequence* $V = (M_i, a)$ of cell M_i is defined by the sequence of values written by terminating tasks T in α for which $M_i \varepsilon R_T$. The set of values contained in the memory cells in M before the first event in any execution sequence starts from the initial state.

A task system C is called *determinate* if any of the intermediate or final results produced by the tasks in C do not depend on the order in which these tasks are executed. That is, for any given initial state I_o, $V(M_i,\alpha) = V(M_i,\alpha')$ for all $i = 1, 2, ., m$.

A task system is *non-determinate* if any of the final results of an execution sequence depend on the order in which the tasks of the sequence are executed. This is the case when parallel tasks read from and write to common domain and range cells. In this case, the uniqueness of the results is not insured. The results depend on the order of read and write sequence. The problem of non-determinacy can be solved by introducing the proper precedence constraints between previously parallel executing tasks (Kowalik et al., [1984]). That is, by executing only those tasks in parallel which are mutually non-interfering. The notion of non-interference is defined as follows:

Two tasks T_i and T_j are *non-interfering* if

(i) $T_i < T_j$ or $T_j < T_i$ or   (3.1)

(ii) $(R_{T_i} \cap R_{T_j}) = (R_{T_i} \cap D_{T_j}) = (D_{T_i} \cap R_{T_j}) = \phi$

Where ϕ denotes the empty set.

A set of tasks $J = (T_1, T_2, \ldots T_n)$ is said to contain *mutually non-interfering* tasks if T_i and T_j are non-interfering for all i and j $= 1, 2, \ldots n$; i $\neq$ j. This implies that a task system consisting of mutually non-interfering tasks is a determinate task system.

A task system C (and its precedence graph) is *maximally parallel* if C is determinate and the removal of any edge (T_i, T_j) from G causes T_i and T_j to become interfering. Thus, if (T_i, T_j) is an edge of a maximally parallel task system (MPTS) then

$$(R_{T_i} \cap R_{T_j}) \cup (R_{T_i} \cap D_{T_j}) \cup (D_{T_i} \cap R_{T_j}) \neq \phi$$

Given a determinate task system $C = (J, <)$ we can derive an equivalent MPTS $C' = (J, <')$ where $<'$ is the transitive closure of the relation set

$$X = \{(T_i, T_j) \mid (R_{T_i} \cap R_{T_j}) \cup (R_{T_i} \cap D_{T_j}) \cup (D_{T_i} \cap R_{T_j}) \neq \phi\} \tag{3.2}$$

The set X is the transitive closure of the mutually interfering tasks.

Applying the foregoing to the task system C_1 (2.1) of figure 1 the domain and range sets of its tasks are given by

$$D_{T_{ji}} = \{\text{MAX}, \text{SUM} (I,J), A (I,L), C (K, J), 1 \leq K \leq N\}$$
$$R_{T_{ji}} = \{\text{SUM} (I, J)\}$$
$$D_{T_j} = \{\text{SUM} (N, J), B (M)\}$$
$$R_{T_j} = \{C(M, J + 1)\}$$

and the set X of (3.2) is given by

$$X = \{(T_j, T_{j+1, i}), (T_{j,n}, T_j), 1 \leq j \leq n-1, 1 \leq i \leq n\}$$

Thus the tasks $T_{i1}, T_{i2}, T_{i3}, \ldots, T_{in}$ are mutually non-interfering and so are the tasks $T_{1j}, T_{2j}, T_{3j}, \ldots, T_{nj}$. For the given value $n = 5$, the MPTS G_1' of the task graph G_1 of figure 2 is shown in figure 3.

Execution Times of G_1 and G_1'

We define the execution time of a task T as the number of time steps (in the sense of section 1) taken for its computation and denote this number as $W (T)$. Thus, for the task system C_1 of (2.1)

$$W (T_{ji}) = 2n$$
$$W (T_j) = n$$

Hence, the serial time to execute C_1 on a uniprocessor machine is given by

$$t_1 = \sum_{j=1}^{n} \sum_{i=1}^{n} W(T_{ji}) + \sum_{j=1}^{n} W(T_j)$$

$$= \quad 2n^3 + n^2 \ \text{time steps} \tag{3.3}$$

$$= \quad 275 \ \text{time steps for} \ n = 5.$$

If $p = n$ processors are available the minimum execution time MET (G_1') to execute the MPTS G_1' is given by

$$MET(G_1') = t_p = \sum_{j=1}^{n} \left(W(T_{jn}) + W(T_j) \right)$$

$$S_p = \frac{2n^3 + n^2}{3n^2} \approx \frac{2n}{3} \quad \text{for large } n$$

$$= 3n^2 \ \text{time steps} \tag{3.4}$$

$$= 75 \ \text{time steps for} \ n = 5.$$

The corresponding speedup and efficiency are given by

$$E_p \approx 2/3$$

for 5 processors

$$S_5 = 275/75 = 3.7 \tag{3.5}$$

$$E_5 = 3.7/5 = .75 \tag{3.6}$$

4.0   SCHEDULING

Scheduling can be done statically by the compiler, dynamically at run-time, or by combining both the compiler and run-time schemes. The requirements for effective static scheduling are restrictive and render purely static schedules impractical for most programs. Given, however, exact timing of the tasks, we can use scheduling algorithms which are polynomially bound in time; e.g., a critical path list scheduling described by Kohler [1975] and Lord [1976]. Of course, such procedures model only make sense in a purely dedicated environment. If these conditions are not satisfied, we have to use dynamic schemes which make decisions at run-time and require overhead processing. We do not pursue this subject further and refer the reader to Polychronopolous [1988] which discusses scheduling and proposes a general framework for an auto-scheduling compiler

5.0   PARTITIONING

Given a serial code, the problem of identifying a set of tasks (code partitioning) remains currently the analyst responsibility. Since, typically, most program parallelism resides in DO loops, it is only natural that loops attract the main attention in the process of partitioning. In addition to loops there are other program constructs that may define tasks; e.g., subroutine calls or groups of source statements that must be executed sequentially. The task granularity is highly architecture dependent and depends on the cost of task creation, synchronization and the number of available processors. The task size may vary from a single source statement for a low cost synchronization and task creation multiprocessor to thousands of statements for a high-cost synchronization and task creation multiprocessor. In the former case, we would need to choose inner nested loops or statements as tasks; in the latter case, we would consider outer loops. For example, if n^2 processors are available with a low cost of task creation and synchronization process, a finer grain parallelism can be implemented by defining each statement of the kth loop and m^{th} loop of figure 1 as separate tasks. That is, by defining the tasks:

$$T_{jik} = \{SUM\,(i, j) = SUM\,(i, j) + A\,(i, l) * C\,(k, j)\}$$

$$T^j{}_m = \{C(m, j + 1) = SUM\,(n, j) + B\,(m)\} \quad i, j, k, m = 1, 2, \ldots, n \qquad (5.1)$$

This task system is then given by

$$C_2 = \{(T_{ji}, T_{jik}, T^j{}_m, T_{start}, T_{end}, i, j, k, m = 1, \ldots, n), < \} \qquad (5.2)$$

The MPTS G'_2 of C_2 for $n = 5$ is shown in figure 4. To execute G'_2 in minimum time $p = O\,(n^{2)}$ processors are required. The MET of G'_2 is then given by

$$MET\,(G'_2) = t_{n^2} = 3n \text{ time steps} \qquad (5.3)$$

The speed up and efficiency are given by

$$S_{n^2} = t_1 / t_{n^2} = \frac{2n^3 + n^2}{3n} \approx \frac{2n^2}{3} \quad \textit{for large } n$$

$$(5.4)$$

$$E_{n^2} = \frac{S_{n^2}}{n^2} \approx \frac{2}{3}$$

The MET and speed up of (5.3) and (5.4) show the extra parallelism achieved. In general, the problem of partitioning is currently an empirical process, but the existing formal knowledge and practical experience in this subject may be sufficient for designing useful tools that could assist the programmer in the task of rapidly testing various code partitioning options.

6.0   AUTOMATIC GENERATION OF MPTS

We propose a scheme for parallelization of Fortran-like serial codes in which a task system C for the given code is defined by the user and then a maximally parallel task system C' equivalent to C is automatically generated without the intervention of the user. The main components of the process of automatic generation of C' are shown in Figure 5. The program PGEN consists of the following steps:

PROGRAM PGEN (C, CRL, C')

Step 1:
Compute the domain and range sets D_T and R_T for each task T in C using the statement numbers defining the task T and the corresponding read and write memory cells (variable names) from the compiler's cross reference listing (CRL). The assumption is that the CRL includes the source program references as desired. For example, the CFT compiler for Cray computers provides CRL containing the information such as:

Address	Variable Name	Variable Type	Main Usage	Source Program References
51370	ALPHA	*R	1 dim array	150S, 11D, 30U, 200R

Where D = defined, S = store, R = read, U = used and the numerals are statement numbers in the source program references column. This column provides the statement number (task ID) in which a particular variable (memory cell) is used as a domain cell (used) or range cell (stored). The output of this step is a Variable Domain and Range Table (VDRT).

Step 2:
Compute the set X of mutually non-interfering (independent) tasks of C.

   Input: VDRT from step 1
   Output: the set

$$X = \{(T_i, T_j) \mid (R_{T_i} \cap R_{T_j}) \cup (D_{T_j} \cap R_{T_i}) \cup (D_{T_i} \cap R_{T_j})\} \neq \Phi$$

Step 3:
Compute C = TRANS (X) the transitive closure of the set X to remove the redundant precedence relations between the tasks.

END PGEN

To illustrate the steps of the program PGEN, assume that the task system C of a given serial code is defined by

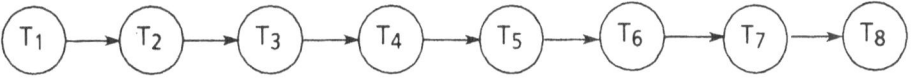

and the variables of the source code are bound with the memory cells M_1, M_2, M_3, M_4, M_5. Assume that the execution of step 1 of PGEN produces the following VDRT table:

The execution of step 2 of PGEN will compute the set X as
$$X = \{(T_1, T_3), (T_1, T_4), (T_1, T_5), (T_1, T_8),$$
$$(T_2, T_3), (T_2, T_4), (T_2, T_5), (T_2, T_7),$$
$$(T_3, T_7), (T_3, T_8)$$
$$(T_4, T_6), (T_4, T_7), (T_4, T_8)$$
$$(T_5, T_7),$$
$$(T_6, T_8)\}$$

Memory Cell (variable name)	In Domain of Tasks	In Range of Tasks
M_1	1, 2, 7, 8	3
M_2	1, 7	5
M_3	3, 4, 8	1
M_4	3, 4, 5, 7	2, 7
M_5	6	4, 6, 8

Step 3 of PGEN computes the maximally parallel task system C′ of C using the transitive closure TRANS (X), where

$$
\begin{aligned}
\text{TRANS (X)} = \{ &(T_1, T_3), (T_1, T_4), (T_1, T_5) \\
&(T_2, T_3), (T_2, T_4), (T_2, T_5) \\
&(T_3, T_7), (T_3, T_8) \\
&(T_4, T_6), (T_4, T_7) \\
&(T_5, T_7) \\
&(T_6, T_8) \}
\end{aligned}
$$

The resulting maximally parallel task graph of C′ is shown in figure 6.

The automatic creation of the maximally parallel procedence graph can be incorporated in some existing portable parallel programming tools; e.g., SCHEDULE by Dongarra et al., [1986], which schedules the execution of tasks on a system of parallel processors for a given precedence graph.

7.0   CONCLUSIONS

For over a decade, extensive efforts have been made at universities and in the data processing industry to design and implement systems that automatically vectorize FORTRAN programs. The result of this work are good quality vectorizing compilers offered now by every commercial manufacturer of vector computers. The next step is extending the current compiler capabilities to handle concurrency by detecting the opportunities for parallel computation and distributing work between multiple processors. But this step is harder. Some experienced researchers caution not to expect too soon robust production quality systems. They also feel that fully automatic techniques will not be able to achieve high performance on asynchrononous parallel systems and the programmer will be needed in the parallelism specification at some level (e.g., Callahan et al. [1988]).

In this paper an assumption is made that the program partitioning is done with the programmer involvement. We show and illustrate how the maximally parallel task system can be obtained automatically at a reasonable computing effort. This system is determinate and has the minimal number of precedence relations (edges). It can be scheduled and executed on a multiprocessor computer system.

8.0   REFERENCES

1. Callahan, C. D., Cooper, K. D., Hood, R. T., Kennedy, K., and Torczon, L.: PARASCOPE: **A Parallel Programming Environment**. The International Journal of Supercomputer Applications, Vol. 2, No. 4, 1988, pp. 84-99.

2. Coffman, Jr., E. G., and Denning, P. J.: **Operating Systems Theory**. Prentice Hall, Englewood Cliffs, NJ, 1973.

3. Dongarra, J. and Sorenson, D.: *SCHEDULE: Tools for Developing and Analyzing Parallel Fortran Programs*. ANL/MSC TM-86, Argonne National Laboratory, Math and Computer Science Division.

4. Kohler, W. H.: *Preliminary Evaluation of the Critical Path Method for Scheduling Tasks on a Multiprocessor System*. IEEE Transaction on Computers, C. 24, 1975, pp. 1235-1238.

5. Kowalik, J. S., Lord, R. E., and Kumar, S. P.: *Design and Performance of Algorithms for MIMD Computers*. **High Speed Computation**, Ed. J. S. Kowalik, Springer-Verlag, 1984.

6. Lord, R. E.: *Scheduling Recurrence Equations for Solution on MIMD Type Computers*. Ph.D. Dissertation, Washington State University, Pullman, WA, 1976.

7. Polychronopoulos, C. D.: *Toward Auto-Scheduling Compilers*. Journal of Supercomputing, No. 2, 1988, pp. 297-330.

PROGRAM SAMPLE

```
REAL A(10,10), B(10), C(10, 11), SUM (10, 10), MAX

    N = 5
    L = 2
    MAX = 10.0                                          } T_start
    READ *, A, B, C

    DO 40 J = 1,N
        DO 20 I = I, N
            SUM (I, J) = MAX
            DO 10 K = 1,N                               } T_ji
                SUM (I, J) = SUM (I, J) + A (I, L) * C (K, J)
10          CONTINUE
20      CONTINUE
        DO 30 M = 1, N
            C(M, J + 1) = SUM (N,J) + B (M)             } T_j
30      CONTINUE
40      CONTINUE
        {PRINT*, 'THE OUTPUT DATA' is, SUM, C       }
        STOP                                               T_end
        END
```

Figure 1: An Example of Defining a Task System for a Serial Program

185

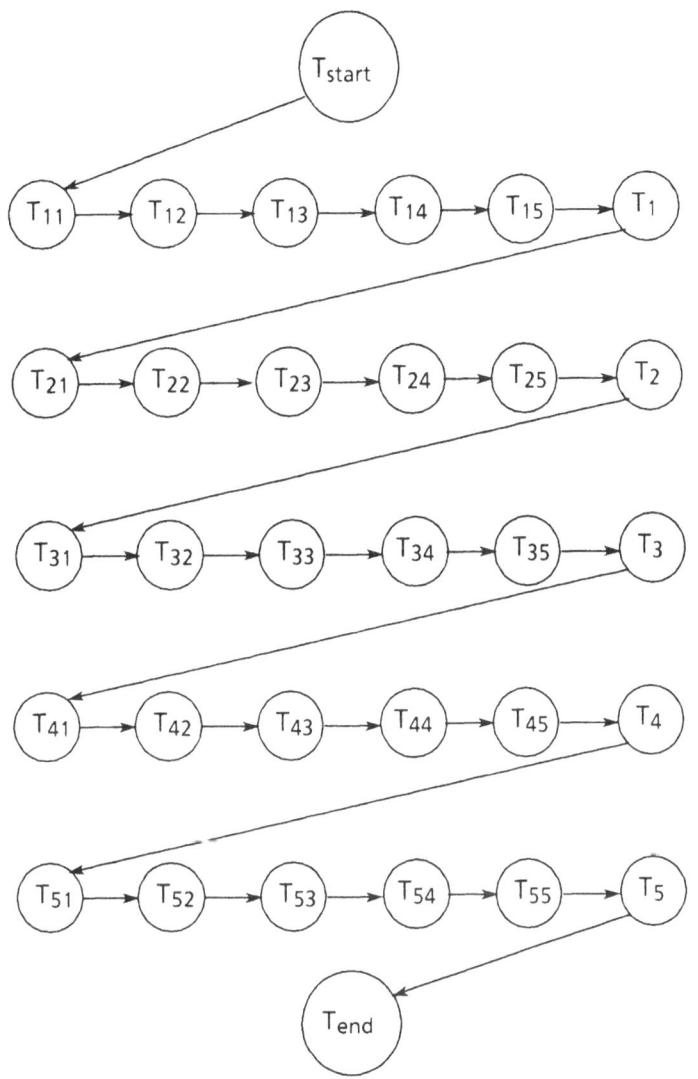

Figure 2: The Task Graph Corresponding to the Task System C

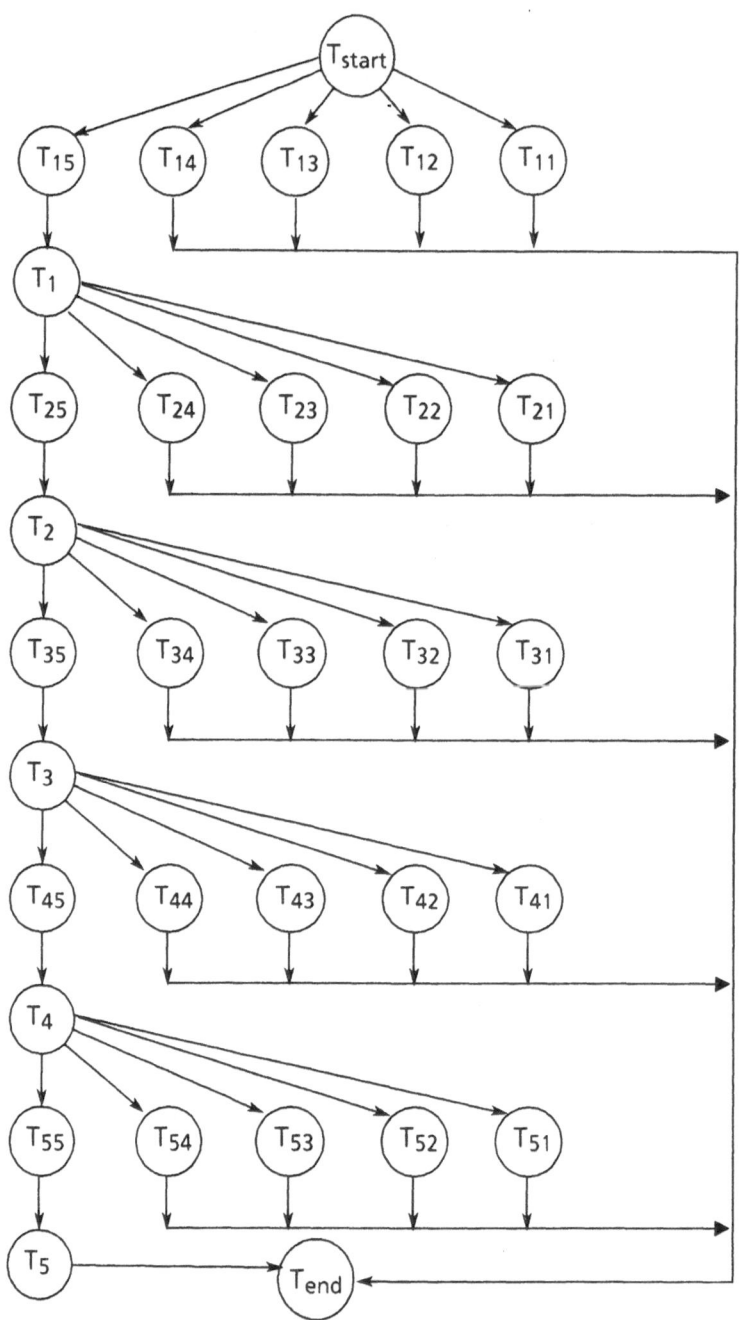

Figure 3: *The MPTS G₁' of the Task Graph G₁*

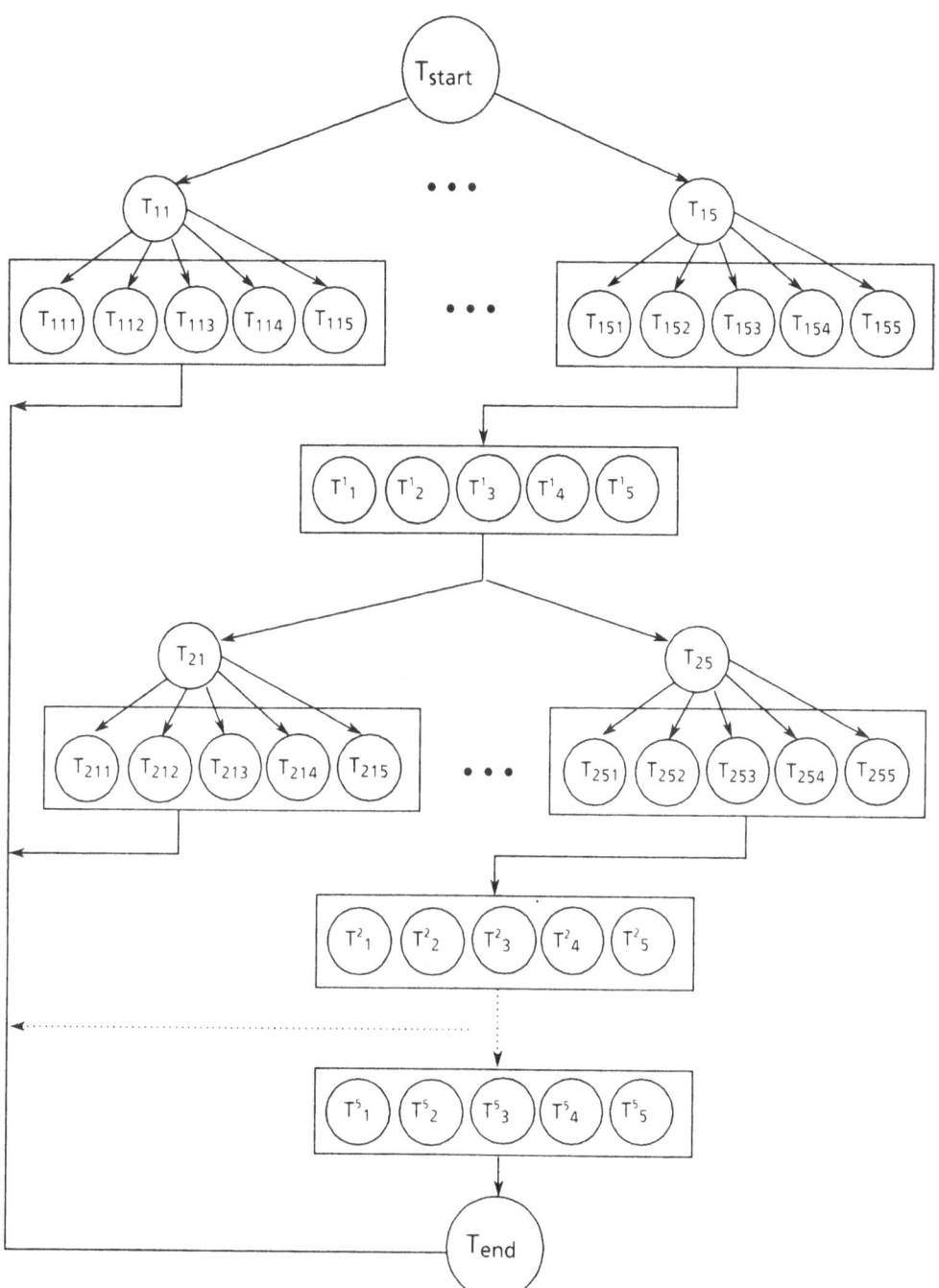

Figure 4: *The MPTS G'₂ of the task system C₂*

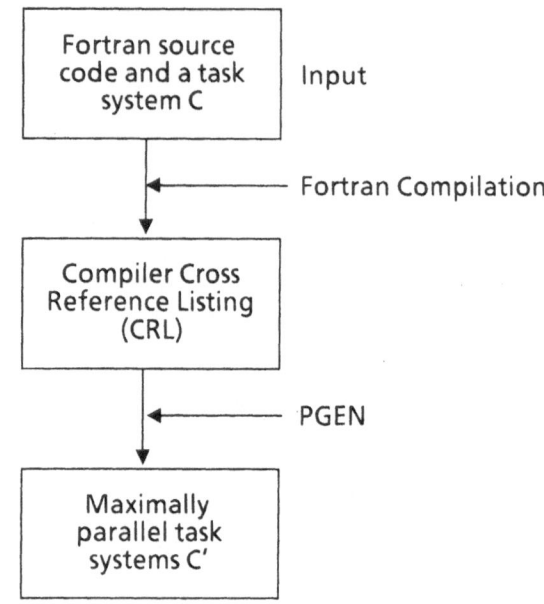

Figure 5: MPTS Generator

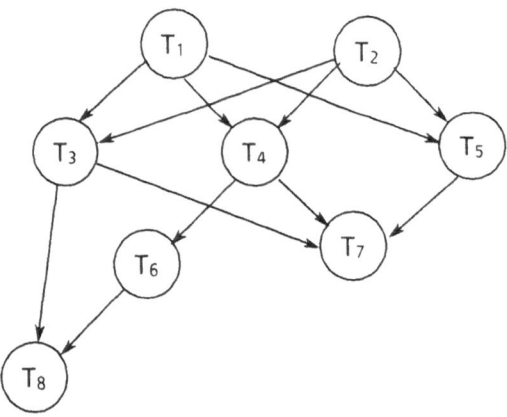

Figure 6: Maximally Parallel Graph G' Generated by PGEN

PART 5

USER ENVIRONMENTS AND

VISUALIZATION

The Role of Graphics Super-Workstations in a Supercomputing Environment

E. Levin

Research Institute for Advanced Computer Science
NASA Ames Research Center, Mail Stop 230/5
Moffett Field, California 94035, USA

KEYWORDS/ABSTRACT: graphics workstations/scientific visualization/supercomputing environments/interactive computing/ distributed processing

A new class of very powerful workstations has recently become available which integrate near-supercomputer computational performance with very powerful and high quality graphics capability. These "graphics super-workstations" are expected to play an increasingly important role in providing an enhanced environment for supercomputer users. Their potential uses include; off-loading the supercomputer (by serving as stand-alone processors, by post-processing of the output of supercomputer calculations, and by distributed or shared processing), scientific visualization (understanding of results, communication of results), and by real time interaction with the supercomputer (to "steer" an iterative computation, to abort a bad run, or to explore and develop new algorithms).

INTRODUCTION

The term "graphics super-workstation" is defined here to refer to a category of workstations introduced in 1988 which combine very powerful computational capability with high quality graphics in a tightly coupled, *architecturally integrated* system. Typically, such workstations provide from $\approx$ 1/10 to 1/100 the floating point speed of the most powerful current supercomputers, they have relatively large main memories plus specialized buffers and caches, and are capable of generating and manipulating realistic, three-dimensional graphics displays. Furthermore, these workstations have been carefully structured to provide for very rapid movement of data between memory resources, computational resources and graphics resources and are designed to enable the efficient, concurrent utilization of these resources. Hardware and software have been incorporated to facilitate networking to supercomputers and to other workstations or computational resources. The extensive system and application software provides users with a powerful and convenient working environment. The primary purpose of this paper is to explore the possible future role and impact of such workstations in a supercomputing environment. This is preceded by an overview of the current status.

CURRENT STATUS

The principal vendors that provide true graphics super-workstations include Ardent Computer Corporation (the Titan), Silicon Graphics Inc.(the high end of the Power Series family), and Stellar Computer Inc. (the GS-1000 and more recently the GS-2000). (The continued presence of the Apollo 10000 Series as a competitor at the very top end is less clear since the acquisition of Apollo by Hewlett-Packard.) Many other workstations have varying degrees

This work supported by Cooperative Agreement NCC 2-387 from the National Aeronautics and Space Administration (NASA) to the Universities Space Research Association (USRA).

of processing and graphics capability (including Sun, DEC, Tektronix, etc.) and there are attached processors using signal processing technology that are capable of generating very impressive graphics approaching the quality of photo-realism (such as the Pixar and the AT&T Pixel machine). There is also a recent trend by vendors of mini-supercomputers (notably Alliant) to add specialized hardware and software to their mainframes to support graphics applications. However, the distinguishing feature of the graphics super-workstation is the efficient, effective and convenient integration of the very powerful computational and graphics capabilities into a single-user system. Typical graphics generation and display manipulation characteristics are summarized in Table 1.

Table 1. Typical Graphics Capabilities
Screen Display
19" Color Monitor
Resolution 1280 x 1024
Image Bit-Planes (+ Control and Overlay Planes)
16-32 Bit Z-Buffer
24 (+) Bit Color Planes
Provision for Double Buffering
Display Speed
500,000 3-D Vectors/Sec
150,000 Gouraud-shaded, Z-buffered Triangles/Sec
30,000 Phong-shaded, Z-buffered Triangles/Sec
Surface Geometry Approximations (Primitives)
Polygons, Triangular Strips, Meshes
NURBS
Lighting/Shading/Rendering
Flat, Phong, Gouraud Shading, Texturing
Transparency, Specular Highlighting, True Ray Tracing
Support for Animation
Playback at 60 frames/sec (Stellar: 74 frames/sec)
NTSC Compatible, RS 170 out
10,000 Gouraud-shaded triangles/frame at 15 frames/sec
Stereo
Full Color
Multiple Viewers

Representative values of the hardware and processing characteristics are shown on Table 2. The range given for the vector processing capabilities reflects both differences between the various vendors as well as the usual obfuscation regarding how the performance is to be measured. (The 64 MegaFlops Peak is for a 4-processor Ardent Titan.) Since current prices range from about $60,000 to $250,000, depending on the vendor and the configuration, the enthusiastic claims of outstanding price/performance seem well justified.

It is interesting to note how much technology has changed the economics of computing. During the 1960's, Grosch's "law" that processing performance was proportional to the square of the cost favored very large central processors. This is no longer the case from the point of view of a single user *provided* there is adequate memory to handle the problem without excessive I/O and runs can be completed in wall clock times that fit human work schedules and limits of patience.

Table 2. Representative Hardware and Processing Characteristics

Number of vector, floating-point processors: 1-4 (8?)

Vector Processing: 5-30 MegaFlops {64 MFlops Peak}

Integer Processing: 10-80 Mips

Main Memory: 16-128 MegaBytes {Access ≈ 300 MegaBytes/sec}

Cache: ≈ 1 MegaByte {Rapid Access ≈ 1 GigaByte/sec}

Input/Output: VME Bus Bandwidth 80-100 Megabits/sec

Disc Storage: 375-2,000 MegaBytes

Typical system software offered by essentially all the vendors is listed in Table 3. In addition to the system software, a very large number of sophisticated application software packages are offered in such areas as:

- Computational Fluid Dynamics
- Computational Chemistry
- Image Processing and Image Synthesis (e.g. Medical)
- Geophysical/Seismic Visualization
- Electrical Computer Aided Design
- Computer-Aided Engineering, Computer-Aided Manufacturing
- Math Libraries (Mathematica, Matlab)
- Animation/Visual Simulation

Table 3. Typical System Software

UNIX
    AT&T System V.3
    Berkeley 4.2/4.3 Extensions
 	
LANGUAGES/COMPILERS
    Fortran 77 + Extensions
    C
    Graphics *"Languages"* (e.g. Phigs+, Dore)
 	
NETWORKING/INTERFACING
    RPC
    NFS, TCP/IP
    Ethernet/Pronet
    FDDI
    Windowing {X-Windows or NEWS (SGI)}

A potentially serious software problem is the lack of agreement on standards for three-dimensional graphics both in terms of the libraries and primitives provided as well as the structure of the "language" to call and concatenate library entities. This lack of agreement limits the portability of code which has the undesirable side-effect of either locking a user into a specific vendor (or set of vendors) or, alternately, requiring time consuming re-write of the code. It also inhibits the communication of graphics between dissimilar workstations.

This is not a new problem. Unfortunately the graphics community has historically failed to reach consensus on such standards. During the 1970's there was the early CORE "standard" that achieved some reasonable degree of acceptance. However, the situation rapidly deteriorated with the adoption of CGI as the ANSI 2-D standard and GKS as the ISO 2-D standard.

Some of the key issues regarding present graphics standards are highlighted in Table 4. PHIGS (the Programmer's Hierarchical Interactive Graphics Standard) has now been formally adopted as the ANSI two dimensional Graphics Standard while GKS continues as the ISO 2-D standard. The evolution of GKS to a three dimensional ISO standard is currently under development and PHIGS+ is out for final comments prior to likely adoption as the ANSI standard. PHIGS+ provides extensions of PHIGS for surfaces, lighting and texturing. There is some need to strengthen the language bindings, however it is very probable that PHIGS+ will be adopted by ANSI.

There are also graphics support systems which might be termed *vendor standards* in that they have been developed by and are maintained by specific vendors rather than defined by national or international committees. To some extent all vendors provide certain unique features such as specialized libraries, customized macros, etc. to provide convenience to the users. For example, Stellar offers AVS (Application Visualization System) as part of the StellarVision^{tm} environment, Silicon Graphics provides an extensive GL^{tm} graphics library and performs most graphics operations in hardware. These vendor packages do not conflict with PHIGS+ which is supported by both Stellar and Silicon Graphics. The RENDERMAN^{tm} system by PIXAR is somewhat different and is proposed as a *general* interface between the geometry and rendering processes. It provides for rendering operations on surfaces defined by NURBS (Non-Uniform Rational B-Splines). NURBS has also become a generic term for surface elements bounded by a broad class of space curves.

DORE^{tm} (Dynamic Object-Rendering Environment) is quite another matter. Ardent is aggressively promoting DORE as a *vendor standard* alternative to PHIGS+. DORE is available on CRAY and SUN computers and Ardent provides relatively easy and inexpensive access to the source code. There are many powerful and convenient attributes incorporated into the concept and design of DORE, yet the incompatibility with PHIGS+ further exacerbates the issue of 3-D graphics standards. (The recent announcement in August of 1989 of the proposed merger of Ardent and Stellar may resolve this incompatibility.)

Table 4. Graphics Standards
GKS {ISO 2D-Standard Exists, 3D Under Development}
PHIGS {ANSI 2D Graphics Standard}
PHIGS+ {*Almost* ANSI 3D Standard}
PHIGS Extension for Surfaces, Lighting, Texturing
Language Bindings Not Well-Defined
PEX {PHIGS+ Extended to X Windows, "3D X-Windows"}
RENDERMAN (Pixar)
Separates Rendering from Geometry (Provides Interface)
Shading/Texturing of NURBS-defined Surfaces
GRAPHICS SUPPORT SYSTEMS (*Vendor Standards*)
GL (Silicon Graphics)
AVS/StellarVision (Stellar Computer)
DORE (Ardent Computer)
Written in C, User Extensible
Easy to Use, Powerful Functionality { >PHIGS+}
Not Compatible with PHIGS+

Ideally, scientific or industrial users should never have to deal with graphics primitives at all nor be concerned with many of the issues discussed above. There should be specialized user environments with appropriate libraries and "natural" languages for each of the various disciplines. If necessary, it should be possible for sophisticated users to modify, extend or customize their libraries and command environments. To some extent this is already taking place but the portability issue between various vendor products continues to be an outstanding problem.

CURRENT ROLE AND A LOOK AHEAD

Supercomputing is currently in the *gigaworld* era. We enjoy GigaFlops of processing power, GigaBytes of main memory, and GigaBits/sec of (local) data communication rates (e.g. UltraBus). Unfortunately, we may also be faced with GigaBytes of output to be digested. This may occur not only from scientific problems that deal with very large amounts of observational input data, but also as output data from solutions to mathematical equations representing some physical process.

A recent study [1] presented at the Third IFIP International Conference on Data Communication Systems and Their Performance reported a stubborn ratio of approximately 10^2 Bytes of output per MegaFlops of calculation. This empirical result was obtained for two quite dissimilar types of large scale calculations: computational fluid dynamics involving repeated iterations over a spatial grid, and *ab initio* computational chemistry involving determination of eigenvalues of very large sparse matrices and multi-dimensional integrations. Furthermore, the time period of the study spanned the transition from Cray 1 or Cyber 205 supercomputing to the Cray XMP era. It may be argued that not all of the output was needed, however the fact remains that it was requested by the users and it is reasonable to believe that such habits will be slow to change. Hence, as we approach processing power of ≈ 10 GigaFlops, this empirical ratio (for a moderate length run of 1000 seconds) would predict a GigaByte of output:

$$(10^2 \text{ Bytes/MFlops}) \times (10^4 \text{ MFlops/Sec}) \times 10^3 \text{ Sec} = 10^9 \text{ Bytes} = 1 \text{ GigaByte}.$$

We are also seeking far greater capabilities to be enable us to use computational methods to solve some of the challenging problems of science and engineering. Figure 1. depicts schematically three of the principal factors that serve as driving forces for increased processing speed and main memory; physics realism, increased dimensionality and data volume. Realistic representations of physical phenomena may increase geometric complexity or eliminate simplifying approximations, e.g., incorporating non-linearities or chaotic behaviour in the mathematical models. The issue of dimensionality is not limited to the three physical dimensions and time but more generally represents the number of degrees of freedom that must be considered as in the number of grid points in a CFD problem or the number of electrons included in an *ab initio* solution of the Schroedinger equation. The driving factor of data volume arises *primarily* from anticipated massive increases in observational data but large output files from computer runs add to the load. The need to process this data in reasonable time places severe demands both on computational speed and memory size.

At a recent conference, Peterson [2] expressed the need in Computational Fluid Dynamics for an ExaFlop processor with TeraWord main memory and noted that NASA has established a high performance computing goal for the next decade of a TeraFlop. At the same conference, Bretherton [3] stated that global monitoring in the 1990's would produce at

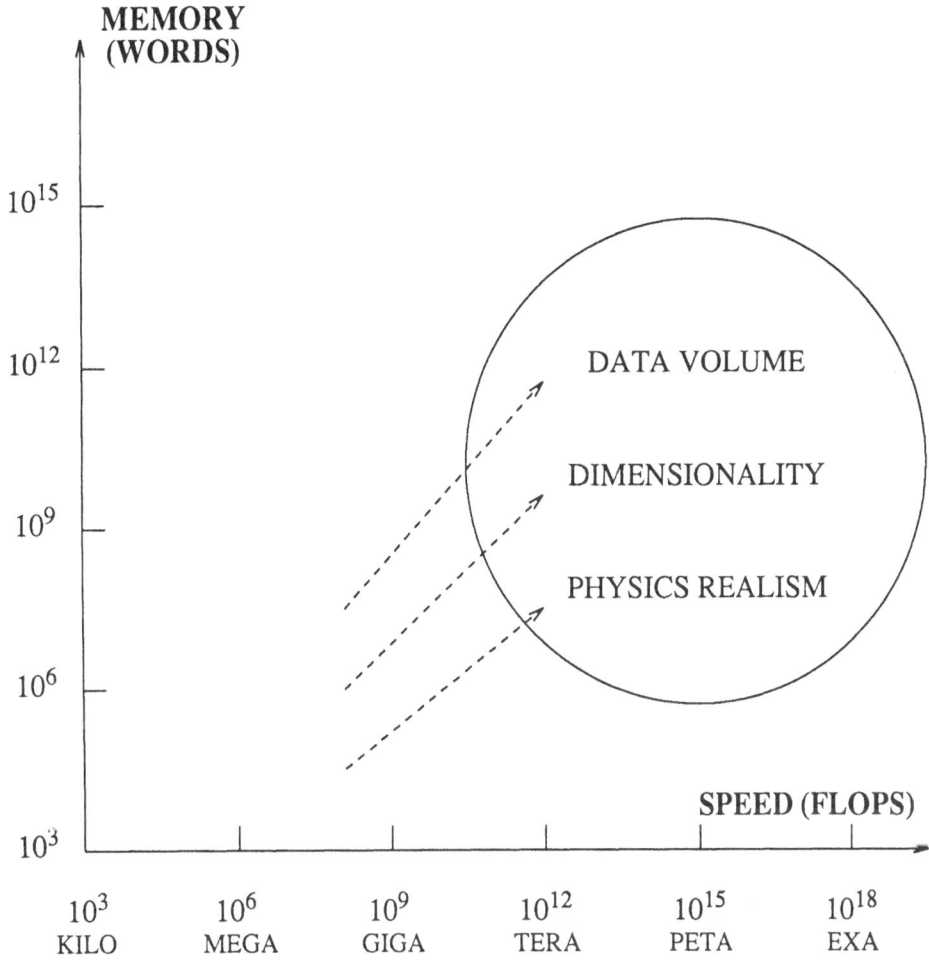

Figure 1. Driving Forces for Increased Computational Capability

least one TeraByte/day of observational data, hence three years of such data gathering would produce a PetaByte of data (if retained). Some of the scientific problems that are driving forces for the indicated increases in computational power are:

- Physics/Chemistry (molecular dynamics, *ab initio* quantum chemistry, surface chemistry, statistical mechanics, relativistic physics, cosmology, astrophysics),
- Computational Fluid Dynamics (including chaos/turbulence and coupling of aerodynamics with structural mechanics and propulsion system behaviour),
- Biology/Pharmacology (genome sequencing, genetic engineering, enzyme activity, cell modelling, drug design),
- Material Science (superconductivity, materials by design),
- Medicine (modelling of human organs and bones, surgical simulator/trainer),
- Planetary Science (global weather, environmental modelling, seismology).

It is a mistake to attribute the need for great increases in computational capability to the aggregate demands of many users each of whom may need only a modest amount of super-computer time. The supercomputers of the future are needed to solve important scientific problems that presently *cannot be done at all* and should be designed and used for this pur-pose. The new generation of powerful workstations provide a logical, cost-effective and *user-time-effective* alternative to shared supercomputers and indeed this is one of their appropriate and important roles in a supercomputing environment. It is anticipated that this role of off-loading the supercomputer for problems of modest size as well as post-processing of supercomputer output will become increasingly important in the future. For graphics pro-cessing they are already a superior alternative to the supercomputer.

Along with processing speed and larger main memories, there is an **implied** concomitant need for increased storage and communication bandwidth. The present limitations on local communication rates of about a Gigabit/second are marginally adequate for now, however long-haul (wide-area) communications are totally inadequate and the problem will be severely exacerbated as computational capability and supercomputer usage increases. Sena-tor Gore [4] has introduced legislation to establish a 3 Gigabit/sec national fiber optic net-work which, if implemented, would provide temporary relief. Fiber-optic links offer a theoretical bandwith of a Terabit/sec but at this transmission rate the bits are spaced about 0.25mm apart and practical implementation beyond 10 Gigabits/sec is questionable. The problem of mass storage is even more limiting and there are fewer promising technological developments on the horizon.

In my view, this system-level problem is being addressed at the wrong end! The only hope for a successful solution is to change the way in which we make use of supercomputers so as to effect an enormous reduction in the amount of data that needs to be stored or communi-cated. This requires a fundamental shift in the approach to handling information in the chain from physical problem to final presentation of results. The emergence of graphics super-workstations offers an opportunity to enable that essential change in methodology.

Figure 2. depicts the steps in the typical solution sequence. The most primitive form of graphics utilization is to prepare a diagram, chart or graph to present the findings after com-pleting the analysis and reduction of the results by computational means. This is denoted as the "old paradigm" on Figure 2. For very large output files, the post-processing itself becomes a major computational problem and a much more powerful approach is to use graphical techniques to search through the data and assist in the analyses by presenting visual representations to the scientist of the content. Only those portions of the data that are "interesting" may then need to be subjected to more detailed analysis and the final prepara-tion of findings. This approach is part of the "new paradigm" indicated on Figure 2. It does not of itself necessarily reduce the volume of data but does facilitate the analysis.

Graphics super-workstations can be utilized effectively further "upstream" in the process as part of the calculation itself. Their specialized hardware and architectural properties can be used effectively in conjunction with the supercomputer in a distributed processing system. Although this augments the computational power available, it does not necessarily alleviate the overall problem of excessive data volume. However, the availability of X-Windows on workstations and on supercomputers not only enables a scientist to view the progress of a computation (and, if necessary, abort a clearly bad run) but permits the user to interact with the computation *in process* (e.g., by modifying parameters such as step size, grid spacing,

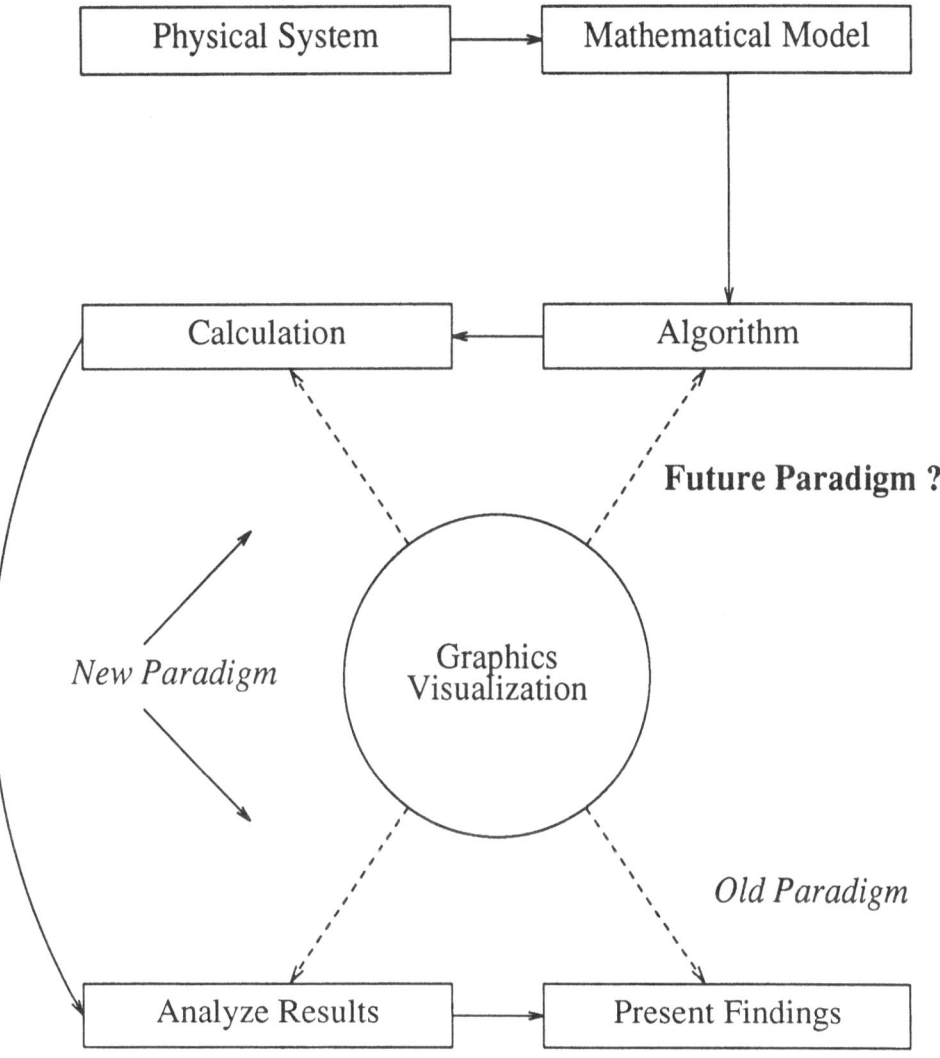

Figure 2. Paradigms for Use of Graphics Visualization

damping terms, etc.). This is not a new idea but the current realization of the concept is indeed new. The possibility was recognized quite early by John von Neumann who said [5], "....if he wishes...he can instruct the machine to present to him the relevant characteristics of the situation, continuously or in discrete succession, as the calculation progresses...then intervene whenever he sees fit." This aspect of the "new paradigm" not only expedites complex computations but has the potential to reduce markedly the volume of output data to be stored and transmitted. Indeed, in many cases, the scientist may have already determined his "answer" (or understanding) and no output at all is requested.

There are entirely new opportunities to explore. It has been estimated [6] that as much as 90% of the supercomputer time used for code development is wasted due to lack of good interactive visual feedback. This is illustrated in Figure 2. as the "future paradigm". Human interaction with the supercomputer by means of powerful graphics workstations will also enable or facilitate the solution of computationally difficult problems where the intervention of a human is a key (or possibly essential) part of the process. Examples include algorithms with "controllable" instabilities, problems involving transition boundaries of unknown location, systems containing multiple extrema where a global extremum is desired, and iterative processes that converge very slowly. In these examples, *the scientist becomes an essential part of the algorithm* and is empowered to act at a very high level by serving as part of a complicated non-linear feedback loop, by using knowledge of physical behaviour not reflected in the computer code or by detecting trends permitting a leap ahead to anticipate the converged state.

Graphics super-workstations also create new opportunities at the terminal end of the process shown in Figure 2., i.e., for the presentation of results in the form of "electronic journals". At present, the delay from completion of a report to publication is often 18 months or more. Advanced workstations could be used to communicate scientific results electronically including both still and animated graphics to enhance the timeliness and quality of information exchange between colleagues. Obviously, sending pixel-level graphics would be a gross misuse of precious communication bandwidth and (hopefully) one of the important attributes of graphics super-workstations of the next decade will be the development of standard or compatible means to communicate (or reconstruct!) graphical information.

This attribute of integration/networking is listed in Table 5. which also presents a prediction of other characteristics of the graphics super-workstation one decade hence. Just as super-computers will achieve great increases in computational power through massive parallelism, so too will workstations. Supercomputer vendors and super-workstation vendors are already adopting many common architectural concepts and technological advances. Massively parallel workstations are a particularly logical development for the graphics processing functions as well as for enhanced computational capability. The predicted 100/1 ratio of processing speed would imply that a TeraFlop supercomputer would co-exist with 10 GigaFlop workstations! This might be achieved by 1000 10MF processors or by 100 processors each of which has a 100MF processing capability. Neither of these models would require a fundamental breakthrough in technology. (It is not unreasonable to expect that the cost of such a workstation might be about the same as present costs as measured in 1989 dollars.)

Note that Table 5. lists ray tracing quality instead of ray tracing since there are already more efficient means to obtain excellent approximations to ray tracing without calculating individual rays. These techniques include radiosity and spatial decomposition. Similarly, there are also relatively simple tools in existence and much more powerful ones under development to facilitate real-time animation.

Of all the attributes predicted, the two most important are integration/networking and ease of use. The integration issue has been addressed previously and Upson [7] clearly captures the essence of the problem of ease of use as follows: "..To date, most software has been developed for programmers and not end-user scientists and engineers.....Designing software for scientists and engineer end-users introduces several new constraints....The most important of these is the need for computational environments that are easy to use and require little documentation....scientists, in general, have exhibited little desire to learn the detail necessary to use graphics.."

Table 5. Attributes of the 1999 Graphics Workstation

WELL-INTEGRATED

> With Supercomputer(s)
>
> With Other Workstations
>
> Into Local and Wide-area Networks

SUPERCOMPUTER/WORKSTATION RELATIONSHIPS

$$\text{Processing Speed} \approx \frac{100}{1}$$

$$\text{Main Memory} \approx \frac{1000}{1}$$

> Similar Parallel Architectures

SOPHISTICATED SOFTWARE

> Powerful Application-Specific Packages
>
> Simple, Real-Time Animation
>
> "Almost" Ray-Tracing Quality
>
> Radiosity
>
> Spatial Decomposition

EASY TO USE BY SCIENTISTS AND ENGINEERS

> "Natural" Human Interface
>
> Discipline-Specific Languages

SUMMARY

An overview of the characteristics and capabilities of currently available graphics super-workstations showed that one of the most important uses in a supercomputing environment was the analysis and interpretation of large masses of complex information. This scientific visualization function was not only important in processing the often overwhelming files of output data or observational data, but could be utilized in a much more powerful mode by interacting with a computation in progress.

As the computational power of supercomputers increases during the next decade, it is predicted that the graphics superworkstations will show a proportionate increase and a TeraFlop supercomputer might co-exist with 10 GigaFlop workstations. Both supercomputers and workstations are expected to use similar technology; in particular, future graphics superworkstations will be based on massively parallel architectures. It is suggested that future supercomputers be reserved for attacking the grand challenges of computational science and the computational needs of scientists with less demanding problems be off-loaded to stand-alone superworkstations.

At least two major improvements are needed to fully exploit the great potential of graphics super-workstations. There is an urgent need to resolve incompatibilities between the various vendor products. These differences limit both the portability of graphics code and, of equal importance, the ability of workstations to communicate efficiently over both local and wide-area networks with supercomputers and with other workstations. Probably the single most important improvement needed is to provide for ease of use by scientists who should not be required to learn a new discipline in order to make use of these valuable tools.

In summary, the role of graphics super-workstations in a supercomputing environment includes:

Off-loading the Supercomputer
- Stand Alone Primary Processing
- Post-Processing of Output
- Graphics Processing
- Program Development
- Distributed (Heterogeneous, Shared) Processing

Scientific Visualization
- Understanding of Results (Scientific Insight)
- Communication of Results

Real-Time Interaction
- "Steering" a Computation, e.g., Convergence
- Aborting a Bad Run

It is the view of this author that *true* real-time interaction with a supercomputer by means of a well-integrated graphics super-workstation has the greatest potential pay-off. Thus

The most important role of the graphics super-workstation in a supercomputing environment is to empower the human as an essential component of the solution process.

REFERENCES

1. Levin, E., Eaton, C.K., and Young, B.: Scaling of Data Communications for an Advanced Supercomputer Network. *Data Communication Systems and Their Performance.* pp. 107-122. North-Holland 1988

2. Peterson, V.: Address at the Conference on Grand Challenges to Computational Science. January 1989. (Proceedings to be published in *Future Generations Computer Science* by Elsevier Publishers, Amsterdam)

3. Bretherton, F.: Address at the Conference on Grand Challenges to Computational Science. January 1989. (Proceedings to be published in *Future Generations Computer Science* by Elsevier Publishers, Amsterdam)

4. Gore, Albert, Jr.: Senate Bill S. 1067, *National High-performance Computer Technology Act of 1989*

5. Goldstine, H., von Neumann, J.: *John von Neumann Collected Works, Design of Computers, Theory of Automata and Numerical Analysis* (A. H. Taub, ed.).: Pergamon Press 1963

6. McCormick, B. H.:Issues in Scientific Visualization. Presented at NATO Advanced Research Workshop on Supercomputing. June 1989. (Proceedings to be published by Springer-Verlag in 1990)

7. Upson, C.:Scientific Visualization Environments for the Computational Sciences. *Proceedings of the 34th IEEE Computer Society International Conference--Compcon Spring 1989*. pp.322-327. IEEE Computer Society Press 1989

ISSUES IN SCIENTIFIC VISUALIZATION

Bruce H. McCormick
Visualization Laboratory
Department of Computer Science
Texas A&M University
College Station, TX 77843-3112, USA

KEYWORDS / ABSTRACT: scientific visualization / supercomputing / medical imaging / graphics / image analysis / geometric modeling / volume visualization / televisualization / scientific visualization environments

Scientific visualization, used in partnership with today's computational tools, whether supercomputer or medical scanner, provides the foremost communications medium in the scientific and engineering world for analyzing and describing phenomena from the atomic, through the anatomic, to the astrophysic.

Several new issues have arisen with the advent of scientific visualization:

- Visually steering computation
- Volume visualization and modeling
- Assembly vs growth of complex structures
- Personal, peer or presentation graphics
- Televisualization: scientific visualization over networks

A framework for scientific visualization environments, responsive to these issues, is presented.

NATO ASI Series, Vol. F 62
Supercomputing
Edited by J. S. Kowalik
© Springer-Verlag Berlin Heidelberg 1990

ROLE OF SCIENTIFIC VISUALIZATION

The report *Visualization in Scientific Computing* (ViSC)* , by the NSF-organized Panel on Graphics, Image Processing and Workstations, recommends and substantiates the need for graphics and image processing hardware and software at research establishments conducting advanced scientific computing. Today's data sources—supercomputers, satellites, spacecraft, geological instrumentation, medical scanners—produce torrents of data, but the human brain still interprets numerical data as poorly as it always has. Scientists and engineers need an alternative to numbers, and that alternative is images.

Much of modern science and engineering cannot be expressed in print at all. Examples include simulated flights through a terrain, visualization of flow fields, molecular models, medical imaging scans, and so on. If poorly communicated, such research cannot stimulate new discoveries and new disciplines. And pragmatically, what is not communicated also is not funded. The old professional adage "publish or perish" is rapidly becoming "visualize or perish."

"The purpose of (scientific) computing is insight, not numbers," said Richard Hamming. The Panel's report concurs and declares that the goal of visualization is to leverage existing scientific methods by providing new scientific insight through visual methods. Visualization of complex computations and simulations is absolutely essential to insure the integrity of analyses, to provoke insights, and to communicate those insights to others. Advanced visualization capabilities may prove as critical as the existence of supercomputers themselves for computational science and engineering.

---

* The ViSC report consists of a printed document and 2-hour videotape, available through the ACM SIGGRAPH organization. "Visualization in Scientific Computing" is published in the organization's journal Computer Graphics, Vol. 21, No. 6, Nov. 1987. The tape is issues 28 and 29 of the SIGGRAPH Video Review. To order, contact: ACM Order Dept., P.O. Box 64145, Baltimore, Maryland 21264; phone: (301)528-4261.

PROBLEMS AND ISSUES

Scientific Visualization is a geometric, or visual, style of computing, as opposed to an exclusively numeric or symbolic approach, and comprises both a network of compatible tools and a subdiscipline of computer science. As a set of tools, it lessens the impedance mismatch between the user and his data, clarifying relationships and exposing unanticipated effects or unforeseen phenomena.

Visually Steering Computation vs Flying Blind

Scientific visualization enables computational scientists and engineers to undertake "human-in-the-loop" problems—a class requiring visualization techniques as different from photorealistic computer graphics as interactive computing is from batch processing.

This goal for scientific visualization is consistent with the vision of the ViSC Report:

> "The process of scientific discovery, however, is essentially one of error recovery and consequent insight. The most exciting potential of wide-spread availability of visualization tools is not the entrancing movies produced, but the insight gained and the mistakes understood by spotting visual anomalies while computing. Visualization will put the scientist into the computing loop and change the way science is done." [9]

Larry Smarr, Director of the National Center for Supercomputing Applications, UIUC, has estimated that during code development for new applications, 90% of allotted supercomputer time is currently wasted because of poor interactive visual feedback. (Personal communication)

Dual Domains of Geometry and Imaging

Scientific visualization is poised with one foot in each of two quite different worlds: the domain of geometry and domain of imaging [12].

The *geometric domain* is the world of the geometric model—analytic, continuous, and readily transformed by coordinate transformations. Some formal description languages exist for geometric modeling: PostScript for illustrations, PHIGS+ for 3D surface graphics, and languages such as GRAMPS or GRANNY for molecular modeling.

The *imaging domain*, in contrast, is the world of discrete sampling (pixels in 2D, voxels in 3D). This is the world of image data sets, as displayed by raster technology on the screen of our workstations.

Visualization involves transformations between these two worlds: objects in the geometric domain are transformed by *rendering* into images in the imaging domain. And conversely, 2D and 3D image data sets in the imaging domain are transformed by *modeling* into objects in the geometry domain (Figure 1).

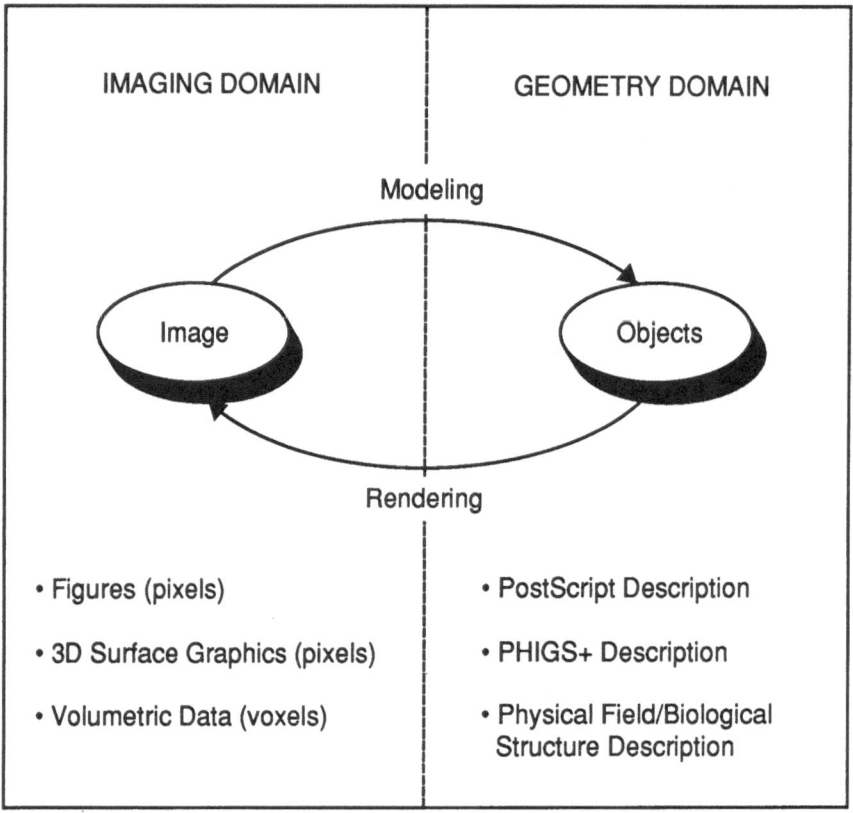

Figure 1. Dual domains of geometry and imaging

For volumetric data, however, standards for geometric description have yet to be defined. Finite element methods, as we show below, serve admirably as an interim formal description language for the fields of classical physics. Biological structure modeling at subcellular, cellular, tissue, and gross anatomical levels of description, on-the-other-hand, has yet to find its optimal expression.

Volume Visualization and Modeling

Constructing a three-dimensional geometric model of a complex object from a volumetric data set is a pervasive problem in many areas of visualization research. A major limitation in the past has been the insufficiency and ambiguity of the visual information in the three-dimensional image data set. Several medical imaging technologies, such as CT and MRI, and laser scanning confocal microscopy, now produce volumetric data sets of sufficient contrast and redundancy. In principle such volumetric data sets can be used to construct a geometric model of a complex object, e.g., the gross anatomical structure of the brain.

To this end, a representation (i.e., trial geometric model) of the object must be provided that can be matched against the sectional image data. Two properties are required for such a representation: a hierarchical organization of the model parts and the ability to dynamically deform these parts. The hierarchical organization guides the focus of attention during the matching process. Dynamic deformation enables the fitting of the model to variable image data.

Such geometric models must be capable of describing complex objects in sufficient detail and yet remain reasonably manageable in terms of their complexity and memory requirements. An attractive approach is to implement dynamic models in an object-oriented scheme. A dynamic description of an object and constraints on its allowable deformations are stored in the frame associated with that object. The hierarchical structure specifies the spatial relationships between objects and the constraints imposed on these constituent objects [4,5].

It is difficult and inefficient to manage the large amount of visual data in the 3D image data set in a single stage interpretation process. Dividing a 3D image into thick slices seems to be a reasonable compromise. Here the amount of image data is manageable, while better contrast and less ambiguity are retained than if individual 2D sections are sequentially examined. The image can be matched thick section by thick section, and a 3D volumetric model can be constructed from the series of matched models. A *rubber-band* model thus stands out as the most appropriate method [6].

The superquadrics model proposed by Barr [1], and adopted by Pentland [10], lacks the capability for dynamic deformation. Again, this model cannot handle complex objects. These shortcomings limit the flexibility in models with image data. The strength of the superquadrics model is that it requires very little memory space to store the description of the object, usually less than 10 bytes per primitive. This efficiency makes this model attractive in televisualization, where communication of visual data is the major concern.

The dynamic model proposed by Terzopoulos et al. [11] lacks the capability of organizing objects into a hierarchy and does not provide for the incorporation of domain knowledge. These shortcomings inhibit the use of domain knowledge in guiding the process of image understanding. Another problem of this model is that when complex objects are encountered, the underlying differtial equations become extremely complicated and hence add to the difficulty of approximating its solution.

Our guided process of image interpretation requires domain-specific knowledge for matching the deformable model with the sectional image data. This presupposes mechanisms exist for knowledge acquisition and for applying this knowledge in guiding the image interpretation process.

For successful image synthesis, the modeling of the sectional volumetric data must then be continually monitored. Therefore the visualization environment must allow interactive three-dimensional viewing of the volume, both with and without juxtaposition of the deformed model.

In summary volumetric visualization and modeling requires:

1. **Volume visualization:** 3D visualization of the volumetric image data set, both before and after construction of the geometric model.
2. **Visual knowledge acquisition:** acquisition of the domain-specific knowledge for guiding the dynamic matching process.
3. **Hierarchical object-oriented descriptions of parts:** this guides the focus-of-attention of the modeling task.
4. **Deformable model construction:** representation of constituent parts by deformable 3D models.
5. **Dynamic matching:** mechanisms to deform models for matching with thick slices of volumetric data.

Assembly vs Growth of Complex Structures

A geometric model, in general, is an artificially constructed object that makes the study and observation of another object easier [8].

A geometric model is normally assembled from a hierarchical object-oriented description of its parts. So predominant is this assembly paradigm that contemporary solid modeling knows virtually no other stratagem. But as we move toward modeling biological structures above the molecular level, we begin to see just how limited this assembly paradigm is. Plants and animals are grown, not assembled. Fractal/graftal algorithms exist to generate workable geometric models of individual plants, shrubs, and trees [7]. However intertwined collections of these cannot, in general, be assembled.

To see the limitation of the assembly paradigm, we consider the following. Imagine one has developed a process model to generate parametric populations of neurons. Now consider how one might go about generating a space-filling arrangement of these 3D neuron models—a model that might give insight into the interpretation of electron micrographs of neural tissue. Conceptually one can no more stomp on a pile of these model neurons to create the space-filling model—than one can stomp on a pile of tree branches to form a volume-filling compacted solid. The developing brain clearly *grows* the volume-filling arrangement of neurons; it is not *assembled* from a collection of fully mature neurons.

Similarly one brain is not an elastic deformation of another of the same species. Rather it is a growth deformation. Growth, because it can involve long, non-local processes, such as nerve fibers, blood vessels, etc., is inappropriately modeled purely by an elastic deformation. A new computational mathematics is needed for the description of biological development, e.g., the study of place and neighbor-hood influences on the developing embryo or the malignant tumor

Topobiological modeling [3] will in time make significant use of supercomputers. The mathematical tools, and visualization technology, are still very new.

Personal, Peer or Presentation Graphics

Scientists and engineers require one type of visualization to present/publish known information; they require another type of visualization to share timely information with colleagues; they require yet a third type to interpret or study phenomena by themselves or with a small group of collaborators. This distinction is discussed in the ViSC report [9]: "Most people see the end result of visualization—reproduced still color photographs or movies. With the exception of flight simulator trainees and video game players, all those not actually in the process of producing visualization see it as one-way and non-interactive. One cannot publish interactive systems in a journal."

DeFanti and Brown [2] provide a concise statement of these different needs:

"Presentation graphics are used to show the end results of scientific discoveries; peer and personal graphics are used to make those scientific discoveries.

Presentation graphics are used for conference presentations and journal publications. Aesthetics are nearly as important as the information being displayed. Image quality affects the way the information is received and the way the speaker/author is perceived by the audience. The more effort put into the preparation of presentation materials, the more persuasive the presentation.

Scientists would most likely need access to powerful equipment, technicians and visualization experts to produce high-resolution, presentation-quality images on 35mm slides or 3/4" or 1" video. If a video animation is the end result, they would also need help with color and composition, editing, soundtracks, narration, and so on.

Peer graphics are used to share timely scientific discoveries with colleagues. The output quality is of some consideration, but more important is the clear visual representation of information. To produce peer graphics, scientists might have something as simple as a consumer video recorder plugged into the back of a graphics workstation, or a 35mm camera on a tripod in front of the computer screen.

Personal graphics, or throwaway graphics, are used for scientific analysis by one individual or a small group of individuals. This is the most important type of visualization as advocated by the ViSC report; here, scientists have the freedom to interact with their data, to observe phenomena in close-to-real-time, and ultimately, to steer their computations and have the effects displayed immediately on their screens in graphical format. Visuals are simply and quickly drawn; images displayed on a computer screen are often all that are needed. The emphasis is on data representation and interaction, not quality."

Our model for a Scientific Visualization Environment, as described below, provides balanced support for all three: personal, peer, and presentation graphics.

Televisualization: Scientific Visualization over Networks

"The application of networks to visualization is termed *televisualization*.

Visualization by shared communication would be much easier if each of us had a CRT in the forehead. We speak—and for 5000 years have preserved our words. But we cannot share vision. To this oversight of evolution we owe the retardation of visual communication compared to language. To overcome the bottleneck, scientists need to improve visual interaction (1) with their data, and (2) with each other.

The United States today faces a formidable challenge—to build the distribution networks, workstations and human interfaces necessary to communicate scientific and engineering knowledge. The magnitude of this task is difficult to see because our scientific and cultural biases in favor of traditional research methodologies would have us think this network is currently in place. Or, we assume that improved networking will soon alleviate the problems of transporting scientific simulations and engineering designs to the scientific/engineering marketplace. However, the problem is no more solved today than the muddy impassable roads of France in the 1880's constituted an adequate transportation system. Modern industrialized France still needed to build government-mandated networks of new roads and railroads in the years prior to World War I.
In the 1980's visual communication in the United States is hobbled by lack of standards, mired in the intellectual frustration of making connections across incompatible media, and held up at the gateways by disparate transmission protocols never designed with visualization in mind. Visual communication cannot be shared among users across a distributed network of incompatible workstations with idiosyncratic interfaces and no common layering of portable software and hardware." [9]

AIDSNET, the proposed National Telecommunications Network for AIDS Research, would address urgent needs for integration of the AIDS biomedical research community, and well illustrates the potential of televisualization to provide this integration.

To combat the AIDS pandemic, a critical need exists to improve investigator-to-investigator information transfer, especially in fields requiring computational and imaging biosciences and technology. Researchers engaged in the computational science of molecular modeling are gaining access to advanced graphical workstations and supercomputers, and are growing dependent upon visualization tools to help them comprehend the volume of data generated. Similarly, those engaged in imaging science (e.g., X-ray crystallography, microscopy and medical imaging) are deluged by torrents of image data, and also increasingly depend upon computational imaging and visualization tools to manage the image data. These AIDS biomedical researchers must be able to communicate their discoveries, their insights, and above all the actual modeling codes they develop to gain those insights.

The long-range goals of the AIDSNET project are:

1. *Make accessible the environment needed to visualize and model complex biomolecular/biological structures at multiple levels of resolution:* from the molecular, through the subcellular and cellular levels, to tissue and gross anatomical levels of description; and

2. *Improve investigator-to-investigator information transfer:* by providing telecommunications-compatible methodologies for visual information display, teleconferencing and modeling. Computational and imaging bioscientists need to communicate their models, visualization tools, and verifying data with their geographically dispersed colleagues without penalty of distance.

Both capabilities are needed to transfer knowledge and foster communication between computational/imaging bioscientists and technologists engaged in AIDS research.

Existing forms of remote visual communication include video teleconferencing, television broadcasting and electronic blackboards, but these are all hopelessly inappropriate for this task. We must use technology to communicate technology, to put scientists back into control of their medium of choice. Interactive visualization and modeling require visualization-based networking as the appropriate mode of scientific communication.

Researchers need to transmit their data and graphical images from their research centers to colleagues at remote sites. However, the sheer scale of graphics and image data sets exceeds the current bandwidth capacity and interaction of networks designed for the transmission of alphanumeric data. Networks that handle screenfuls of textual information exist and work well; network nodes are simply gateways that neither add nor detract from the quality of the message. But a 2-dimensional visual image of 512 x 512 pixels x 8 bits/pixel contains approximately 100 times more information than a screen of text with 25 rows x 80 characters/row.

Volumetric images, approaching 1024 x 1024 x 1024 x 4 bytes/voxel, present approximately 16,000 times more information than a single 512 x 512 pixel image.

Telecommunication costs for the added bandwidth needed to transmit visual imagery, however, must be balanced against local computational costs. For the foreseeable future this balance dictates the following strategy of "value-added processing" over visualization-compatible networks: (1) transmit only graphic models over common carriers (for example, transmit only molecular models, not the screen images themselves), and (2) use local graphics workstations to animate, render, and display the transmitted spatial data sets. In a televisualization network, unlike existing computer networks, cost effectiveness requires that value is added at multiple nodes of the network, to provide for the semantic compression and subsequent reconstruction, rendering, and display of the image.

In summary, televisualization is the study of real-time, interactive visualization and modeling, conducted over computer networks that link geographically dispersed investigators.

The panel convened by the National Science Foundation, in its report *Visualization in Scientific Computing* of November 1987 [9], concluded that televisualization technology could greatly improve the quality of scientific and technical communication and discovery. When examining the ways scientists use visualization today, the Panel observed that visualization tasks are best handled when distributed across several tiers of machine types. Visual telecommunication among geographically dispersed research centers would serve as the driving force to integrate these visualization and modeling capabilities.

FRAMEWORK FOR SCIENTIFIC VISUALIZATION ENVIRONMENTS

Objectives of the Research Environment

We are developing a highly interactive *ScientificVisualization Environment* to aid in the interpretation of massive volumetric data sets, increasingly common to scientific computing. The Scientific Visualization Environment is designed for applications requiring intensive visualization and geometric modeling: in the interpretation of seismic data in the geosciences, space exploration and astrophysics, medical imaging, brain mapping, computational fluid dynamics, and microelectronic device field modeling. Additional applications to the analysis of time-varying imagery may be also possible.

The Scientific Visualization Environment will (1) provide a portable highly interactive visualization interface to aid in the visualization and geometric modeling of volume data sets, (2) develop a general model for representing visual knowledge and apply the model to create a protoype knowledge base in a specific visually-oriented domain, and (3) develop an imaging/graphical interface coupling the domain-specific knowledge bases and the workstation environment.

Three critical aspects of volume visualization and modeling require definition and are being given pilot evaluation: (1) navigation in 3D image data sets, (2) constraint-based assembly/growth of complex 3D models from a set of basic parametrized objects, and (3) finite element modeling of 3D imagery. The pilot Scientific Visualization Environment will serve as a testbed for the development and evaluation of these key visualization tools.

In summary, the Scientific Visualization Environment aims to provide computational scientists and engineers with the ability:

1. to visualize and steer computation

2. to navigate through, visually explore, and model volumetric data sets

3. to assemble and/or grow geometric models of complex structures

4. to play back segments of simulations in real time, with the option of producing videotapes and slides of research results

5. to interact with colleagues across visual telecommunications networks

Organization of the Research Environment

The general organization for a scientific visualization and modeling environment is described below (Figure 2).

Scientific visualization support for contemporary computational science and engineering is fractured into subdisciplines (such as geometric modeling, finite element analysis, graphics, and image processing) which have developed independently and can scarcely communicate with one another. This disjointed environment makes attainment of the goals for scientific visualization enunciated above, such as interactive visualization and steering of computation, currently unattainable. To ameliorate this situation we propose a scientific visualization environment described below.

This model of the Scientific Visualization Environment, consistent with a broader view of computational science and engineering, is shown in Figure 2. Underlying this view is the critical need to provide a consistent geometric modeling environment for scientific visualization and modeling. The visualization components identified in Figure 2 are defined operationally below.

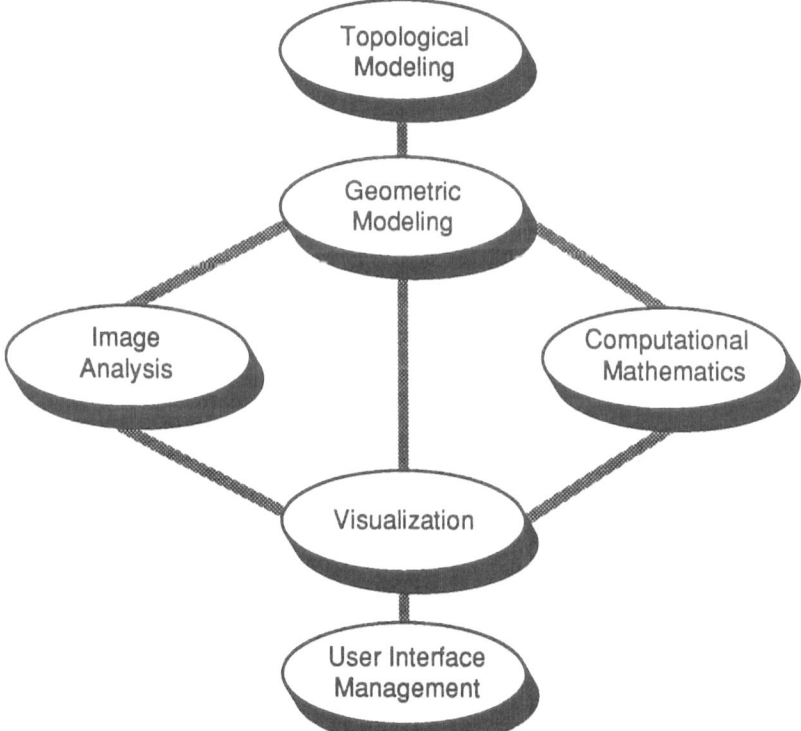

Figure 2.   Scientific visualization environment for computational science and engineering

Figure 2.   Scientific visualization environment for computational science and engineering

Topological Modeling

Topological modeling allows for the interconnecting, assembly and/or growth of a complex structure from a hierarchical object-oriented description of its parts.

Assembly-based modeling allows for the construction of a complex three-dimensional geometric model from a hierarchical object-oriented description of its parts. Computer-aided design (e.g., of an automobile engine) typically uses this paradigm.

Growth-based modeling allows for the growth over time of a three-dimensional model from seed objects which preserve their topological connectivity throughout the growth process. This type of modeling is appropriate for the simulations of developmental biology/neurobiology (e.g., visualization and modeling of gene expression). Meteorological modeling of the growth of a storm can also use this paradigm.

Network-based modeling facilitates the construction and simulation of networks embedded in a geometrical model. Examples in electrical engineering include printed circuit and VLSI layout. In cognitive neuroscience, network-based modeling is used to describe topological neural networks. Here neighboring units behave similarly and exhibit the type of coordinated behavior seen in cortical maps.

Geometric Modeling

The representation of a 3-dimensional object or shape is referred to as a *geometric model*. Geometric modeling provides the geometric model, such as a molecular configuration, or the container of the classical field (e.g.,fluid dynamics) to be computed.

Molecular Modeling provides open-form polyhedral descriptions of molecules, viruses, and the space-filling tessellations of mathematical crystallography. Direct extensions enable the modeling of robotic manipulators.

Solid Modeling provides symbolic, numerical, and graphical techniques for (1) the construction of a designed object as a rigid solid, (2) data structures to represent solids, and (3) methods to compute the intersection of two solids. Typical solid modeling primitives include regularized polyhedrons, generalized cylinders, and superquadrics.

Fractal Modeling provides geometric models using fractal, graftal, and grammar-based construction algorithms. Most commonly these so-called process models are used to model natural phenomena, such as undeveloped terrain, plants, and trees.

Physically-based Modeling extends geometric models by including not just spatial information, but also physical parameters associated with the models. Once these physical characteristics (e.g., mass, moments, force, torque, stress, and energy) are incorporated into the model, the behavior of the model is then simulated according to well-known natural physical laws. Physical models, by incorporating rules into the model which define the physical behavior of an object over time, allow us to use a generic representation scheme for the description of both rigid and deformable objects.

Topobiological Modeling allows dynamic description of biological and neurobiological development, including the following:

- primary processes of development (cell division, cell motion, and cell death)
- regulatory processes (cell adhesion, differentiation and induction, and hence the transformation between epithelia and mesenchymal states)
- the parallel concurrent execution of these processes to develop space-filling configurations of cells

Computational Mathematics

Computational mathematics brings to bear the classical field descriptions, the symbolic and algebraic manipulation, and the numerical analysis algorithms (e.g., finite element analysis) typically required to compute the classical field at all spacetime points compatible with or contained by the geometrically-defined solid model. For example, a boundary representation model of the Space Shuttle can define the geometric context for a fluid flow computation. In this example geometric modeling must also provide the cell decomposition employed by a finite element analysis of the Navier-Stokes equation.

Classical Field Description provides a geometrical description of classical fields in flat spacetime, including symbolic, numerical, and graphical representations for the following classical geometric objects:

- scalars, vectors, and tensors
- differential forms
- gradients and directional derivatives
- coordinate representations of geometric objects
- exterior derivatives and closed forms
- stress-energy tensor and conservation laws

Symbolic and Algebraic Manipulation provides symbolic and algebraic manipulation of the formulae of computational mathematics. For instance, the symbolic and algebraic manipulation package, *Mathematica* [13], is built around a static surface graphics model. An extension to volume visualization has yet to be defined. In the interim it is undoubtedly appropriate to build *Mathematica* into the dynamic geometric modeling and scientific visualization environment.

Numerical Analysis provides symbolic, numerical, and graphical algorithms to compute and analyze functions defined over the geometric models constructed above.

Geometrical linkages must be made to the matrices for the solution of partial differential equations by finite element and finite difference methods, as in such applications as computational fluid dynamics, and the computation of electrostatic/magnetostatic fields.

For example, numerical algorithms for finite element analysis require an environment with sufficient symbolic computational power and visualization support to design the finite element grid (typically in 3D), carry out the modeling of the individual elements and their nodal linkages, and establish the mapping between the elements of the geometric description and the banded matrix required for the relaxation computation. Facilities of course are also needed for the visualization of the computed results.

In addition, geometrical linkages need be made for numerical analysis by:

- differential equations and dynamic systems modeling
- analytic function expansions
- stochastic processes
- lumped parameter modeling

We view numerical analysis itself as lying outside the intended scope of the Scientific Visualization Environment research program. However we must provide geometric linkages to those computational forms commonly employed by numerical analysis algorithms.

Image Analysis

Image analysis provides dynamic geometric modeling support for:

- remote sensing and terrain modeling
- computer vision
- geometric modeling of volumetric data

Image analysis starts with an intensity function defined everywhere in a 2- or 3-dimensional world. Image analysis typically attempts to find a geometric model of some phenomena in the image data set; increasingly often these are 3-dimensional. For instance, in the computational fluid dynamics of the Space Shuttle in flight, turbulence eddies can be visualized in the computed data set. This turbulence does not appear explicitly in the Navier-Stokes equation, but can be modeled geometrically after appropriate image analysis. In brief, *computational mathematics* provides the tools to compute a volumetric data set constrained by a geometric model; *image analysis* proceeds in the opposite direction: for example, beginning with a volumetric data set, image analysis computes an inferred geometric model.

Visualization

Visualization provides the interactive viewing transformations, surface- and volume-based rendering techniques, and animation to visualize the computed trajectories, surfaces, and fields in the image data set.

User Interface Management

User Interface Management addresses the design and implementation of application-dependent user interfaces.

The Scientific Visualization Environment proposed above would provide for the symbolic, numerical, and graphical representation of objects in spacetime. Nonetheless each class of applications calls for a facile computer-human interface. The goals for user interface management are:

1. to manage the diverse modes of visual, tactile, and auditory interaction of the client with the computer
2. to provide the tools to build these interfaces

This assemblage of tools includes windowing systems, video editors, and interfaces to special-purpose 3-dimensional pick-and-place input devices. Such groups of interface tools have recently been referred to as User Interface Management Systems (UIMS). Unfortunately UIMS technology for the support of supercomputing is currently in an abysmal state of underdevelopment.

Frontend and Kernel Subsystems

The Scientific Visualization Environment must provide both frontend and kernel subsystems. The frontend system would typically run on advanced visualization workstations, as exemplified by the Silicon Graphics Personal Iris graphics workstation.

More computationally intensive applications require that the kernel run on a minisupercomputer or supercomputer. To facilitate system integration, both the frontend and the kernel subsystems of the Scientific Visualization Environment are being developed in UNIX.

Fragments of this Scientific Visualization Environment currently exist commercially, but much tool building and system integration remains to be done. The Visualization Laboratory at Texas A&M University is focusing much of its research program on designing this integrated environment for scientific visualization and modeling. We build upon standards and ensure compatibility through systems integration and on-going support.

INTERDISCIPLINARY RESEARCH PROGRAM IN SCIENTIFIC VISUALIZATION

In summary, we propose that the scientific and engineering community commit its resources in the following visualization initiatives:

1. Drive world class research by scientists and engineers who need both leading edge graphics/image processing and the unique capabilities of advanced supercomputers to achieve interactive steering of computation and/or realistic animation.
2. Introduce visualization capabilities into computational science and engineering across multiple disciplines.
3. Build a three-tiered graphical support environment for presentation, peer and personal graphics. In particular, help users who have direct, even prosaic, requirements for visual output, emphasizing its research and development usability.
4. Develop advanced capabilities for image and graphics support.
5. Build upon standards and ensure compatibility through systems integration and ongoing support.
6. Through televisualization, support collaborators increasingly distributed across networks.
7. Train users and staff in visualization techniques.

Our research program at Texas A&M University is building the tools and human interfaces for human-in-the-loop environments for selected computational research areas, in order to help computer scientists and computational scientists alloy their strengths.

To be successfully utilized, however, a Scientific Visualization Environment must be backed by a computation center which will provide practical support such as documentation, distribution, consulting, feedback from real users, and migration of the tools and concepts to a broad range of disciplines. Support of this nature is not generally provided by universities and industry.

REFERENCES

1. Barr, A. H., Global and local deformations of solid primitives. *Computer Graphics*, 18, 3, July 1984, pp. 21-30.

2. DeFanti, T. A., and M. D. Brown, Scientific animation workstations: Creating an environment for remote research, education, and communication. *Academic Computing*, 3, 6, February 1989.

3. Edelman, G. M., *Topobiology: An introduction to molecular embryoloby.* NY: Basic Books, 1988.

4. Koons, D. B., *A model for the representation and extraction of visual knowledge from illustrated texts.* MS Thesis, Report TAMU 88-010, Department of Computer Science, Texas A&M University, August, 1988.

5. Koons, D. B., and B. H. McCormick, A model of visual knowledge representation. Proc. First International Conf. on Computer Vision, 1987, pp. 365-373.

6. Liaw, J. S., *Visualization environment for editing and modeling volumetric data.* MS Thesis, Department of Computer Science, Texas A&M University, August, 1989.

7. Mandelbrot, B., R. Voss, D. Saupe, P. Prusinkiewicsz, and H.-O. Peitgen, Fractals: Introduction, basics and applications. *SIGGRAPH '88 Course Notes No. 8*, 1988.

8. Mantyla, M., *An introduction to solid modeling.* Computer Science Press, 1988.

9. McCormick, B. H., T. A. DeFanti, and M. D. Brown (eds.), Visualization in scientific computing. *Computer Graphics*, 21, 6, November 1987.

10. Pentland, A. P., Perceptual orgnanization and the representation of natural form. *Artificial Intelligence*, 28, 1986, pp. 293-331.

11. Terzopoulos, D., J. Platt, A. Barr, and K. Fleischer, Elastical deformable models. *Computer Graphics*, 21, 4, July 1987, pp. 205-214.

12. Smith, A. R., Geometry and imaging: Clarifying the major distinctions between the two domains of graphics. *Computer Graphics World*, November 1988.

13. Wolfram, S., *Mathematica: A system for doing mathematics by computer.* CA: Addison-Wesley, 1988.

PART 6

REQUIREMENTS AND

PERFORMANCE

REQUIREMENTS FOR MULTIDISCIPLINARY DESIGN OF AEROSPACE VEHICLES ON HIGH PERFORMANCE COMPUTERS

Robert G. Voigt[1]

ICASE

NASA Langley Research Center

Hampton, VA 23665

Abstract

The design of aerospace vehicles is becoming increasingly complex as the various contributing disciplines and physical components become more tightly coupled. This coupling leads to computational problems that will be tractable only if significant advances in high performance computing systems are made. In this paper we discuss some of the modeling, algorithmic and software requirements generated by the design problem.

Introduction

The classical scientific method is undergoing a fundamental change that has the numerical experiment or simulation taking its place alongside the more traditional laboratory experiment. More detailed experiments require more sophisticated models, and in turn, more powerful computational systems. Thus it is becoming increasingly clear that a pacing technology for advances in many areas of science and engineering is high performance computing.

In 1987 the Federal Coordinating Council for Science, Engineering and Technology (FCCSET) Committee on Computer Research and Applications conducted a review of high performance computing issues and opportunities. The result of this study was a report issued by the Executive Office of the President, Office of Science and Technology Policy on November 20, 1987, entitled "A Research and Development Strategy for High Performance Computing." The report concluded that maintaining leadership in the development and application of high performance com-

---

[1]This research was supported by the National Aeronautics and Space Administration under NASA Contract No. NAS1-18605 while the author was in residence at the Institute for Computer Applications in Science and Engineering (ICASE), NASA Langley Research Center, Hampton, VA 23665.

puting is crucial to continued preeminence in science and engineering and that this leadership position is being challenged by advances in Europe and Japan. Four areas were singled out as focal points to address this challenge: development of high performance computing systems including parallel systems, development of algorithms and software to bring the power of such systems to bear efficiently on complex problems, development of the networking technology required to make the systems readily accessible for collaboration among scientists who are geographically dispersed and finally support of the research infrastructure to assure that the trained personnel will be available to make effective use of the systems resulting from the first three areas. To stimulate the desired development and assure its relevance, the FCCSET report suggests pursuing "Grand Challenges." As defined in the report, a Grand Challenge "... is a fundamental problem in science and engineering, with broad application, whose solution will be enabled by the application of the high performance computing resources that could become available in the near future."

Various federal agencies responsible for support of research and development in the U.S. are selecting and refining Grand Challenges to provide a focus for the research programs they are developing to respond to the FCCSET report. In the remainder of this paper, one such challenge put forth by the National Aeronautics and Space Administration (NASA) will be described, and some of the issues arising in its pursuit will be discussed.

A Grand Challenge

The NASA Grand Challenge in aerosciences as first put forth in [2], and subsequently refined, is the integrated multidisciplinary design of aerospace vehicles and their numerical simulation throughout a mission profile. The goal is to demonstrate the utility of advanced parallel computer systems, including hardware, software and algorithms, capable of delivering teraflop performance for the design of new generations of aerospace vehicles. Such a demonstration requires separate developments within a number of disciplines as well as the tight integration of those disciplines.

The integration of multiple disciplines arises in at least three different ways. First there are the various components of the vehicle that must function in a tightly coupled fashion. These include the airframe, the propulsion system, the control systems, etc. Second there are the scientific disciplines required for the basic understanding and modelling of the components. Here one must involve aerodynamics, chemistry, combustion, structural dynamics, solid mechanics and control theory to name a few. Finally there are the disciplines such as applied mathematics, numerical analysis and computer science that must come together for the successful numerical simulation of a complex physical phenomena on a parallel computer.

Most of these disciplines have been involved in traditional aerospace vehicle design and analysis, but present trends toward improved performance are leading to a tighter coupling of these disciplines. The impact of this trend on the design process will be discussed in the next section.

The Design Problem

The traditional approach to the design of aerospace vehicles is to treat each discipline separately and in turn. A simplified and exaggerated example may serve to illustrate the point. The aerodynamics team working under the requirement to keep weight to a minimum while providing a specified range completes a clean aerodynamic design and passes it to the structure team. Their analysis reveals that the juncture between the wing and the fuselage may fail under extreme load conditions. They correct for this weakness by increasing the thickness and thus the weight of the root of the wing. This design is passed to the propulsion team who must now provide for more power than originally intended to overcome the added weight. This requires a larger engine which adds additional weight and changes the flow characteristics of the original aerodynamic design. Finally, to improve maneuverability, a controls team includes an active device to increase lift at takeoff. This device adds weight and changes the flow characteristics. At this point there is a design dilemma: accept the reduced range made necessary by the increased weight or return the design to the aerodynamics team to improve the efficiency of the design. The difficulty with the latter approach is that the process may not produce a solution to the original design objectives; that is, it may not converge to an optimal design.

As designs become more sophisticated and approach finer and finer tolerances, the various disciplines involved become more tightly coupled.

This tight coupling inhibits convergence of the design resulting in higher design costs or compromises in the design objectives. It may also lead to modifications in the design based on prototype performance resulting invariably in higher costs and decreased performance.

An Optimization Approach

One obvious way to overcome the design dilemma described in the previous section is to approach the problem as a coupled, multidisciplinary optimization problem. Thus one might seek to minimize the gross take off weight of a vehicle subject to the requirement or constraint that the range be a certain number of miles, or in general

$$\left. \begin{array}{c} \text{minimize } G(y^{(1)}, \cdots, y^{(p)}, x) \\[2em] \text{subject to } c(y^{(1)}, x) = 0 \end{array} \right\}. \tag{1}$$

The complication in this deceptively simple formulation is that the dependent variables in G and c may be specified implicitly through a complex exquation

$$F(y^{(1)}, x) = 0.$$

For example, $y^{(1)}$ might represent the pressure distribution over the vehicle and $F(y^{(1)}, x)$ could be the full Navier-Stokes equations. Optimization algorithms for solving (1) require repeated evaluations of G and c, and thus in the example, repeated solutions to the Navier-Stokes equations. Now if we consider that there are many other factors contributing to the weight of a vehicle that may be tightly coupled, we see that p, the number of dependent variables, may be quite large; furthermore, each one may be specified by a complex system of partial differential equations and hence be very expensive to obtain.

Figures 1 and 2, reproduced from [2], provide some indication of the computational complexity and the present state of the art for two disciplines: aerodynamics and structural analysis. The underlying assumption is that a single simulation must be completed in 15 minutes. Figure 1 shows a range of configuration complexities from an airfoil through a wing to a full aircraft. The underlying models also increase in complexity beginning with a greatly simplified version of the Navier-Stokes equations which includes nonlinear effects but neglects viscous terms. The Reynolds-averaged Navier-Stokes equations include all terms but averages over time are taken and turbulence models are required. Finally,

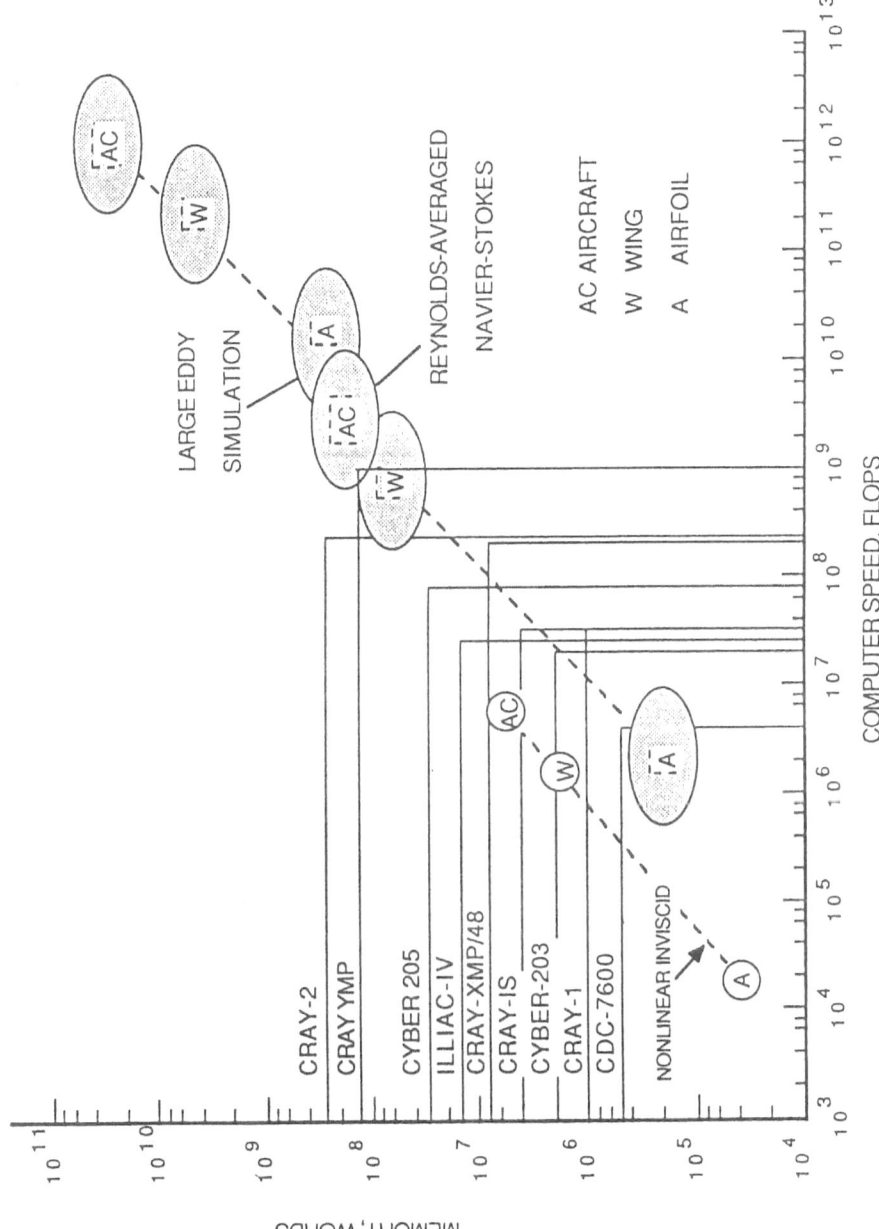

Figure 1. Memory and Speed Requirements for Computational Aerodynamics

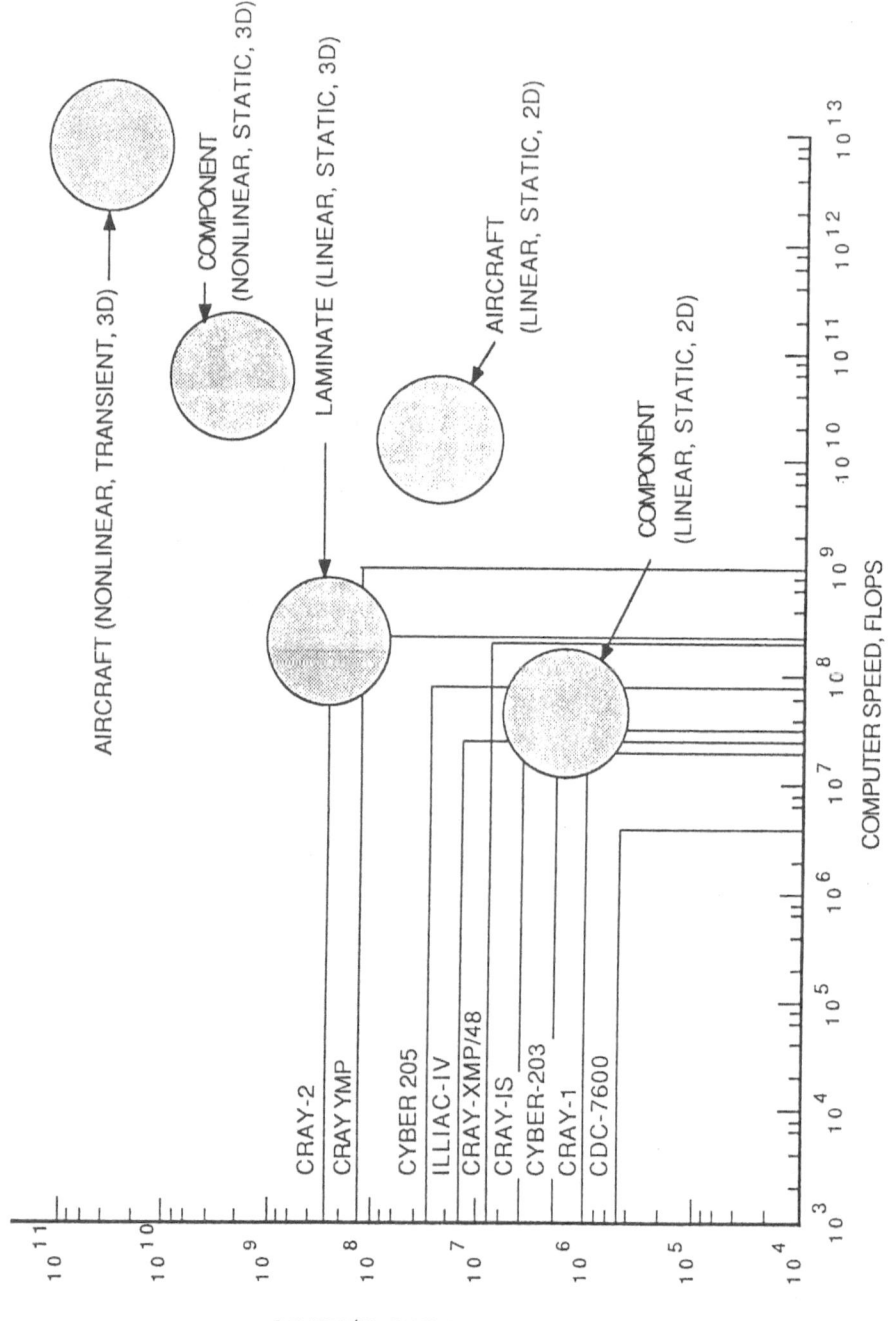

Figure 2. Memory and Speed Requirements for Computational Structural Analysis

large eddy simulation involves the direct numerical simulation of turbulent eddies over a large range of scales but still requires turbulence models for the smallest scales.

Figure 2 also shows a range of computational requirements relative to past and present high performance computers. Again the configuration complexity moves from a simple laminated material through a component to a full aircraft. The models range from simple linear two dimensional models for static analysis to nonlinear three dimensional models appropriate for studying transient behavior.

The computational requirements implied by these figures are severe in their own right. When one thinks of coupling these and other disciplines that are equally computationally demanding through an optimization formulation that requires repeated evaluation of these models the "challenge" is truly "grand."

Alternative Approaches

We have seen that attacking the aerospace vehicle design problem with traditional optimization techniques leads to potentially enormous computational requirements and may be only slightly more attractive than the brute force approach involving parametric studies. One alternative that has received increased attention is based on sensitivity analysis (see, for example, [1] and [7]). The idea is that if the designer knew how sensitive the dependent variable was to changes in the independent variable he could use that information to guide the design process to an optimal solution.

For example, if $F(y, x) = 0$ represents the Navier-Stokes equations with y the pressure and x the vehicle geometry, the designer would like to know the sensitivity of y to changes in x, that is $\partial y/\partial x$. This partial is readily available through the implicit function theorem:

$$dF/dy = \partial F/\partial y\ \partial y/\partial x + \partial F/\partial x.$$

But since $F(y, x) = 0$, a change in x must be compensated for by a change in y. Setting $dF/dy = 0$ yields

$$\partial F/\partial y\ \partial y/\partial x = -\partial F/\partial x. \qquad (2)$$

For some disciplines, $\partial F/\partial y$ and $\partial F/\partial x$ are available analytically, for others they must be obtained numerically. In either event if y is a vector of length n, equation (1) is an $n \times n$ matrix equation for the vector of unknown derivatives of y with respect to a single x. Thus $\partial y / \partial x$ is available for a number of different x's by factoring $\partial F / \partial y$ and solving (1) with different right-hand sides. It is important to note that if $\partial F / \partial y$ and $\partial F / \partial x$ are not available analytically, they can be obtained numerically by computing differences with respect to x; this only requires evaluation of $F(y, x)$ not the solution of $F(y, x) = 0$. Of course, this approach yields only an approximation to $\partial F / \partial y$ and $\partial F / \partial x$.

Other disciplines may be coupled through a large block matrix equation similar to Equation (2). Thus if the sensitivity information can be incorporated into an automatic design process, the computational burden of solving repeatedly the equivalent of $F(y, x) = 0$ will be removed.

Recently Jameson, [3] and [4], has suggested a design procedure in which the design problem is treated as a control problem with the control chosen as some appropriate design objective. This approach has been demonstrated for a three dimensional wing design using the Euler equations [3]. By considering the design problem as a problem in control, a variety of formulations are available based on the theory for control of systems governed by partial differential equations. It remains to be seen if this approach can be extended to solve efficiently the more complex multidisciplinary design problem.

Implementation Issues

A number of interrelated issues arise in the implementation of a multidisciplinary design problem on a high performance parallel computer. The first of these is the choice of the appropriate models for the physical phenomena of interest. When conducting an analysis within a single discipline the model choice is normally based on the need to resolve the phenomena of interest; for multidisciplinary analysis the demands of one discipline on the quality of the results from another may influence the model choice as well. For example, a simple panel method might be adequate to provide data on the pressure distribution at cruise conditions, where as the Navier-Stokes equations would be required if a prediction of pressure at high angle of attack were needed. The issue of model choice also arises in the need to maintain a particular level of realism from discipline to discipline. For example, there might be little point in using the Navier-Stokes equations for aerodynamics if linear thin shell theory was to be used for structural analysis.

Another factor influencing the model selection, and to an even greater extent the numerical algorithm, is the computer system and its utilization of parallelism. It is a well established fact that different algorithms exhibit different degrees of parallelism, but this is also true of models. For example, the cellular automata model for fluid dynamics contains a high degree of parallelism that is easier to exploit on most parallel systems than the parallelism available in differential equation models.

For multidisciplinary design problems to execute efficiently on high performance parallel computers, advances will have to be made in several software areas. Providing a programming environment including compilers, debuggers and performance monitors is crucial to achieving good utilization of the hardware and productive use of the scientists trying to use the system. Special language constructs and data structures may be appropriate for different disciplines and then there must be an efficient linking of these across disciplines. For example, a rectilinear grid may be appropriate for an aerodynamic calculation involving a wing-engine configuration whereas an unstructured triangular mesh may be required by the finite element analysis of the structural properties of that same configuration.

System software must also be developed to automate some of the tedious yet crucial aspects of implementing a large simulation on a parallel computer. Areas of particular importance include communication and synchronization constructs, mapping the data and program onto distributed processors, and dynamically load balancing the processors when computational changes cause some processors to be overloaded while others are idle.

Programming communication and synchronization for a distributed memory system is both time consuming and error prone; furthermore, it clutters up a program and makes it hard to read and modify in the future. Several research efforts are underway to provide compilers which generate the necessary code automatically. One such compiler, Kali Fortran 1, uses sequential Fortran annotated with a "distribution clause" appended to array declarations, [5] and [6]. Using this distribution the compiler is able to generate the necessary message passing code to handle communication. Such compiler concepts must mature if implementation of the multidisciplinary design problem on a parallel system is to become feasible.

The various data structures and changing computational requirements of the design problem are going to make dynamic load balancing an essential software tool. Static load balancing or the mapping problem must be accomplished with heuristics as it is known that there are no polynomial-time algorithms. When the load changes during execution, the problem becomes even worse. First a mechanism must exist to detect and evaluate the imbalance. Then a new mapping must be computed. The cost of carrying out the remapping must be calculated and the improvement resulting from the remapping must be estimated. The improvement and its cost must be compared with the strategy of continuing the computation without remapping. Finally, if deemed appropriate, the remapping is carried out. Obviously, very efficient algorithms must be found and the process must be automated if all the processors in a parallel system are to be used effectively.

Conclusions

The NASA aerosciences Grand Challenge is the integrated multidisciplinary design of aerospace vehicles and their numerical simulation throughout a mission profile. This challenge will require the interaction and cooperation of scientists and engineers from a wide range of disciplines. In addition, it will require advances in the numerical simulation of a wide range of physical phenomena and the close integration of a number of these. Finally, a number of advances in parallel high performance computing hardware and software will be required. The wide range of expertise required may well necessitate forming research teams whose members are geographically distributed. This "institute without walls" will put an extra burden on national networks, but if it is successful, it may represent a new way of doing science that is as important as the results themselves.

References

1. Adelman, H. and Haftka, R.: Sensitivity Analysis of Discrete Structural Systems. AIAA Journal, Vol. 24, No. 5. pp. 823-832. (1986).

2. Bailey, F., Chow, L., Denning, P., Halem, M., Smith, P., South, J., Stevens, K., Szuch, J., Voigt, R., and Weber, W.: Concept Document

for a NASA High Performance Computing Initiative. National Aeronautics and Space Administration, Office of Aeronautics and Space Technology. January 31, 1989.

3. Jameson, A.: Aerodynamic Design via Control Theory. ICASE Report No. 88-64. NASA Langley Research Center. November 21, 1988.

4. Jameson, A.: Computational Aerodynamics for Aircraft Design. Science, Vol. 245. pp. 361-371. (1989).

5. Mehrotra, P. and Van Rosendale, J.: Compiling High Level Constructs to Distributed Memory Architectures. ICASE Report No. 89-20. NASA Langley Research Center. March 7, 1989.

6. Mehrotra, P. and Van Rosendale, J.: Parallel Language Constructs for Tensor Product Computations on Loosely Coupled Architectures. ICASE Report No. 89-41. NASA Langley Research Center. June 2, 1989.

7. Sobieszczanski-Sobieski, J.: On the Sensitivity of Complex, Internally Coupled Systems. AIAA Paper No. 88-2378. (1988).

Supercomputer Performance Evaluation: The PERFECT Benchmarks

Joanne L. Martin
IBM T.J. Watson Research Center
Yorktown Heights, New York 10598   U.S.A.

1. Introduction

Supercomputers, designed to solve problems that would be otherwise intractable, have received considerable attention in recent years. It has been claimed that they are critical to national defense, that the economic well-being of companies and nations depends on their exploitation, and that the grand challenges facing today's scientists -- global change phenomena and human genome definition, e.g. -- would remain as open challenges without super-computer access [1].

Designing systems to meet the extreme requirements of rapidly-advancing computational technology, and therefore purchasing and using supercomputer systems, is a very costly proposition. Because supercomputers have become such a significant commodity with large associated costs, the measurement and evaluation of their performance has emerged as an area that must be el-evated from art or magic to science.

Evaluating performance is more than a clocking function. Careful evaluations explain why performance is what it is and thus provide insights into how it can be improved. Subjective components of the process are at best difficult to contain. Furthermore, how well a system performs depends on the perspective from which the evaluation is viewed as well as on the tools employed in the process. Very seldom, if ever, is it possible to reduce a thorough evaluation of supercomputer performance to a single figure of merit.

Three perspectives from which to view performance evaluation are presented in Section 2, as are three techniques for evaluating performance. In each case, these categories are neither all-inclusive nor mutually exclusive. After a general description of the problem, the interested parties, and the possible techniques, we will focus on the specific case of benchmarking as it applies to system designers and applications developers. The question of procurement-

NATO ASI Series, Vol. F 62
Supercomputing
Edited by J. S. Kowalik
© Springer-Verlag Berlin Heidelberg 1990

related evaluation is beyond the scope of this work because any procurement must be considered a unique situation requiring specific tuning of the needs of the site in question to the systems under consideration. Section 3 presents the history and development of a group of researchers known as the PERFECT Club, where the acronym represents PERFormance Evaluation for Cost-effective Transformations. In Section 4 we consider future directions and provide some conclusions based in part on the discussion that followed the presentation of this work at the NATO workshop from which these Proceedings were developed.

2. The Performance Evaluation Problem

Performance evaluation is an ongoing process, performed as part of the development, purchase, and use of supercomputing systems. Architects need to be able to quantify the costs and benefits that are associated with modifications to their designs. Groups writing compilers or operating systems need to measure the contributions made by enhanced optimization algorithms or new scheduling techniques. Users, eager to make optimal use of limited resources, endeavor to conform their applications to the underlying hardware/software system. What is common in all of these areas is the need to understand and quantify the workload associated with the evaluation task.

System architects, both hardware and software, set out to develop a system that will accommodate most effectively the requirements of a particular workload. In the case of supercomputing, the workload is generically described as a set of scientific or engineering applications for which the memory requirements are extremely large and the computational characteristics can grow almost without bound. More specific characterizations depend on details that pertain to individual workload components, and therefore can be generalized only by classes. For example, some applications are dominated by their scalar components, others have considerable scalar content but are parallelizable, while still others are conveniently structured in a manner that permits the exploitation of both vector and parallel systems. Further definition of these classes is needed at the design stage.

Application developers, or users, are interested in exploiting a given system as fully as possible. The quest for effective use of a given system often results in what Jack Dongarra has termed "bottleneckology". Users modify algorithms based on their understanding of a system's structure and capabilities. Having a stable and well-understood workload fosters the development of algorithmic changes that map to features introduced into new and evolving systems.

People responsible for procuring systems want to ensure an appropriate match between the needs of their installation and the system that they acquire. As noted, supercomputing systems are very costly to purchase and to operate. Mistakes made because of a lack of understanding of an installation's workload and its requirements are long remembered.

Three common techniques for evaluating computer performance are simulation, benchmarking, and analytic modeling. Simulation is typically the most expensive technique, but it can be quite accurate. At the design stage, especially before there is hardware available for testing, simulation is the method of choice. Analytic modeling is at the other end of the spectrum. It is typically applied in situations where there is an understood workload that has been carefully parameterized running on an existing system. Analytic models can then predict quite reasonably how changes in operating systems will effect system throughput, or how system enhancements -- often additional DASD or an extra CPU, for example -- will increase the overall operation of a system.

Benchmarking fits in between the other techniques both in its cost and its accuracy. In fact, good analytic models often require substantial system benchmarking for the development of appropriate input parameters. Unfortunately, it is not easy to construct a worthwhile set of benchmarks that has significance in any general context. Some of the more well-known examples of benchmarks make this point well. The Livermore Fortran Kernels [2] are related to the workload at the Lawrence Livermore National Laboratory in a way that is understood by scientists there. Thus, it is possible in that context to extrapolate performance on the loops to performance on a Laboratory computing system. Outside of Livermore, it is much less clear how the loops can be applied effectively to evaluate performance. Similarly, the LINPACK benchmark [3] is able to measure performance across a wide range of systems, but only in terms of specific matrix manipulation routines. Unless a workload is comprised in large part of this type of computation (and few workloads are), the ability to predict overall performance based on the LINPACK results is non-existent.

In 1986, the Department of Energy and the Office of Naval Research recognized the need for a scientific approach to the evaluation of supercomputer systems and commissioned a National Academy of Sciences Committee to evaluate the state-of-the-art and make recommendations on how to improve this field. The Committee found that existing techniques for benchmarking supercomputers tended to be ad hoc and experiential and that although there were individual examples of structured approached to the benchmarking problem, there was no well-developed method for enhancing, generalizing, or sharing results obtained. Furthermore, despite the body of performance evaluation literature concerning queueing models and related issue, there was no well-defined effort to develop existing models for use in supercomputing systems. One of the major recommendations of the report, *An Agenda for Improved Evaluation of Supercomputer Systems* [4], was a 5-stage methodology for better supercomputer performance evaluation (Table 1).

Table 1.   5-Stage Methodology (NAS report, 1986)

Determine major applications areas and predominant solution techniques

Select a collection of representative programs and a set of target architectures

Define appropriate parameters of the applications and architectures to allow models to be developed

Define metrics and measure performance in a controlled environment

Assess relationship between application and architecture models

---

3. The PERFECT Club

3.1 Background

Partially in response to the NAS report, and also because of their interest in supercomputer development (hardware, system software, applications, and algorithms), David Kuck and Ahmed Sameh of the Center for Supercomputer Research and Development (CSRD) established the PERFECT Club. Beginning with the premise that applications drive the development of supercomputer systems, Kuck and Sameh collected a group of individuals who represented various application areas as well as a number of different architectures (Table 2).

Table 2.   Senior members of the PERFECT Club

CSRD (Kuck, Sameh, Berry, and others)

California Institute of Technology (Fox, Messina)

Houston Area Research Center (Johnson)

IBM (Clementi, Martin)

Institute for Supercomputing Research (Mendez)

Princeton (Orszag)

Their goal was to attack the performance evaluation problem by creating a portable benchmark set and defining the methods and parameters that would be used for measurements. This limited first step was aimed at building tools with which to perform detailed analyses in subsequent studies of architectures, compilers, systems, algorithms, and the matches or mismatches across them.

3.2 The Benchmarks

Thirteen programs comprise the initial PERFECT benchmark suite (Table 3). Although there is no claim that the programs represent all of computational science, they do represent a collection of applications that are indicative of what a particular set of scientists believe is executing in some of today's high-speed computing environments. With the exception of ARC3D (which has been removed from the collection) the programs are in the public domain, have been executed on a small number of diverse systems, and have provided the base for developing a methodology for improved benchmarking techniques. Considerable effort was expended to ensure portability of the PERFECT programs. Furthermore, guidelines were developed, using the base programs and the porting experiences of the study participants, for Fortran standards to promote portability across systems. Result verification also was built in to all of the programs based on significant output variables, setting a reasonable range for errors that might result from program execution on systems with different standards for floating point arithmetic.

Table 3. The PERFECT Club Benchmarks

Program	Type of Calculation	Lines of Code
ADM	Air Pollution	6142
ARC3D	Comp. Fluid Dynamics	3605
BDNA	Nucleic Acid Simulation	3962
DYFESYM	Structural Dynamics	7599
FLO52Q	Comp. Fluid Dynamics	2250
MDG	Liquid Water Simulation	1231
MG3D	Seismic Migration	2754
OCEAN	Comp. Fluid Dynamics	4215
QCD	Quantum Chromodynamics	2342
SPEC77	Weather Simulation	3880
SPICE	Circuit Simulation	18504
TRACK	Signal Processing	3770
TRFD	Quantum Mechanics	479

Perhaps the most important contribution of the PERFECT methodology is the emphasis placed on program modification and the recording of the performance enhancement achieved by various software optimizations. In each benchmarking endeavor, one of the primary questions involves how extensively the benchmark programs should be modified during the evaluation. Because any application is an implementation of a problem on a target architecture, choices are made concerning algorithmic approach and numerical method that would often be quite different if the implementation were targeted for a different architecture. Certainly methods that might be chosen for a tightly-coupled multi-processor with a small number of vector processors would be unlikely choices for a massively parallel computer, for example.

In procurement situations, the disparity in programming methods and algorithmic requirements often has the effect of maintaining the status quo rather than encouraging a shift to a radically different system. Because there is generally a large investment in existing software, users are reticent to re-write programs that are working. Thus, when benchmarking for procurement, one of the rules typically is to permit only very minor program modifications. This approach minimizes conversion costs on new systems, but it impedes progress in the development of new systems as well as new algorithms.

The approach taken by the PERFECT Club was to make baseline measurements of program execution with no modifications to the portable code to establish a common initial point for future measurements. Subsequently, no limits were placed on what modifications were performed provided each optimization was recorded and classified (see Table 4) and its resulting performance change was measured. Because there is a subjective component to measurements of the level of effort expended on program modification, an attempt was made to calibrate the various modifications. All of this information was recorded and maintained in "diaries" by application and system.

Analyzing the diaries provides insights concerning what general techniques are most applicable to specific systems and therefore which systems naturally belong to classes that can be defined. In further studies, compiler designers could benefit from analyzing the diaries to determine which of the optimizations were most cost-effective, which could be automated, or which should be encouraged or assisted by an interactive compiler at the user level. Similarly, hardware designers analyzing the diaries could gain insight about which architectural features are being exploited by these programs and where there is room for improvement. In the user domain, the diaries chart paths for program improvement based on generalizing the measurements and selectively applying the various modifications that are likely to produce the best results.

3.3 Unresolved Issues

Despite the significance of the diaries, the collection of a stable and portable benchmark suite, and the methodology being developed by the PERFECT Club,

Table 4. Software Transformation - Optimization Diaries

Compiler directives: concurrent loop (microtask)

Compiler directives: concurrent subroutine call (macrotask)

Compiler directives: ignore data dependence

Vectorize loop

Remove data dependence

Common subexpression elimination

Modify for concurrent subroutine call

Removal of redundant operations

Modify for stride one memory access

Loop fusion

Loop interchange

Loop unrolling

Cache management

Register management

Replace Fortran statements by assembly language code

Library subroutine

Change of algorithm

Loop restructuring: move loop inside/outside of subroutine

Subroutine elimination: contents moved inline

I/O modification: formatted to unformatted I/O

Unknown Diary modification

the work completed and published to date must be seen as a beginning rather than as a completed product [5]. Major contributors to system performance were avoided in the first phase of the group's efforts.

Problem sizes were selected to permit timely execution of all programs on all of the systems in the first set of tests. Although this permitted the establishment of a single baseline measurement, it limits the validity of the results, especially for the larger systems. More work needs to be performed to develop problem sets by class of system measured so that the requirements placed on system resources are consistent with what is reasonable for the problem being solved. It is important to maintain an appropriate balance between set-up costs and the main computations of a program so that the overheads associated with program bookkeeping functions are properly amortized over problem calculation times.

Including larger problem sizes may have the effect of introducing I/O measurements into the PERFECT benchmarking effort. Because most user problems/workloads have some dependence on the various levels of a system's memory hierarchy, this is a necessary inclusion. I/O was intentionally excluded from the first phase of PERFECT activity because standardization is difficult and measurements in this area introduce multiple variables into the evaluation process. Nevertheless, measurement and prediction is inaccurate without I/O components, implying that this is an area that must be addressed in future studies.

Finally, very few supercomputers are designed or used as single-user systems. Thus, benchmarking them only in standalone mode with a single program thread masks the issue of total system performance. Throughput depends not only on the computational speed of the central processor(s) of a system, but also on the efficiency with which the system exploits its resources. Although throughput measures are best if based on a specific workload, the PERFECT Club should address the construction of a general framework for developing throughput results.

4. Conclusions and Directions

When considering the contribution of the PERFECT effort, it is fair to ask if this suite of benchmarks is representative of general scientific workloads. Further, it is reasonable to ask if it binds us to yesterday's solutions rather that forging a path for the future.

The real requirement concerning representativeness is that the benchmarks be comprised of a spectrum of applications with a reasonably wide variation in their exploitation of various system attributes. Although future iterations may indicate a need to enhance the set, the original PERFECT codes certainly cover a good range of problem classes and are therefore a solid foundation for future work.

The problem of binding to old architectures as a result of employing existing algorithms is answered mainly by the use of the optimization diaries. Because any program modifications are permitted, and encouraged, there is considerable flexibility available to any target architecture. That programs require total reconstructions in some cases is valuable information, and need not be viewed as condemnation. There is a school of thought that real progress can be achieved only through this type of reconsideration and re-programming at the algorithmic levels of program development.

The general goals of benchmarking are to understand the limits of computer systems, to predict the performance of classes of problems on given systems, and to understand system performance relative to individual workloads. With the PERFECT codes and the ongoing efforts based on these programs, there is a concerted effort to address these goals and to move beyond existing ad hoc and experiential approaches to supercomputer performance evaluation.

REFERENCES

[1]   *The Federal High Performance Computing Program*, 53pp. Executive Office of the President, Office of Science and Technology Policy, September 8, 1989.

[2]   McMahon, F.H., *The Livermore Fortran Kernels Test of the Numerical Performance Range.* in **Performance Evaluation of Supercomputers.** Martin, J.L. editor, North-Holland, pp. 143-186 Amsterdam, 1988

[3]   Dongarra, J., Martin, J.L, and Worlton, J, *Computer benchmarking: paths and pitfalls.* **IEEE SPECTRUM**, July. pp. 38-43. July, 1987

[4]   Infante, E.F. (committee chairman), *An Agenda for Improved Evaluation of Supercomputer Performance.* 58pp. 1986 Available from: Energy Engineering Board, Commission on Engineering and Technical Systems, National Research Council 2101 Constitution Avenue, N.W., Washington, DC 20418 USA.

[5]   Berry, M. et al. *The PERFECT Club Benchmarks: Effective Performance Evaluation of Supercomputers*, **The International Journal of Supercomputer Applications**, 3.3, 5-40. 1989

MEASUREMENTS OF

PROBLEM-RELATED PERFORMANCE PARAMETERS

Roger W. Hockney

Computer Science Department

Reading University, P.O. Box 220, Whiteknights

Reading, Berks., England, U.K., RG6 2AX

Abstract - Measurements are presented of the problem-related performance parameters $r_\infty, n_{1/2}, s_{1/2}, f_{1/2}$ on the Cray-2, ETA-10, IBM 3090/VF and the INMOS T800 Transputer.

1. INTRODUCTION

The problem related performance parameters $r_\infty, n_{1/2}, s_{1/2}, f_{1/2}$ have been defined fully elsewhere ([1] chapter 1, [2]). Their purpose is to relate performance to some aspect of a users problem or program, such as average vector length or grain size of parallelism. The purpose of this paper is to present some recent measurements of these parameters on the Cray-2, ETA10, IBM 3090 and the INMOS Transputer.

2. $(r_\infty, n_{1/2})$ CRAY-2 and ETA-10

The performance of vector computers is characterized by the performance of their vector pipelines. If we assume a linear relationship between the time, t, of the vector operation and the length of the vector, n, we are at liberty to express this using the two parameters $(r_\infty, n_{1/2})$ as follows

$$t = r_\infty^{-1} (n + n_{1/2}) \tag{1}$$

Then the performance, r, is given by

$$r = \frac{n}{t} = \frac{r_\infty}{(1 + n_{1/2}/n)} \tag{2}$$

NATO ASI Series, Vol. F 62
Supercomputing
Edited by J.S. Kowalik
© Springer-Verlag Berlin Heidelberg 1990

Showing that r_∞ is the asymptotic performance approached as the vector length goes to infinity, and $n_{1/2}$ is the vector length needed to achieve half the asymptotic performance. Although $n_{1/2}$ is defined in terms of half the asymptotic performance, its value gives the performance for all vector lengths by use of Eqn. (2).

The values of $(r_\infty, n_{1/2})$ were obtained by measuring the time for the following simple loop as a function of the loop length N

$$\text{DO} \ 10 \ \ I = 1, N \tag{3}$$
$$10 \quad A(I) = B(I)*C(I)$$

This loop is replaced by a vector instruction by the vectorizing compiler. A best least squares straight line is fitted to the measurements of time against vector length. The inverse slope of this line is the value of r_∞, and the negative of the intercept with the N-axis is the value of $n_{1/2}$. Such measurements have already been reported for the Cray X-MP [3], and we give below measurements for the Cray-2.

The technique used for timing is crucial for obtaining accurate values for the time of loop (3), particularly when the computer system used is busy. On the Cray-2 we used the cycle timer and took the minimum time observed for about 1000 trials conducted in the early hours of the morning. The effect of this strategy is to eliminate almost entirely the effect of interference from other users of the machine. Such interference arises from competition for memory cycles from other programs running on other CPUs of the four CPU Cray-2. We do not recommend the common practice of taking the average of many trials because such a measures includes the arbitrary effect of other users of the system. We tend to use the above vector dyadic kernel as a standard of comparison between computers because experience shows that its performance is usually similar to that to be expected from average Fortran code, as opposed to carefully optimised code. However measurements were also made with various other kernels in statement 10 of Eqn. (3).

Performance measurements on the Cray-2 were obtained in 1987/8 on the machine at UKAEA Harwell. This was serial number 8 with 256 Mword of DRAM as common memory with a cycle time of 234ns. The numbers given in this paper do not therefore apply to the later model Cray-2S which has a faster SRAM memory. Figure 1 shows the timing line obtained as described above on the Cray-2. The solid line follows the measured times and the dotted line is the least squares fit used to determine the performance parameters. We see immediately that the true timing line is far from linear, and the above two parameter characterization can only give an approximate description. The quality of the approximation is shown in Fig. 2 where the same data is shown in terms of the performance as a function of vector length. We see that although the performance predicted by Eqn. (2) may differ from the actual performance be as much as 30% for very short vectors, the two-parameter model gives a reasonable average description of the variation of performance with vector length that becomes increasingly good as the vector length increases.

In Fig. 1 the vertical upward steps in time at intervals of 64 elements (the capacity of a vector register) arise from the need to refill the vector registers with a new segment of the arrays involved, and is a feature of all machines with vector registers. The drop in time at intervals of 256 in vector length was unexpected, and is a feature of the compiler used (here CFT (1.3)) and not the computer hardware. An examination of the code generated by the compiler shows that an unrolled loop of four consecutive vector multiply instructions is generated capable of computing 256 elements, as well as a rolled loop containing a single vector multiply operation. The unrolled loop is used to compute integral multiples of 256 elements, and the rolled loop is used for the remainder. Consequently there is a reduction in loop-control overhead every time the unrolled loop can be used, that is to say at intervals of 256 elements. Curiously the drop in time is seen to occur between a vector length of 256 and 257, and not as one would expect between 255 and 256.

A complete description of the Cray-2 measurements is to be found in a UKAEA Harwell report [4], but we show in Tables 1 and 2 a summary of the results. The value of 66 Mflop/s obtained for the memory-to-memory dyadic multiply kernel (A=B*C) is approximately 1/4 of the peak rate of 244 Mflop/s for a single pipeline of this machine (one result per clock period of 4.1ns). This is because the existence of only one memory access pipeline and the absence of chaining make it necessary to pass through the elements of the vector four times in sequence to produce the result, giving a maximum rate of only one result every four clock periods. These passes (or chimes) are respectively: read B, read C, multiply, store A, none of which can be overlapped. A higher performance is observed for the all vector triad (A=B*C+D) and the SAXPY kernel (A=B*s+D), and for the inner-product. But even the best of these, the inner-product, only reaches 1/4 of the maximum performance of the add and multiply pipelines working simultaneously (488 Mflop/s). In this case the $n_{1/2}$ is quite large and it is necessary to have vectors of $4n_{1/2}$ (i.e. greater than 1000) to obtain 80% of the r_∞ value given. These experiments with simple kernels on the Cray-2 show that this machine, like the original Cray-1, has a severe memory bottleneck arising from the small memory bandwidth. Performance approaching the theoretical peak can only be achieved by ensuring that much arithmetic work is performed in the vector registers between references for more data from main memory. That is to say the computational intensity is high (see section 3 below). This is difficult to achieve through Fortran even on suitable problems, so that the performance rates quoted in Table 1 may be quite typical for anything but very carefully organised codes probably using assembler.

Table 1 also shows the corresponding measurements made on the ETA-10, both the model E with a 10.5ns clock, and the model G with a 7ns clock. The ETA-10, like its parent the Cyber 205, has three pipelines to memory. These allow the two input vectors and one output vector of a dyadic vector operation to be transferred from or to the main memory in parallel, and permit the arithmetic pipelines to run at the maximum theoretical rate when the pipelines and full. Each CPU of the ETA-10 has two 64-bit arithmetic pipelines, thus in 64-bit arithmetic this maximum dyadic rate is two operation per clock period, or 190 Mflop/s on the model E and 285 Mflop/s on the model G. These maximum rates are seen as the r_∞ values obtained for the vector dyadic operation. The $n_{1/2}$ for both models is about 200, similar to that of the Cyber 205, and about three times that of the Cray-2. Thus the ETA-10, like the Cyber 205 is a long vector machine, requiring vector lengths over 1000 to get a reasonable fraction of its maximum performance (see Eqn. 2). This was something of a surprise because it had been reported that the $n_{1/2}$ of the ETA-10 had been significantly reduced from that of the Cyber 205. If there have been some hardware improvements these are not evident when the ETA-10 is programmed in Fortran. The use of descriptor notation reduces the $n_{1/2}$ significantly by a factor of three. However putting up to 10 descriptor operations in sequence, in order to allow some overlap between successive vector operations, only increased $n_{1/2}$. In Fortran, 10 successive vector operations reduced the $n_{1/2}$ to 63, approximately the value seen using descriptor notation.

The scalar performance of the ETA-10(E) is slightly less than the Cray-2, and the ETA-10(G) is slightly greater. However the DO-loop overhead, as expressed by the value of $n_{1/2}$, is significantly smaller on the ETA-10. The principal differences in performance between the Cray-2 and ETA-10 are seen when triadic and SAXPY operations are considered. The ETA-10's ability to perform two 32-bit operations in the time of one 64-bit operation, together with the linked triad operation which allows the multiply and add pipelines to run in parallel, gives an asymptotic rate of 1.14 Gflop/s in 32-bit mode for the SAXPY operation. This is slightly more than 10 times the rate achieved by the Cray-2, which has no facility for 32-bit arithmetic or automatic parallel operation of the add and multiply operations (i.e. unlike the Cray X-MP and Y-MP, there is no ability to chain these operations together).

All the measurements given in Table 1 are for contiguous vectors which have a stride in memory address of unity between successive elements. The performance for other strides in memory address are shown in Table 2. The Cray-2 shows a decrease in performance due to memory quadrant conflict when the stride modulo four equals zero or two. Further decreases occur if the stride is a power of two due to memory bank conflicts. The performance reaches a minimum of 1.3 Mflop/s when the stride is a multiple of the number of pseudo-banks (256) in the memory. Thus strides equal to any power of two are to be avoided at all cost on this computer. On the other hand, a stride of any odd number performs approximately the same as a contiguous vector at about 60 Mflop/s for a vector dyad (see e.g. stride=1025). The behaviour of the ETA-10 is much simpler in that any stride other than unity performs at about 1/10 of the contiguous rate, namely about 15 Mflop/s on the model E and 22 Mflop/s on the model G. It is worth noting that in Fortran, if the stride is set as the third parameter of a DO-loop as a variable, the slow rate is observed even if the variable is previously set to unity. The maximum rate is only seen if the third parameter is absent and taken to be unity by default.

3. $(r_\infty, f_{1/2})$ IBM 3090/VF

Suppose a memory access pipeline characterised by the parameters $(r_\infty^m, n_{1/2}^m)$ feeds data to the local memory or cache of a vector unit with an arithmetic pipeline characterised by the parameters $(r_\infty^a, n_{1/2}^a)$, and that the vector unit performs f floating-point operations for every word (i.e. number) transferred from memory, then we call f the *computational intensity* of the calculation. A little algebra [5] shows that the combined memory/arithmetic unit acts like a pipeline with the average parameters

$$r_\infty = \frac{\hat{r}_\infty}{(1+f_{1/2}/f)} \tag{4}$$

$$n_{1/2} = \frac{n_{1/2}^m + n_{1/2}^a x}{(1+x)} \tag{5}$$

where $x = f/f_{1/2}$ and $f_{1/2} = r_\infty^a/r_\infty^m$

and $\hat{r}_\infty = 1/r_\infty^a$

Equation (4) shows how the peak performance of the arithmetic pipeline $\hat{r}_\infty$ is degraded by inadequate memory bandwidth. The computational intensity, f, must be large compared to the hardware parameter $f_{\frac{1}{2}}$ (called the *half –performance intensity*) if a high proportion of the vector pipeline performance is to be realised. The above formulae assume the worst case that memory access and arithmetic are not overlapped. It is shown in reference [5] that if overlap takes place, the value of $f_{\frac{1}{2}}$ is halved, and this has been confirmed by measurements on the FPS 5000 [6].

For a computer such as the IBM 3090VF which has a cache memory between the main memory and the vector unit, a high value of computational intensity corresponds to the array variables being found *in–cache* (of course for vector lengths that will fit into the cache), whereas a value of $f = 1/3$ for a dyadic operation corresponds to a measurement taken when the array variables were *out –of –cache* , and have to be retrieved from main memory. Hence an in-cache and out-of-cache measurement enable one to measure the value of $f_{\frac{1}{2}}$ and f_{∞} from

$$ f_{\frac{1}{2}} = f\,(out) \left[\frac{r_{\infty}(in)}{r_{\infty}(out)} - 1 \right] \qquad (6) $$

where $f\,(out) = 1/3$ and

$$ f_{\infty} = r_{\infty}(in) $$

Table 3 gives some values for a number of kernels on the IBM 3090VF. Two cases are distinguished according to whether the data used is in-cache or out-of-cache. In scalar mode a performance between 5 and 8 Mflop/s is seen if the data is in-cache. Out-of-cache values are reduced by about 20% for contiguous vectors and down to 1.6 Mflop/s for a stride of 8, and for scatter/gather. The value of $n_{\frac{1}{2}}$ is essentially zero for in-cache scalar work, but increases to 10 to 20 if the data is out-of-cache. This increase represents the time taken to retrieve data from main memory. The above scalar performance is comparable to that seen on the Cray-2 and ETA-10. In vector mode, however, the IBM 3090/VF is not comparable to the Cray-2 or ETA-10. The best performance of 26 Mflop/s is achieved on unit stride inner product, compared to 122 Mflop/s and 286 Mflop/s seen on the Cray-2 and ETA-10(G). However the $n_{\frac{1}{2}}$ values are generally less on the 3090 except for out-of-cache scatter/gather.

The difference between in-cache and out-of-cache values can be used to calculate a value of $f_{\frac{1}{2}}$ using Eqns. (5) & (6). The values for a vector dyad give :

$$ f_{\frac{1}{2}} = 0.26, \quad n^{a}_{\frac{1}{2}} = 34, \quad n^{m}_{\frac{1}{2}} = 120 $$

4. $(\hat{r}_\infty, \hat{s}_{1/2}, f_{1/2})$ INMOS Transputer

A simple model of an MIMD program is as a sequence of parallel sections separated by synchronization points. Each parallel section has a total amount of arithmetic (or grain size) s which is given to p processors to perform, and this requires the communication of m words of data. The synchronization requires that all work is completed in one parallel section before any work starts in the next.

The simplest model for the time behaviour of a parallel section assumes the worst case of no overlap between synchronisation, communication and arithmetic. In this case, the time for a parallel section is

$$t = t_s(p) + t_c(p)m + t_a(p)s \tag{7}$$

where the three terms are identified respectively with synchronization, communication, and arithmetic, and

$t_s(p)$ - time for synchronizing p processors

$t_c(p)$ - time per data word communicated on average

$t_a(p)$ - time per floating point operation on average

Introducing the *computational intensity* $f = s/m$ as a new problem variable describing the ratio of arithmetic to communication in a parallel section, the performance of a parallel section can be written

$$r = \frac{s}{t} = \frac{\hat{r}_\infty}{\left[1 + \dfrac{\hat{s}_{1/2}}{s} + \dfrac{f_{1/2}}{f}\right]} \tag{8}$$

where $\hat{r}_\infty = t_a^{-1}$, $f_{1/2} = t_c/t_a$ and $\hat{s}_{1/2} = t_s/t_a$ \tag{9}

are hardware parameters. Alternatively if we consider f fixed and wish to study the variation with grain size s, we can define the following values corresponding to infinite grain size

$$r_\infty = \frac{\hat{r}_\infty}{(1 + f_{1/2}/f)} \tag{10}$$

$$s_{1/2} = \frac{\hat{s}_{1/2}}{(1 + f_{1/2}/f)} \tag{11}$$

then

$$r = \frac{r_\infty}{(1+s_{1/2}/s)} \qquad (12)$$

and substituting Eqn.(10) in Eqn.(12) we obtain

$$r = \frac{\hat{r}_\infty}{(1+s_{1/2}/s)(1+f_{1/2}/f)} . \qquad (13)$$

where the first factor in the denominator gives the degradation of performance due to synchronization, and the second factor the degradation due to communication delays. In equation (13) it must be remembered that $s_{1/2}$ is itself a function of f via equation (11). Thus the ratio of f (a program variable) to $f_{1/2}$ (a ratio of hardware variables) determines the fraction of the peak performance $\hat{r}_\infty$ that is obtained. The peak performance being that obtained for large grain size ($s = \infty$) in the absence of any communication delays ($f_{1/2}=0$), and reflects the bare performance of the arithmetic hardware.

Given a set of values of r_∞ for different values of f, the values of $f_{1/2}$ and $\hat{r}_\infty$ are obtained by plotting (f/r_∞) versus f and fitting the best straight line. From equation (10) we have

$$(f/r_\infty) = \hat{r}_\infty^{-1}(f + f_{1/2}) \qquad (14)$$

Then by analogy with equation (1), we find that $\hat{r}_\infty$ is the inverse slope of the straight line, and $f_{1/2}$ is the negative intercept of the line with the f-axis.

The variation of computer performance with the ratio f of arithmetic to memory references, through equation (10), suggests that problems might usefully be classified according to the value of the problem variable f. For example, the following cases can immediately be distinguished:

(1) VECTOR PROBLEMS - $f = O(1)$

Dyadic operation	Z=X*Y	$f=1/3$
All vector triad	Z=W+X*Y	$f=1/2$
BLAS 1, or SAXPY	Z=a*X+Y	$f=2/3$
Tridiag. solve: flop=$5n$, mrefs=$5n$		$f=1$
BLAS 2		$f=2$

(2) LOGARITHMIC PROBLEMS - $f = O(logn)$

FFT: flop=$2.5n\log_2 n$, mrefs=$2n$ $f=1.25\log_2 n$

(3) MATRIX PROBLEMS - $f = O(n)$

Full matrix solve: flop=$1/3n$, mrefs=n^2 $f = n/3$

Matrix Multiply: flop=$2n^3$, mrefs=$3n^2$ $f = 2n/3$

BLAS 3   flop=$2n^3$, mrefs=$4n^2$ $f = n/2$

The above table shows that vector and logarithmic problems are unlikely to be computed efficiently on computer systems with $f_{1/2}>0.1$, whereas matrix problems can be computed efficiently even on computers with quite large $f_{1/2}$ provided the problem is large enough. The improvement in performance of the basic linear algebra subroutines reported by Dongarra in this workshop, can also be attributed to the increase in the value of computational intensity f as one goes from BLAS1 to BLAS2 to BLAS3.

Measurements of the parameters r^∞, $f_{1/2}$, $s_{1/2}$ have previously been reported for the IBM LCAP computer system [7,8], and we give here the results obtained on the INMOS Transputer. The INMOS T800 Transputer is a 10 Mips 32-bit microprocessor with an approximately 1 Mflop/s floating-point coprocessor, sharing a single chip with 4KBytes of RAM and four bidirectional 20 Mhz 1-bit channels or links. The links of different transputers can be joined directly together to form any network that is no more than four valent (in the chemical sense). To measure the problem-related parameters for a network of transputers we consider the dyadic multiply problem in equation (3) with the work performed by a number of transputers. One transputer is used as a master and transfers portions of the input vectors B(I) and C(I) to a number of slave transputers. Each slave computes its portion of the dyadic multiply operation and returns the result A(I) to the master, which assembles the complete result vector. In order to vary the value of computational intensity the arithmetic is repeated $3f$ times in each slave transputer. Although this repeat makes no computational sense, it does vary the ratio of arithmetic to communication correctly according to the definition of f. Note also that when $f=1/3$ the arithmetic is done once, giving the correct behavior for a dyadic operation. The time for this complete operation is measured and fitted to the timing formula (7).

The results for 1 slave and 3 slave transputers are shown in Table-4. For each value of computational intensity, f, the asymptotic performance, r_∞, and half-performance grain size, $s_{1/2}$, are measured by fitting a straight line to the variation of time with grain size, $s=3fN$. Next the best value of $r_\infty, s_{1/2}, f_{1/2}$ are obtained in the formulae (10) and (11). Columns 2 & 3, and 5 & 6 show the agreement obtained between calculated and measured values. For one transputer $s_{1/2}$ is negligible and taken to be zero, and the agreement between the measured and calculated figures is exact to three decimals. For three slave transputers we have a small value of $s_{1/2}$ and fair agreement with the theory. The value of r_∞ is three times that for a single transputer. In both cases, the value of $f_{1/2}$ is approximately unity.

5. CONCLUSIONS

We have compared the performance of the Cray-2, ETA-10 and IBM 3090/VF supercomputers using the problem-related parameters $r_\infty, n_{1/2}, s_{1/2}, f_{1/2}$. These parameters successfully show up numerically the main weaknesses and strengths of the machines. Although it has a high theoretical register-to-register performance, the observed performance of the Cray-2 using Fortran is severely limited by memory bottleneck and lack of chaining. The ETA-10 does achieve its advertised performance for long vectors in Fortran, however the large value of $n_{1/2}$ means that, like the Cyber 205, special long-vector algorithms must be used. Whilst the IBM 3090/VF has excellent scalar performance equal to and sometimes surpassing that of the Cray-2 and ETA-10, it single CPU vector performance is not in the supercomputer range. In order to achieve a performance comparable to a single CPU of the Cray-2 or ETA-10, it is necessary to use multiple CPUs on the 30090/VF, of which six are available on the largest model. Many supercomputers are planned based on assemblies of INMOS Transputers. The measurements presented here suggest that 1/3 Mflop/s per T800 Transputer should be realisable from virtually any code, and that the basic synchronisation cost measured by $s_{1/2}$ is small. Much higher performance up to 1 or even 1.5 Mflop/s can be obtained on the T800 for suitable problems that are carefully coded, probably in the Occam language.

Acknowledgements

The measurements reported here have been made at a number of different computer centres, and the author would like to thank the management and staff at the following centres for their assistance and hospitality: IBM European Center for Scientific and Engineering Computing, Rome; UKAEA Harwell Laboratory; Supercomputing Research Institute, Florida State University, Tallahassee. The author is indebted to Ahmed Sharieh of the Computer Science Department at FSU for data on the 7ns ETA-10 (model G).

References

1   Hockney R. W. and Jesshope C. R.: *Parallel Computers 2 - Architecture, Programming and Algorithms* Bristol and New York: Adam Hilger 1988. Distributed in USA by American Institute of Physics (attn. Dale Cunningham) 335 East 45th Street, New York 10017

2   Hockney R. W.: Problem-related performance parameters for supercomputers. In: *Performance Evaluation of Supercomputers* (J. L. Martin. ed.). pp 215-235. Amsterdam: Elsevier Science BV (North Holland) 1988

3   Hockney R. W.: $(r_\infty, n_{1/2}, s_{1/2})$ measurements on the 2-CPU CRAY X-MP *Parallel Computing* **2** 1-14 (1985)

4   Hockney R. W.: *Cray-2 Performance Measurements - I: Simple Kernels on One CPU*, UKAEA Harwell Report CSS215 May 1988

5   Hockney R. W. and Curington I. J.: $f_{1/2}$: a parameter to characterise memory and communication bottlenecks *Parallel Computing* **10** 277-286 (1989)

6   Curington I. J. and Hockney R. W.: Synchronisation and pipeline overhead measurements on the FPS-5000. In: *Parallel Computing85* (M. Feilmeier, G. Joubert, U. Schendel eds.) pp 469-476 Amsterdam: Elsevier Science BV (North Holland) 1986

7   Hockney R. W.: Characterising overheads on VM-EPEX and multiple FPS164 processors. In: *Parallel Systems and Computation* (G. Paul and G. S. Almasi eds.) pp 255-271 Amsterdam: Elsevier Science BV (North Holland) 1988

8   Hockney R. W.: Synchronization and communication overheads on the LCAP multiple FPS-164 computer system *Parallel Computing* **9** 279-290 (1988/9)

TABLE 1   Measurements of $(r_\infty, n_{1/2})$ on a single CPU of the
Cray-2 and ETA-10. All Fortran code with maximum
scalar and vector optimisation.

KERNEL	bits	CRAY-2 4.1ns, DRAM		ETA-10(E) 10.5ns		ETA-10(G) 7ns	
		r_∞ Mflop/s	$n_{1/2}$ vlen	r_∞ Mflop/s	$n_{1/2}$ vlen	r_∞ Mflop/s	$n_{1/2}$ vlen
SCALAR	64	5.2	30.7	4.6	3.7	6.9	3.6
A=B*C	32						
VECTOR	64	66	117	190	193	285	212
A=B*C	32						
10 VECOPS	64			173	63	285	63
A=B*C	32						
DESCRIPTOR	64			190	71	285	126
DA=DB*DC	32						
10 DESCRIPT.	64			187	443	281	440
DA=DB*DC	32						
TRIADIC	64	83	39	190	132	285	148
A=B*C+D	32			380	274	570	261
SAXPY	64	96	47	380	252	571	257
A=s*C+D	32			763	518	1141	517
INPROD	64	122	270	190	173	286	176
s=s+B*C	32			190	178	285	178

TABLE 2   Measurements of $(r_\infty, n_{1/2})$ on a single CPU of the
Cray-2 and ETA-10. Effect of non-unit stride on the
performance of the dyadic vector multiply operation.
All Fortran code with maximum scalar and vector
optimisation.

STRIDE	CRAY-2 4.1ns, DRAM		ETA-10(E) 10.5ns		ETA-10(G) 7ns	
	r_∞ Mflop/s	$n_{1/2}$ vlen	r_∞ Mflop/s	$n_{1/2}$ vlen	r_∞ Mflop/s	$n_{1/2}$ vlen
$1^{(a)}$			190	193	285	212
$1^{(b)}$	66	117	16	67	24	65
2	21	18	15	59	23	60
3	66	115	15	56	23	58
4	12	1.5	15	53	22	54
5	66	115	15	52	22	51
6	21	18	14	49	21	48
7	66	118	14	47	21	45
8	12	1.4	14	46	20	43
16	12	-2.0	14	62	21	60
32	7.8	-10	14	104	21	102
64	4.0	-16	14	169	21	166
128	2.0	-18	14	199	21	197
256	1.3	-17	14	204	21	202
512	1.4	-16	14	206	21	204
1024	1.4	-13	14	210	21	208
1025	52	62				

(a) using: DO I=1,N
(b) using: ISTRID=1; DO I=1,N,ISTRID

TABLE 3   Measurements of $(r_\infty, n_{1/2})$ on a single CPU of the IBM 3090/VF. All Fortran code with maximum scalar and vector optimisation. IN and OUT refer to whether the data used is in-cache or out-of-cache respectively.

KERNEL	SCALAR				VECTOR			
	r_∞ Mflop/s		$n_{1/2}$ vlen		r_∞ Mflop/s		$n_{1/2}$ vlen	
	IN	OUT	IN	OUT	IN	OUT	IN	OUT
A(i)=B(i)*C(i) Stride=1	5.1	4.1	0.6	14	13	7.3	34	71
A(i)=B(i)*C(i) Stride=8	5.1	1.6	0.0	4	13	1.7	36	14
A(i)=B(i)*C(i)+D(i) Stride=1	8.0	6.5	0.1	21	27	11	56	36
Inner Product s=s+A(i)*B(i)	10	8.9	0.2	14	26	16	52	95
Recurrence A(i)=B(i)*A(i-1)+D(i)	8.0	6.8	0.1	21	NV	NV	NV	NV
Matrix Mult Compiler choice	6.8	6.1	1.6	0.9	16	15	12	12
Scatter+Gather A(i)<=>B(j(i))	5.1	1.6	0.6	5	11	4.7	30	435
Indirect Update A(j(i))=A(j(i))+s	4.1	2.0	0.2	19	NV	NV	NV	NV
Transpose A(i,j)=A(j,i)	7.2	2.9	6.4	6.5	NV	NV	NV	NV

TABLE 4   Measurements of $(r_\infty, s_{1/2}, f_{1/2})$ on one and three INMOS T800 Transputers. Programmed in Occam.

| f | 1-SLAVE | | 3-SLAVES | | |
| | r_∞ | $r_\infty^{(a)}$ | r_∞ | $s_{1/2}$ | $s_{1/2}^{(b)}$ |
flop/mref	Mflop/s	Mflop/s	Mflop/s	flop	flop
1/3	0.093	0.093	0.275	12.3	11.6
2/3	0.147	0.147	0.438	19.5	18.5
1	0.183	0.183	0.544	23.8	23.0
2	0.242	0.242	0.723	31.1	30.5
5	0.300	0.300	0.900	38.0	38.0
10	0.326	0.326	0.978	39.4	41.3
20	0.341	0.341	1.023	34.7	43.2
$\hat{r}_\infty$	-->	0.357	-->	-->	1.072
$\hat{f}_{1/2}$	-->	0.947	-->	-->	0.966
$\hat{s}_{1/2}$	-->	-->	-->	-->	45.3

(a) calculated from Eqn. (10): $r_\infty = \dfrac{\hat{r}_\infty}{(1 + f_{1/2}/f)}$

(b) calculated from Eqn. (11): $s_{1/2} = \dfrac{\hat{s}_{1/2}}{(1 + f_{1/2}/f)}$

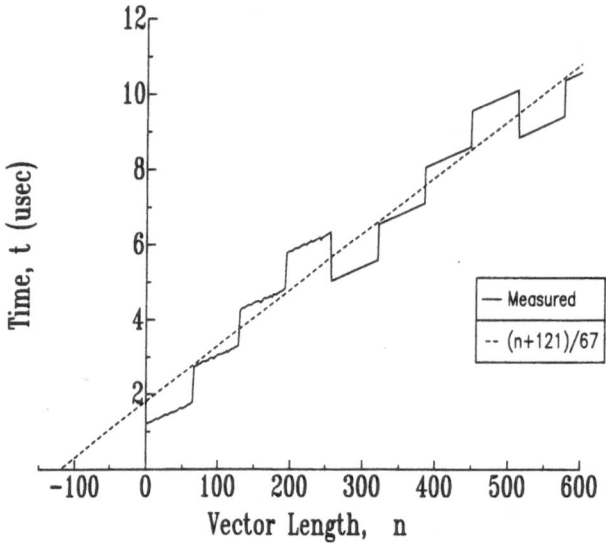

Figure-1 Time as a function of vector length for the dyadic multiply benchmark using contiguous vectors. Each measurement is the minimum time for 900 trials. The dotted line is the best least-squares straight-line fit to all data $n \geq 4096$

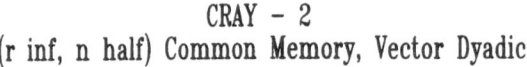

CRAY – 2
(r inf, n half) Common Memory, Vector Dyadic

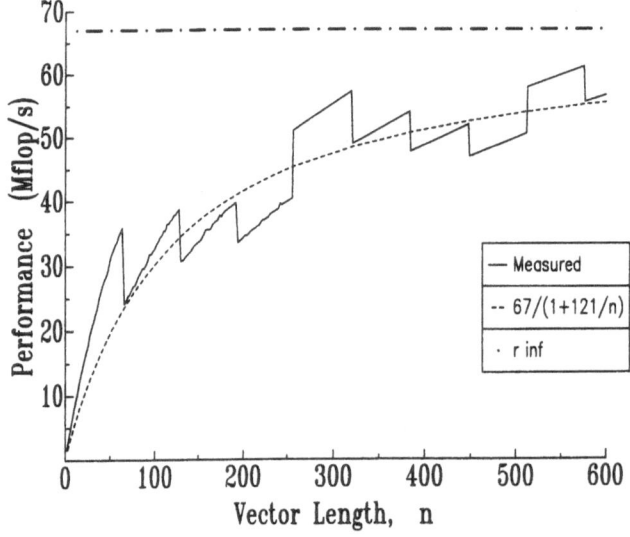

Figure-2 The performance as a function of vector length using the same data as Figure-1.

PART 7

METHODS AND

ALGORITHMS

A Parallel Iterative Method for Solving 3-D Elliptic Partial Differential Equations*

Anne Greenbaum
New York University
Courant Institute of Mathematical Sciences
251 Mercer Street
New York, NY 10012

Edna Reiter
Department of Mathematics and Computer Science
California State University at Hayward
Hayward, California,94542

Garry Rodrigue
Department of Applied Science
University of California, Davis
P.O. Box 808
Livermore, California, 94550

1 Domain Decomposition

In this paper, we will consider the parallel solution of the Dirichlet problem for a second-order uniformly elliptic equation in two and three dimensions. Specifically, we shall consider the problem

$$
\begin{aligned}
L(u) &= f \text{ in } \Omega \\
u &= g \text{ on } \Gamma
\end{aligned}
\tag{1}
$$

where

$$
L(u) = -\sum_{j=1}^{3} \frac{\partial}{\partial x_i} \left(a_i \frac{\partial u}{\partial x_i} \right)
$$

*This work was performed under the auspices of the U.S. Department of Energy at the Lawrence Livermore National Laboratory under Contract W-7405-Eng-48 and at the Courant Institute of Mathematical Sciences under Contract DE-AC02-76ER03077

NATO ASI Series, Vol. F 62
Supercomputing
Edited by J. S. Kowalik
© Springer-Verlag Berlin Heidelberg 1990

with a_i positive, bounded, and piecewise smooth on bounded Ω with boundary Γ. For sake of exposition, we will assume the equation is 2-dimensional, however, the extension to 3-dimensions of the numerical methods to be described will be obvious.

The solution of elliptic problems such as those given by (1) arise quite often on supercomputers when they are used to numerically simulate problems arising in mechanical systems. They are often solved when the steady state of dissipative systems as in gas dynamics is desired. These equations are particularly difficult to solve on a computer because their discrete representation on a computational grid yields a linear system of equations where the number of unknowns is equal to the number of grid points on the lattice and, in 3-dimensions, the number of unknowns can be very large, (e.g. 10^6).

A general framework for developing parallel algorithms for solving the elliptic equation (1) on a multiprocessing supercomputer is provided by a techinique that has come to be known as *domain decomposition*. For time-independent problems such as (1), the technique of domain decomposition begins by subdividing the *global* domain Ω into a union of subdomains:

$$\Omega = \bigcup_{i=1}^{n} \Omega_i.$$

Again, for ease of exposition, we will assume $\Omega = [0,1] \times [0,1]$ is the unit cube and the subdomains are themselves cubes, $\Omega_i = [\alpha_i, \beta_i] \times [\gamma_i, \delta_i]$. Examples of such decompositions are given in Figure 1.

For a given decomposition, numerical methods are constructed by solving equation (1) on each of the subdomains and then piecing together the sub- solutions in some manner to provide an approximation to the global solution,[3]. This seems quite straightforward provided there is some mechanism for defining appropriate boundary conditions on each of the subdomain problems. The first attempts at providing subdomain boundary conditions were based around the classical Schwarz alternating method. Here, an initial guess is provided for the boundary conditions and then each sub-problem is solved numerically in some sequential order. Boundary conditions are updated whenever sub-solutions are computed on subdomains containing neighboring subdomain boundaries. As an example, let us consider the

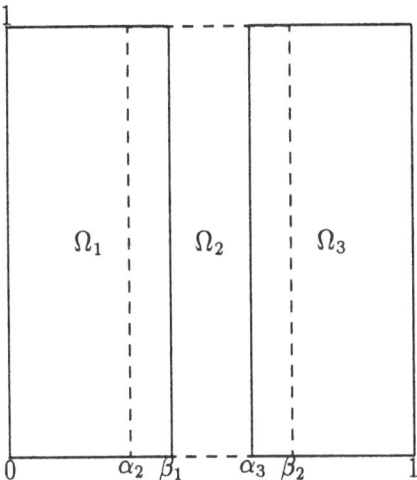

Figure 1: Three Subdomains

solution of (1) using a decomposition such as the one given in Figure 1. Let $u_i^{(0)}$ be given initial guesses for the solution u on each of the subdomains Ω_i. Then for $k = 1, 2, \ldots$, define the sequence of sub-solutions $\{u_i^{(k)}\}$ as follows:

$$Lu_i^{(k)} = f \Big|_{\Omega_1} \quad \text{on } \Omega_1,$$

$$u_i^{(k)} = g \text{ on } \Gamma \bigcap \Gamma_i,$$

$$
u_i^{(k)} = \begin{cases} u_2^{(k-1)} \Big|_{\Gamma} & \text{if } i = 1, \\[20pt] u_{i-1}^{(k-1)} \Big|_{\Gamma_{i-1}} & \text{if } 1 < i < n, \\[20pt] u_{i+1}^{(k-1)} \Big|_{\Gamma_{i+1}} & \text{if } 1 < i < n, \\[20pt] u_{n-1}^{(k-1)} \Big|_{\Gamma_n} & \text{if } i = n. \end{cases}
$$

Convergence of the above procedure to the global solution follows from the max-imum principle. The numerical analog of the previous algorithm is obvious and has been shown to be equivalent to a block-Jacobi matrix iterative method, and con-sequentely, often results in slow convergence, [3]. Specifically, the matrix iteration takes the form

$$
\hat{x}^{(k+1)} = M^{-1} N \, \hat{x}^{(k)} + \hat{q} \tag{2}
$$

where M is a block-diagonal matrix and N is a non-symmetric matrix. Conse-quentely, a major effort over the past few years has been to develop iterative meth-ods based around domain decomposition that will converge faster than the Schwarz methods.

2   Conjugate Gradient Acceleration

The easiest and, probably, the most popular way of accelerating an iteration of the form (2) is by the method of conjugate gradients (or, Krylov subspace methods),[2]. Basically, an additive splitting $A = M - N$ of a symmetric ,positive-definite, matrix is given. Then, the following iteration is used to solve the matrix equation $A\hat{x} = \hat{b}$.

Let $\hat{x}^{(0)}$ be an initial guess and $\hat{r}^{(0)} = \hat{b} - A\hat{x}^{(0)}$. Then, for $k = 0, 1, \ldots,$ define

$$i) \quad M\hat{z}^{(k)} = \hat{b} - A\hat{x}^{(k)};$$

$$ii) \quad \beta_k = \begin{cases} 0, & k = 0; \\[2ex] \dfrac{\hat{z}^{(k)^t} M\hat{z}^{(k)}}{\hat{z}^{(k-1)^t} M\hat{z}^{(k-1)}}, & k \geq 1; \end{cases} \tag{3}$$

$$iii) \quad \hat{p}^{(k)} = \hat{z}^{(k)} + \beta_k \hat{p}^{(k-1)};$$

$$iv) \quad \alpha_k = \frac{\hat{z}^{(k)^t} M\hat{z}^{(k)}}{\hat{p}^{(k)^t} A\hat{p}^{(k)}};$$

$$v) \quad \hat{x}^{(k+1)} = \hat{x}^{(k)} + \alpha_k \hat{p}^{(k)}.$$

The convergence of this iteration in a finite number of steps is guaranteed if the matrix M is symmetric and positive-definite. In fact, it is known that

$$E(x^{(k)}) \leq 2 \left(\frac{\sqrt{\kappa} - 1}{\sqrt{\kappa} + 1} \right)^k E(x^{(0)}) \tag{4}$$

where

$$\kappa = \frac{\lambda_{max}}{\lambda_{min}},$$

$\lambda_{max} = $ largest absolute eigenvalue of $M^{-1}A$,

$\lambda_{min} = $ smallest absolute eigenvalue of $M^{-1}A$,

$E(x) = (x^t Ax)^{1/2}.$

Consequentely, convergence will be rapid if the matrix M is *spectrally* close to A in the sense that κ should not be large. The literature on the choice of M is vast especially for the case when the matrix A represents a discrete approximation to (1). However, it is not until recently that reasonable choices of M have been provided that make efficient use of a decomposed domain. The difficulty of using domain decomposition methods in conjugate gradient lies in the origin of each of the two ideas. The conjugate gradient method has its foundation in linear algebra whereas domain decomposition is embedded in the theory of differential equations, and at the present time a unified mathematical connection between the conjugate gradient method and differential equations has not been established.

3 Additive Schwarz Method

Because the convergence rate of the conjugate gradient method depends on the relationship between the matrix A that approximates equation (1) and the preconditioning matrix M, it may be possible to build a matrix M from a domain decomposed differential equation that is related to equation (1). We do so by first splitting the global solution of (1) into the sum of a *homogeneous* solution and a *particular* solution. That is, $u = u^{(H)} + u^{(P)}$ where

$$L(u^{(H)}) = 0 \ \text{on} \ \Omega,$$

$$u^{(H)} = g \ \text{on} \ \Gamma,$$

$$(5)$$

and

$$L(u^{(P)}) = f \ \text{on} \ \Omega,$$

$$u^{(P)} = 0 \ \text{on} \ \Gamma.$$

$$(6)$$

Now, let

$$\Omega = \bigcup_{i=1}^{n} \Omega_i \tag{7}$$

be a decomposition of Ω and define the subproblems

$$L(u_i) = f \Big|_{\Omega_i} \quad \text{on} \ \Omega_i,$$

$$u_i = \begin{cases} g & \text{on} \quad \Gamma \cap \Gamma_i, \\ u & \text{on} \quad \Gamma_i - \Gamma. \end{cases}$$

$$(8)$$

Clearly, $u = u_i$ on Ω_i. Again, $u_i = u_i^{(H)} + u_i^{(P)}$ where

$$L(u_i^{(H)}) = 0 \text{ on } \Omega_i,$$

$$u_i^{(H)} = \begin{cases} g & \text{on} \quad \Gamma_i \cap \Gamma, \\ u & \text{on} \quad \Gamma_i - \Gamma, \end{cases} \tag{9}$$

and

$$L(u_i^{(P)}) = f \Big|_{\Omega_i} \quad \text{on } \Omega_i, \tag{10}$$

$$u_i^{(P)} = 0 \text{ on } \Gamma_i.$$

Clearly, each of the "particular" problems defined by (10) can be solved in parallel. Thus, we take as an approximation to $u^{(P)}$ the function $v^{(P)}$ where

$$v^{(P)} = \sum_{i=1}^{n} v_i^{(P)} \tag{11}$$

and

$$v_i^{(P)} = \begin{cases} u_i^{(P)} & \text{in} \quad \Omega_i, \\ \\ 0 & \text{in} \quad \Omega - \Omega_i. \end{cases} \tag{12}$$

Then,

$$u \simeq u^{(H)} + v^{(P)} \tag{13}$$

and

$$\mathcal{L}(u) \simeq L(u) \tag{14}$$

where

$$\mathcal{L}(u) = L(u^{(H)}) + \left(\sum_{i=1}^{n} L(v_i^{(P)}) \right). \tag{15}$$

We now carry out the numerical equivalent of the above construction. To do so, we overlay the global domain Ω by a grid G such that the boundaries Γ_i of the subdomains are contained in G. Now, suppose the grid $G^{(H)}$ also overlays the domain Ω with $G^{(H)} \subset G$. Let

$$G_i = G \cap \overline{\Omega}_i$$

where $\overline{\Omega}_i$ is the closure of Ω_i, $\hat{u}$ the discrete representation of the function u on grid G and $\hat{v}_i$ the restriction of $\hat{u}$ on G_i. If A_i is the matrix representation of the equation (10) on G_i, then define the matrix $\mathcal{A}_i$ on G by

$$\mathcal{A}_i \hat{u} = \begin{cases} A_i \hat{v}_i, & \text{on} \quad G_i \\ \\ 0, & \text{otherwise.} \end{cases}$$

Now let P be an operator that maps grid functions on $G^{(H)}$ into grid functions defined on G by bilinear interpolation. If $A^{(H)}$ is the matrix approximation to equation (5) on $G^{(H)}$, then define

$$\mathcal{A}^{(H)} \hat{u} = A^{(H)} P \hat{u}, \text{ on } G^{(H)}.$$

Now, if M is the preconditioning matrix to be used in the conjugate gradient method, then we carry out the computation of step $(3i)$ by

$$\begin{aligned} z^{(k)} &= M^{-1} r^{(k)} \\ &= \mathcal{A}^{(H)^*} r^{(k)} + \left(\sum_{i=1}^{n} \mathcal{A}_i^{\dagger} r^{(k)} \right) \end{aligned}$$

where the matrices $\mathcal{A}_i^{\dagger}$ are the pseudo-inverses of $\mathcal{A}_i$, i.e.,

$$\mathcal{A}_i^{\dagger} r^{(k)} = \begin{cases} A_i^{-1} r^{(k)} & \text{on} \quad G_i, \\ \\ 0 & \text{otherwise} \quad . \end{cases}$$

and

$$\mathcal{A}^{(H)^*} r^{(k)} = P A^{(H)-1} P^t r^{(k)}.$$

Again, we see that M is symmetric and positive-definite. The above method of constructing M is called the additive Schwarz method,[1],[4],[5].

4 Computational Example

We apply the above the previously described to compute a numerical solution to the 3-dimensional Poisson equation

$$L(u) = \frac{\partial^2 u}{\partial x^2} + \frac{\partial^2 u}{\partial y^2} + \frac{\partial^2 u}{\partial z^2} = f \tag{16}$$

where

$$f(x,y,z) \;=\; -2((y-y^2)(z-z^2) + (x-x^2)(z-z^2) + (x-x^2)(y-y^2));$$

$$(x,y,z) \;\in\; \Omega = [0,1] \times [0,1] \times [0,1];$$

$$u \;=\; 0 \text{ on } \Gamma.$$

On a uniform grid with meshsize Δ that is overlayed on Ω, the operator L in (16) is central differenced to yield a matrix approximation A where

$$A = \begin{bmatrix}
E & F & G & & & & \\
F & E & F & G & & & \\
G & F & E & F & G & & \\
& \ddots & \ddots & \ddots & \ddots & \ddots & \\
& & G & F & E & F & G \\
& & & G & F & E & F \\
& & & & G & F & E
\end{bmatrix}$$

where

$$E = \begin{bmatrix}
6 & -1 & & & \\
-1 & 6 & -1 & & \\
& \ddots & \ddots & \ddots & \\
& & -1 & 6 & -1 \\
& & & -1 & 6
\end{bmatrix}$$

and

$$G = F = - \begin{bmatrix}
1 & & & \\
& \ddots & & \\
& & \ddots & \\
& & & 1
\end{bmatrix}.$$

Let Δ_c be the mesh-size of a grid $G^{(H)}$ and $A^{(H)}$ the matrix approximation of the homogeneous equation (5). The grid lines of $G^{(H)}$ define boundaries of a decomposition

Δ_c	Δ_f	t_s	t_p
1/3	1/16	.243	.199
1/3	1/24	.76	.51
1/3	1/32	2.33	1.25

Table 1: Timings on NMFECC Cray-II

of Ω. Let this decomposition be given by

$$\Omega = \bigcup_{i=1}^{n} R_i.$$

On the closure of each subdomain R_i define a grid G_f of mesh-size Δ_f. Extend the internal boundaries of the regions R_i by Δ_f in each coordinate direction to obtain a larger sub-domain Ω_i. Clearly, $\Omega = \bigcup_{i=1}^{n} \Omega_i$. We use the conjugate gradient method given in (3) where the preconditioner M is defined by the additive Schwarz method. The calculations were executed on the Cray-II located at the NMFECC in the Lawrence Livermore National Laboratory. t_s refers to time in minutes without multiprocessing and t_p refers to the time in minutes with multiprocessing. The banded version of the LU-factorization algorithm was used to solve the linear systems involving A_i and $A^{(H)}$ in step $(3i)$ of the conjugate gradient method.

5   Summary

We have described an iterative algorithm for solving linear systems arising from the discretization of elliptic partial differential equation. The algorithm combines domain decomposition techiniques and the conjugate gradient method in order to achieve rapid convergence and parallelism. Subdividing the domain allows us to solve 3-dimensional problems in an efficient manner. Table 1 indicates the timings that were achieved on the Cray-II located at the NMFECC. Although the maximum speedup we achieved was 1.9, the parallel performance of the algorithm was disappointing. The idealized timing which assume we receive all four processors simultaneously should produce speedups in the range of 3.5-3.8. We are continuing to research the cause of this poor performance and will be the subject of a future paper.

References

[1] P.E.Bjorstad, Multiplicative and Additive Schwarz Methods: Convergence in the Two-Domain Case, Proc. of the 2nd Int. Symp. on Domain Decomposition Methods, Los Angeles, Calif., Jan. 14-16,1988, SIAM Pulications, Philadelphia, Pa., pp.147-159.

[2] G.Golub and C.Van Loan, Matrix Computations, Johns Hopkins University Press, Baltimore, Maryland, 1983.

[3] G. Rodrigue, Inner/Outer iterative methods and numerical Schwarz methods, Journal of Parallel Computing, 2, 1985, pp. 205-218.

[4] O. Widlund, Optimal iterative refinement methods, Proc. of the 2nd Int. Symp. on Domain Decomposition Methods, Los Angeles, Calif., Jan. 14-16, 1988, SIAM Publications, Philadelphia, Pa., pp. 114 - 126.

[5] A. Greenbaum, C. Li, and H.Z. Chao, Parallelizing Preconditioned Conjugate Gradient Algorithms, Computer Physics Communications, 53, 1989, pp. 295-309

Solving Linear Equations by Extrapolation

Walter Gander[†] , Gene H. Golub[‡] and Dominik Gruntz[†]

Abstract

This is a survey paper on extrapolation methods for vector sequences. We have simplified some derivations and we give some numerical results which illustrate the theory.

1   Introduction

Many iterative methods have been proposed for solving large systems of equations which are either sparse or structured or both. Unfortunately, the iteration procedures converge rather slowly and consequently acceleration methods are required. If the original matrix is symmetric and positive definite, then the conjugate gradient method can be used in combination with the Jacobi or SSOR iteration method. Such procedures often have excellent convergence properties.

For non-symmetric problems there is a great need to develop, analyze and experiment with a variety of methods. The solution time is not only a function of the convergence properties but of the architectural structure of the computer being used. To the best of our knowledge, there is no acceleration procedure which is uniformly best for all architectures.

In [2], a system of equations was given; the efficiency of the ε-algorithm was demonstrated in [4] in extrapolating the solution from approximations obtained from Block Gauss-Seidel iterations. Though the Gauss-Seidel method is convergent for diagonally dominant matrices cf. [18], convergence may be very slow.

In this paper, we continue our study and show to the relevance of this and other methods. All methods make use of *the minimal polynomial of a matrix with respect to a vector*. Since this polynomial is not known in practice, there are several ways to approximate it. This leads to the algorithms *minimal polynomial extrapolation* related to the *method of Arnoldi, reduced rank extrapolation* related to *GCR* or *orthomin* and *topological ε-algorithm* related to *Lanczos*. We also mention the algorithms *modified minimal polynomial extrapolation, scalar-* and *vector ε-algorithm* which are not related to known Krylov subspace methods. Finally we report some numerical results from experiments with large linear systems obtained from the discretisation of an elliptic differential equation.

Much of the theoretical work on extrapolation methods has been done by Brezinski [1] and his students [3] and by Avram Sidi et al. An extensive list of publications are given in the **References**.

2   Notation and general comments to the methods

Given a matrix A and a vector b, we consider the vector sequence generated in real or complex n-space from a starting vector x_0 by

$$x_{j+1} = Ax_j + b, \qquad j = 0, 1, \dots \tag{1}$$

It is assumed that 1 is not an eigenvalue of A and therefore that the iteration has a unique fixed point

$$s = (I - A)^{-1}b, \tag{2}$$

which is the solution of the linear system $Cx = b$ with $C = I - A$.

If all the eigenvalues of A are less than one in magnitude, then $s = \lim_{j \to \infty} x_j$ for all x_0.

If the sequence diverges, s is called the *anti-limit* and one may still try to determine s from a finite number of terms of the sequence.

Let $e_j := x_j - s$ denote the error and assume $A = QDQ^{-1}$ where Q is the matrix of the eigenvectors q_j and $D = diag\{\lambda_1, \dots, \lambda_n\}$ the eigenvalues of A. Then from iteration (1), we have

$$e_j = A^j e_0 = QD^j \underbrace{Q^{-1}e_0}_{z} = \sum_{l=1}^{n} q_l \lambda_l^j z_l \tag{3}$$

[†]Institut für wissenschaftliches Rechnen, Eidgenössische Technische Hochschule, CH-8092 Zürich, Switzerland, e-mail : gander@inf.ethz.ch respectively gruntz@inf.ethz.ch.

[‡]Department of Computer Science, Stanford University, Stanford, CA 94305, e-mail : na.golub@na-net.stanford.edu. The work of this author was supported in part by the National Science Foundation under grand DCR-8412314.

and therefore

$$x_j = s + \sum_{l=1}^{n} q_l \lambda_l^j z_l. \tag{4}$$

The idea of extrapolation is to form a *linear combination of the* x_j, so that error terms in (4) are cancelled. We wish to find $\gamma_0, \ldots, \gamma_k$ such that

$$\begin{aligned} \sum_{j=0}^{k} \gamma_j &= 1 \\ \sum_{j=0}^{k} \gamma_j x_j &= \sum_{j=0}^{k} \gamma_j s = s. \end{aligned} \tag{5}$$

Now taking the same linear combination with the e_j in (3) gives

$$0 = \sum_{j=0}^{k} \gamma_j e_j = \sum_{j=0}^{k} \gamma_j A^j e_0 = P_k(A) e_0.$$

P_k is some polynomial with coefficients γ_j such that $P_k(A)e_0 = 0$. Note that (5) implies $P_k(I) = I$. Of course, one wishes to have k as small as possible. Therefore we *choose as* γ_j *the coefficients of the minimal polynomial of A with respect to* e_0 *with the normalization* $\sum_{j=0}^{k} \gamma_j = 1$, $\gamma_k \neq 0$.

If $P_k(A)e_0 = 0$ then by multiplying with A^m one obtains

$$A^m P_k(A) e_0 = P_k(A) A^m e_0 = P_k(A) e_m = 0.$$

Therefore if A is non singular then P_k is also minimal polynomial of A with respect to e_m for all m (cf. lemma 1 in [15]).

It is not possible to compute P_k since the error e_m is unknown. However let

$$u_j = \Delta x_j = x_{j+1} - x_j \tag{6}$$

denote the difference of two successive iterates. Then we have the important relation

$$u_j = (A - I) e_j = -C e_j. \tag{7}$$

Note that for the *residual vector*,

$$r_j = b - C x_j = A x_j + b - x_j = x_{j+1} - x_j = u_j \tag{8}$$

therefore the differences of successive iterates are also the residuals.

Consider now the minimal polynomial of A with respect to u_m:

$$P_k(A) u_m = 0. \tag{9}$$

Because $u_m = (A - I)e_m$ and $(A - I)$ is nonsingular and commutes with $P_k(A)$, the minimal polynomial with respect to e_m is also P_k, since

$$P_k(A)(A - I)e_m = (A - I)P_k(A)e_m = 0 \iff P_k(A)e_m = 0.$$

Let $P_k(\lambda) = c_0 + c_1 \lambda + \cdots + c_k \lambda^k$ with $c_k \neq 0$. Then

$$\sum_{j=0}^{k} c_j (x_{m+j} - s) = \sum_{j=0}^{k} c_j x_{m+j} - \sum_{j=0}^{k} c_j s = 0,$$

and we get the barycentric representation

$$s = \frac{\sum_{j=0}^{k} c_j x_j}{\sum_{j=0}^{k} c_j}. \tag{10}$$

Since we assumed that $A - I$ is non singular, $\lambda = 1$ is not an eigenvalue of A and therefore $P_k(1) = \sum_{j=0}^{k} c_j \neq 0$. *We can therefore extrapolate* s *from* $x_m, \ldots, x_{m+k}$ *by using the coefficients of the minimal polynomial of A with respect to* u_m.

If we define the coefficients

$$\gamma_j := c_j / \sum_{i=0}^{k} c_i,$$ (11)

we can also write the barycentric representation as given in equation (5).

We conclude this section with some definitions which will be used in the following.

$$w_j = \Delta^2 x_j = \Delta u_j = u_{j+1} - u_j.$$ (12)

For some fixed k, we introduce the matrices

$$U = [u_m, u_{m+1}, \ldots, u_{m+k-1}]$$ (13)

$$W = [w_m, w_{m+1}, \ldots, w_{m+k-1}].$$ (14)

We note that for each j,

$$u_j = A u_{j-1} = A^j u_0.$$ (15)

The same relation holds between w_j and u_j as between u_j and e_j:

$$w_j = (A - I) u_j.$$ (16)

Writing this relation, using the matrices U and W, we get

$$W = (A - I) U.$$ (17)

3 Minimal Polynomial Extrapolation (MPE)

Let P_k be the (unique) minimal polynomial of A of degree k with respect to u_m, i.e.

$$P_k(A) u_m = 0.$$ (18)

Let $P_k(\lambda) = c_0 + c_1 \lambda + \cdots + c_k \lambda^k$ with $c_k = 1$. Using (15), equation (18) becomes

$$\sum_{j=0}^{k-1} c_j u_{m+j} = -u_{m+k}.$$ (19)

In [15] it is proposed to solve (19) as linear least squares problem for c_j (numerically this may not be a good way to go, see p. 371 in [19]). The normal equations are

$$U^T U c = -U^T u_{m+k}$$

$$\iff \sum_{j=0}^{k-1} (u_{m+i}, u_{m+j}) c_j = -(u_{m+i}, u_{m+k}) \quad i = 0, \ldots, k-1.$$

Using the fact that $c_k = 1$, we divide the equations by $\sum_{j=0}^{k} c_j$ and obtain with γ_j from (11)

$$\begin{aligned} \sum_{j=0}^{k} (u_{m+i}, u_{m+j}) \gamma_j &= 0 \quad i = 0, \ldots, k-1 \\ \sum_{j=0}^{k} \gamma_j &= 1. \end{aligned}$$ (20)

Since k, the degree of the minimal polynomial, is in general unknown using the coefficients computed by (20) will give only an approximation to s:

$$s_{mk} = \sum_{j=0}^{k} \gamma_j x_{m+j}.$$ (21)

One can show (cf. [11]) that for k values smaller than the degree of the minimal polynomial, solutions of (20) exist if the matrix $U^T C U$ is non singular, that is, if the symmetric part of C is positive definite. In the same paper a connection to the method of Arnoldi is established.

In the method of Arnoldi an orthogonal basis is constructed from the Krylov subspace

$$\mathcal{K}_k(C, r_0) := \operatorname{span}\{r_0, Cr_0, \ldots, C^{k-1}r_0\}.$$

Using Gram-Schmidt, one computes the QR-decomposition of the matrix

$$[r_0, Cr_0, \ldots, C^{k-1}r_0] = P_k R.$$

Let p_j denote the j-th column of the orthogonal matrix P_k. Then $p_0 = r_0/\|r_0\|$ and

$$h_{j+1,j}p_{j+1} = Cp_j - \sum_{i=0}^{j} h_{ij} p_i \tag{22}$$

with $h_{ij} = p_i^T C p_j$ and $h_{j+1,j}$ such that $\|p_{j+1}\| = 1$. From equation (22) we conclude that

$$CP_k = P_k H_k + p_{k+1} e_k^T h_{k+1,k} \tag{23}$$

where H_k is a Hessenberg matrix. To compute an approximate solution to the linear system $Cx = b$, one makes the *ansatz*

$$x_k = x_0 + P_k y \quad \Rightarrow r_k = r_0 - CP_k y \in \mathcal{K}_{k+1}(C, r_0)$$

and determines y such that

$$r_k \perp \mathcal{K}_k(C, r_0) \iff H_k y = \|r_0\| e_1.$$

On the other hand, if we compute the residual of (21) with $m = 0$ then

$$r(s_{0k}) = \sum_{j=0}^{k} \gamma_j r(x_j) = \sum_{j=0}^{k} \gamma_j A^j u_0 \in \mathcal{K}_{k+1}(A, r_0).$$

Now $\mathcal{K}_{k+1}(A, r_0) = \mathcal{K}_{k+1}(C, r_0)$ since $I - A = C$. Furthermore from the normal equation (20) we see that

$$r(s_{mk}) \perp \mathcal{K}_k(A, r_0) = \mathcal{K}_k(C, r_0).$$

Therefore,

Theorem 1 *The residuals of the approximations of the method of Arnoldi and of MPE are in the same subspaces and thus the two methods are equivalent.*

4 Reduced rank extrapolation (RRE)

In the method of Arnoldi, the approximate solution x_k is formed as a linear combination of columns the orthogonal matrix P_k, which form an orthogonal basis of the Krylov-subspace $\mathcal{K}_k(C, r_m)$

$$s_{mk} = x_m + P_k y \in \mathcal{K}_k(C, r_m) = \mathcal{K}_k(A, u_m).$$

Also the columns of $U = [u_m, \ldots, u_{m+k-1}]$ are a basis of $\mathcal{K}_k(A, u_m)$ and therefore one can also express the approximate solution as

$$s_{mk} = x_m + U\xi. \tag{24}$$

If k is the degree of the minimal polynomial of A with respect to u_m then $s_{mk} = s$. Therefore we have to find ξ such that

$$U\xi = -e_m = s - x_m. \tag{25}$$

However since e_m is not known, this equation is not practical. By multiplying (25) from the left by $A - I$, we get a computable right hand side:

$$W\xi = (A - I)U\xi = -(A - I)e_m = -u_m. \tag{26}$$

This equation defines the

Reduced rank extrapolation (RRE) method

- generate $x_m, x_{m+1}, \ldots, x_{m+k+1}$

- compute $U = [u_m, u_{m+1}, \ldots, u_{m+k-1}]$ and $u_{m+k} = x_{m+k+1} - x_{m+k}$

- compute $W = [w_m, w_{m+1}, \ldots, w_{m+k-1}]$ where $w_j = u_{j+1} - u_j$

- Solve the least squares problem $W\xi = -u_m$ so that $\xi = -W^+ u_m$

- compute $s_{mk} = x_m + U\xi$

Writing ΔX for U and $\Delta^2 X$ for W, equation (24) becomes

$$s_{mk} = x_m - \Delta X (\Delta^2 X)^+ \Delta x_m$$

and the analogy of reduced rank extrapolation and Aitken's Δ^2 method becomes visible.
The normal equations of (26) are

$$W^T W \xi = -W^T u_m$$

$$\iff \sum_{j=0}^{k-1} (w_{m+i}, w_{m+j}) \xi_j = -(w_{m+i}, u_m) \quad i = 0, \ldots, k-1.$$

If we replace $w_{m+j} = u_{m+j+1} - u_{m+j}$, we obtain

$$(w_{m+i}, u_m)(1 - \xi_0) + \sum_{j=1}^{k-1} (w_{m+i}, u_{m+j})(\xi_{j-1} - \xi_j) + (w_{m+i}, u_{m+k})\xi_{k-1} = 0.$$

We now introduce the new variables

$$\begin{aligned}
\gamma_0 &= 1 - \xi_0 \\
\gamma_j &= \xi_{j-1} - \xi_j \quad j = 1, \ldots, k-1 \\
\gamma_k &= \xi_{k-1}.
\end{aligned} \tag{27}$$

Notice there are $k+1$ variables γ_j and only k variables ξ_j, however the γ_j are related by the relationship $\sum_{j=0}^k \gamma_j = 1$. Expressed in this new variables the normal equations become

$$\begin{aligned}
\sum_{j=0}^k (w_{m+i}, u_{m+j}) \gamma_j &= 0 \\
\sum_{j=0}^k \gamma_j &= 1.
\end{aligned} \tag{28}$$

Furthermore

$$\begin{aligned}
s_{mk} &= x_m + \sum_{j=0}^{k-1} u_{m+j} \xi_j \\
&= x_m + \sum_{j=0}^{k-1} (x_{m+j+1} - x_{m+j}) \xi_j \\
&= x_m(1 - \xi_0) + \sum_{j=1}^{k-1} x_{m+j}(\xi_{j-1} - \xi_j) + x_{m+k} \xi_{k-1}
\end{aligned}$$

and therefore again,

$$s_{mk} = \sum_{j=0}^k x_{m+j} \gamma_j \quad \text{with} \quad \sum_{j=0}^k \gamma_j = 1.$$

If k is smaller than the degree of the minimal polynomial, then in contrast to MPE the intermediate solutions always exists for RRE [11]. RRE is connected to GCR or ORTHOMIN.

ORTHOMIN or GCR algorithm:

$$\begin{aligned}
r_0 &= b - Cx_0; \quad p_0 = r_0 \\
\alpha_j &= \frac{(r_j, Cp_j)}{(Cp_j, Cp_j)} &&\Rightarrow r_{j+1} \perp Cr_j \\
x_{j+1} &= x_j + \alpha_j p_j \\
r_{j+1} &= r_j - \alpha_j Cp_j \\
\beta_{ij} &= \frac{(Cr_{j+1}, Cp_i)}{(Cp_i, Cp_i)}, \quad i = 0, \ldots, j &&\Rightarrow Cp_{j+1} \perp Cp_i \\
p_{j+1} &= r_{j+1} - \sum_{i=0}^j \beta_{ij} p_i.
\end{aligned}$$

This algorithm has the following properties:

1. $r_k \in \mathcal{K}_{k+1}(C, r_0)$

2. $r_k \perp Cr_j, \quad j < k$ ("conjugate residuals")

3. $r_k \perp \mathcal{K}_k(C, Cr_0)$.

For the residuals in RRE we have

$$r(s_{mk}) = \sum_{j=0}^{k} \underbrace{r(x_{m+j})}_{u_{m+j}} \gamma_j \in \mathcal{K}_{k+1}(A, u_m) = \mathcal{K}_{k+1}(C, r_m).$$

It follows from the normal equations (28) that

$$r(s_{mk}) \perp w_{m+j}, \quad \text{for} \quad i = 1, \ldots, k-1.$$

Since $w_{m+j} = A^j w_m$ this is equivalent to $r(s_{mk}) \perp \mathcal{K}_k(A, w_m)$. Finally since $w_m = (A - I)u_m = -Cu_m$, we have

$$\mathcal{K}_k(A, w_m) = \mathcal{K}_k(A, Cu_m) = \mathcal{K}_k(C, Cu_m) = \mathcal{K}_k(C, Cr_m).$$

Now for $m = 0$, $r(s_{0k}) \in \mathcal{K}_{k+1}(C, r_0)$ and $r(s_{0k}) \perp \mathcal{K}_k(C, Cr_0)$. Hence

Theorem 2 *The residuals of RRE are in the same subspaces as those of GCR. The methods are therefore equivalent.*

5 Modified minimal polynomial extrapolation (MMPE)

Instead of solving the system (26) $W\xi = -u_m$ in the least squares sense, an arbitrary matrix $Q \in R^{n \times k}$ with full rank k is chosen and the solution ξ is computed from the $k \times k$ system

$$Q^T W\xi = -Q^T u_m.$$

Writing $Q = [q_0, \ldots, q_{k-1}]$, these equations are

$$\sum_{j=0}^{k-1} (q_i, w_{m+j})\xi_j = -(q_i, w_m), \quad i = 0, \ldots, k-1.$$

Introducing the new variables γ_j we obtain by the same manipulation as in the previous section the equations

$$
\begin{aligned}
\sum_{j=0}^{k} (q_i, u_{m+j})\gamma_j &= 0 \\
\sum_{j=0}^{k} \gamma_j &= 1.
\end{aligned}
\tag{29}
$$

Note that the subspace spanned by the columns of Q can be interpreted as a Krylov subspace (with some matrix G). We construct G as follows: Augment linear independent column vectors $\tilde{q}_j$ to form a $n \times n$ non singular matrix

$$\tilde{Q} = [q_0, \ldots, q_{k-1}, \tilde{q}_k, \ldots, \tilde{q}_n].$$

Let us denote $\tilde{q}_j = q_j$, $j = 0, \ldots, k-1$. We wish to find G such that $\tilde{q}_{j+1} = G\tilde{q}_j$ i.e.

$$[\tilde{q}_1, \ldots, \tilde{q}_n, x] = G\tilde{Q}$$

where x is an arbitrary vector. From this equation we get

$$G = [\tilde{q}_1, \ldots, \tilde{q}_n, x]\tilde{Q}^{-1}.$$

Therefore we have an analogous result as with RRE:

$$r(s_{mk}) \in \mathcal{K}_{k+1}(A, u_m) \quad \text{and} \quad r(s_{mk}) \perp span\{q_0, \ldots, q_{k-1}\} = \mathcal{K}_k(G, q_0).$$

Taking $q_j = e_i$ with randomly chosen unit vectors such that rank $Q = k$ (i.e. picking simply k equations out of $W\xi = -u_m$) minimizes the computational work. If k is smaller than the degree of the minimal polynomial then it is shown in [11], that the intermediate values s_{mk} exists if $Q^T CU$ is non singular.

6 The topological ε-algorithm (TEA)

Here again one replaces the linear system (26) $W\xi = -u_m$ by $Q^T W\xi = -Q^T u_m$. However now

$$Q = [q, A^T q, \ldots, (A^T)^{k-1}q] \tag{30}$$

and q is some fix chosen vector. By the same calculation as before we obtain the equations for γ_j, namely,

$$0 = \sum_{j=0}^{k}((A^T)^i q, u_{m+j})\gamma_j = \sum_{j=0}^{k}(q, A^i u_{m+j})\gamma_j = \sum_{j=0}^{k}(q, u_{m+j+i})\gamma_j, \quad (i = 0, \ldots, k-1).$$

We note that for this extrapolation, we need the iterates $x_m, \ldots, x_{m+2k}$ whereas for MPE,RRE and MMPE we only need $x_m, \ldots, x_{m+k+1}$. Originally TEA was developed by Brezinski in analogy to Wynn's ε-algorithm for extrapolation of sequences in a topological vector space. The approximations can be computed recursively without using the normal equations cf. section 9. It is interesting that TEA is connected [11] to the non-symmetric Lanczos algorithm [7].

Non-symmetric Lanczos

Choose y_0 and z_0 such that $y_0^T z_0 = 1$
for $j = 0, 1, \ldots, k$ do

$$\begin{aligned}
\alpha_j &= (Cy_j, z_j) \\
\hat{y}_{j+1} &= Cy_j - \alpha_j y_j - \beta_j y_{j-1} \quad (\beta_0 y_{-1} = 0) \\
\hat{z}_{j+1} &= C^T z_j - \alpha_j z_j - \delta_j z_{j-1} \quad (\delta_0 z_{-1} = 0)
\end{aligned}$$

Choose δ_{j+1} and β_{j+1} such that $\delta_{j+1}\beta_{j+1} = \hat{y}_{j+1}^T \hat{z}_{j+1}$

e.g. $\delta_{j+1} = \sqrt{|\hat{y}_{j+1}^T \hat{z}_{j+1}|}$ and $\beta_{j+1} = sign(\hat{y}_{j+1}^T \hat{z}_{j+1})\delta_{j+1}$

$$\begin{aligned}
y_{j+1} &= \hat{y}_j/\delta_{j+1} \\
z_{j+1} &= \hat{z}_j/\beta_{j+1}.
\end{aligned}$$

This algorithm has the following properties:

1. $Y_k^T Z_k = I$ (the vector y_j and z_i are *biorthogonal*)

2. $\hat{y}_{j+1} = P_j(C)y_0$ and $\hat{z}_{j+1} = P_j(C^T)z_0$, i.e. $\hat{y}_j \in \mathcal{K}_j(C, y_0)$ and $\hat{z}_j \in \mathcal{K}_j(C^T, z_0)$

3.

$$Z_k^T CY_k = T_k := \begin{pmatrix} \alpha_0 & \beta_1 & & \\ \delta_1 & \ddots & \ddots & \\ & \ddots & \ddots & \beta_k \\ & & \delta_k & \alpha_k \end{pmatrix}$$

4. $CY_k = Y_k T_k + \underbrace{\delta_{k+1}y_{k+1}}_{\hat{y}_{k+1}} e_k^T$.

To solve linear equations with Lanczos one makes the *ansatz* $x_k = x_0 + Y_k y$. Then

$$Cx_k = Cx_0 + CY_k y = Cx_0 + Y_k T_k y + \hat{y}_{k+1} e_k^T y = b$$

$$\Longleftrightarrow Y_k T_k y + \hat{y}_{k+1} e_k^T y = r_0.$$

Multiplying from the left with Z_k^T gives

$$\underbrace{Z_k^T Y_k}_{I} T_k y + \underbrace{Z_k^T \hat{y}_{k+1}}_{0} e_k^T y = Z_k^T r_0.$$

Therefore by choosing $y_0 = r_0/\|r_0\|$ and putting $\beta = \|r_0\|$ one has to compute y as the solution of

$$T_k y = \beta e_1.$$

Note that for the residual $r_k = b - Cx_k$ we have by construction $r_k \perp \mathcal{K}_k(C^T, z_0)$ and furthermore

$$r_k = b - Cx_0 - CY_k y = \underbrace{r_0 - Y_k T_k y}_{r_0} - \hat{y}_{k+1} e_k^T y = -\hat{y}_{k+1} e_k^T y \in \mathcal{K}_{k+1}(C, r_0).$$

For the residual in TEA we have for $m = 0$

$$r(s_{0k}) \in \mathcal{K}_{k+1}(A, r_0) = \mathcal{K}_k(C, r_0).$$

From the normal equations (with Z defined by (30)) we conclude

$$Z^T r(s_{0k}) = 0 \iff r(s_{0k}) \perp \mathcal{K}_k(A^T, q) = \mathcal{K}_k(C^T, q).$$

Therefore again

Theorem 3 *The residuals of Lanczos and TEA lie for $q = z_0$ in the same subspaces and so both algorithms are equivalent.*

An important difference of both algorithms for practical computations is the fact, that the Lanczos algorithm needs both operators for A and A^T while it is sufficient for TEA to have A.

7 Determinantal representation and Shanks transformation

All four extrapolation methods (MPE,RRE,MMPE,TEA) have as we have seen a similar structure. The extrapolated value is expressed as a linear combination of $k + 1$ successive iterates

$$s_{mk} = \sum_{j=0}^{k} \gamma_j x_{m+j},$$

where the coefficients γ_j are determined as solution of the linear system

$$
\begin{array}{ccccccc}
\gamma_0 & + & \gamma_1 & + & \cdots & + & \gamma_k & = & 1 \\
\mu_{00}\gamma_0 & + & \mu_{01}\gamma_1 & + & \cdots & + & \mu_{0k}\gamma_k & = & 0 \\
\vdots & & \vdots & & \cdots & & \vdots & & \vdots \\
\mu_{k-1,0}\gamma_0 & + & \mu_{k-1,1}\gamma_1 & + & \cdots & + & \mu_{k-1,k}\gamma_k & = & 0
\end{array}
\tag{31}
$$

The coefficients μ_{ij} of these "normal equations" are

$$
\begin{array}{llll}
\mu_{ij} & = & (u_{m+i}, u_{m+j}) & \text{for MPE;} \\
\mu_{ij} & = & (w_{m+i}, u_{m+j}) & \text{for RRE;} \\
\mu_{ij} & = & (q_i, u_{m+j}) & \text{for MMPE;} \\
\mu_{ij} & = & (q, u_{m+j+i}) & \text{for TEA.}
\end{array}
$$

Let R denote the matrix of (31). If R_j is R with its j-th column replaced by the right hand side e_1 then using Cramer's rule we have

$$\gamma_j = \frac{\det R_j}{\det R}$$

and therefore

$$s_{mk} = \sum_{j=0}^{k} \gamma_j x_{m+j} = \frac{1}{\det R} \sum_{j=0}^{k} (\det R_j) x_{m+j}. \tag{32}$$

If we now introduce the determinants,

$$D(\sigma_0, \ldots, \sigma_k) := \begin{vmatrix} \sigma_0, & \cdots & \sigma_k \\ \mu_{00}, & \cdots & \mu_{0k} \\ \vdots & \cdots & \vdots \\ \mu_{k-1,0}, & \cdots & \mu_{k-1,k} \end{vmatrix} \tag{33}$$

then *formally* the last sum in (32) can be written as the Laplace expansion of the first row of the determinant $D(x_m, \ldots, x_{m+k})$. Thus we obtain the following expression for s_{mk}, which is a *generalized Shank transformation*

$$s_{mk} = \frac{D(x_m, \ldots, x_{m+k})}{D(1, \ldots, 1)}. \tag{34}$$

8 The scalar ε-algorithm (SEA)

Let $\{x_i\}$ now be a *scalar* sequence with $x_i \to s$ as $i \to \infty$ and with for which the error has an expansion like (4) or more generally

$$\sum_{i=0}^{k} c_i(x_{m+i} - s) = 0, \quad \text{for all} \quad m > N \quad \text{and} \quad \sum_{i=0}^{k} c_i \neq 0.$$

If we consider these equations for $m, m+1, \ldots, m+k$ then for the existence of non trivial solutions the determinant of the matrix has to vanish. Introducing the *Hankel determinant*

$$H_{k+1}(x_m) := \begin{vmatrix} x_m & x_{m+1} & \cdots & x_{m+k} \\ x_{m+1} & x_{m+2} & \cdots & x_{m+k+1} \\ \vdots & \vdots & \cdots & \vdots \\ x_{m+k} & x_{m+k+1} & \cdots & x_{m+2k} \end{vmatrix}$$

it is shown in [1], using some manipulations with determinants, that

$$s = \frac{H_{k+1}(x_m)}{H_k(\Delta^2 x_m)}. \tag{35}$$

Since k is generally unknown, the right hand side of (35) is considered for various values of m and k and is called the *Shanks transformation* of the sequence x_j. It is often denoted by s_{mk} or $e_k(x_m)$.

Wynn discovered a simple recursion *that allows the computation of $e_k(x_m)$ without solving linear systems or computing determinants: the ε-algorithm.*

Let $\varepsilon_{-1}^{(n)} = 0$ and $\varepsilon_0^{(n)} = x_n$ for $n = 0, 1, 2, \ldots$. From these values, the following table using the recurrence relation

$$\varepsilon_{k+1}^{(n)} = \varepsilon_{k-1}^{(n+1)} + \frac{1}{\varepsilon_k^{(n+1)} - \varepsilon_k^{(n)}} \tag{36}$$

is constructed:

$$
\begin{array}{ccccccc}
\varepsilon_{-1}^{(0)} \\
 & \varepsilon_0^{(0)} \\
\varepsilon_{-1}^{(1)} & & \varepsilon_1^{(0)} \\
 & \varepsilon_0^{(1)} & & \varepsilon_2^{(0)} \\
\varepsilon_{-1}^{(2)} & & \varepsilon_1^{(1)} & & \varepsilon_3^{(0)} \\
 & \varepsilon_0^{(2)} & & \varepsilon_2^{(1)} & & \\
\varepsilon_{-1}^{(3)} & & \varepsilon_1^{(2)} & & \cdots \\
 & \varepsilon_0^{(3)} & & \cdots \\
\end{array}
\tag{37}
$$

The theorem of Wynn then states that

$$\varepsilon_{2k}^{(n)} = e_k(x_n) \quad \text{and} \quad \varepsilon_{2k+1}^{(n)} = \frac{1}{e_k(\Delta x_n)}.$$

For a vector sequence we can apply the ε-algorithms to each component. We call the resulting algorithm SEA. The main disadvantage of SEA is, *that it ignores the linkage between the scalar sequences in different components, and it also runs the risk that, in one or more components the required reciprocals frequently will either fail to exists or be quite large numerically* [15]. In [1], modified recursions are mentioned that eliminates that problem.

9 The vector ε-algorithms VEA and TEA

In order to generalize the ε-scheme for a vector sequence $\{x_j\}$ we have to give a meaning to the "inverse of a vector" used in the recursion (36). One possibility is to use the pseudoinverse (Samelson inverse), defined by

$$y^{-1} := \frac{1}{||y||^2} \bar{y}^T. \tag{38}$$

It has been conjectured by Wynn and proved by McLeod that if the sequence $\{x_j\}$ has the property that

$$\sum_{i=0}^{k} c_i(x_{m+i} - s) = 0, \quad m > N$$

then $\varepsilon_{2k}^{(m)} = s$. This is however all that ones knows about VEA. No determinantal representations of $\varepsilon_i^{(n)}$ are known.

A second generalization for vector sequences is based on a different interpretation of "inverse" [1]. Brezinski defines the inverse of a ordered pair (a, b) of vectors such that $a^T b \neq 0$ to be the ordered pair (b^{-1}, a^{-1}), where

$$b^{-1} = \frac{a}{b^T a}, \quad a^{-1} = \frac{b}{b^T a}.$$

For real sequences $\{x_j\}$ two algorithms result with this interpretation [17], where the first refers to TEA defined in section 6 (note that the difference in $\Delta \varepsilon_m^{(n)}$ is with respect to n).

Choose an arbitrary vector q and set

$$\varepsilon_{-1}^{(n)} = 0, \quad \varepsilon_0^{(n)} = x_n, \quad n = 0, 1, 2, \ldots$$

Algorithm 1: (TEA1)

$$\varepsilon_{2m+1}^{(n)} = \varepsilon_{2m-1}^{(n+1)} + \frac{q}{q^T \Delta \varepsilon_{2m}^{(n)}}, \varepsilon_{2m+2}^{(n)} = \varepsilon_{2m}^{(n+1)} + \frac{\Delta \varepsilon_{2m}^{(n)}}{\Delta \varepsilon_{2m+1}^{(n)T} \Delta \varepsilon_{2m}^{(n)}} \quad m, n = 0, 1 \ldots \quad (39)$$

Algorithm 2: (TEA2)

$$\varepsilon_{2m+1}^{(n)} = \varepsilon_{2m-1}^{(n+1)} + \frac{q}{q^T \Delta \varepsilon_{2m}^{(n)}}, \varepsilon_{2m+2}^{(n)} = \varepsilon_{2m}^{(n+1)} + \frac{\Delta \varepsilon_{2m}^{(n+1)}}{\Delta \varepsilon_{2m+1}^{(n)T} \Delta \varepsilon_{2m}^{(n+1)}} \quad m, n = 0, 1 \ldots \quad (40)$$

As it is shown in [1] TEA1 computes

$$\varepsilon_{2k}^{(n)} = e_k(x_n)$$

and yields the same extrapolated values as if it is computed using the "normal equations" (31). It is possible that the recursive computation will be numerically more stable, since then the coefficients of the minimal polynomial are not computed.

10   Numerical experiments

In this section we report a number of numerical experiments executed on a CRAY X-MP/2 using 64-bit real arithmetic. We have compared the Lanczos and different ε-algorithms using the following Model Problem (found in [8]):

We solve the elliptic partial differential equation in the region $\Omega = (0, 1) \times (0, 1)$

$$-\Delta u + \gamma(x \frac{\partial u}{\partial x} + y \frac{\partial u}{\partial y}) + \beta u = g \quad \text{in } \Omega$$

with Dirichlet boundary conditions on $\partial \Omega$. This problem is discretized using centered differences for both the first and second order derivatives on a uniform mesh with red-black ordering. The mesh size $h = 1/32$ leads to 961 unknowns. The boundary conditions (i.e. the right hand side vector) is chosen such that the solution u to the *discrete* system is one everywhere. This allows an easy verification of the result. The nonsymmetrical linear system has now the form $G(\gamma, \beta)u = f$.

The parameters γ and β can be varied to make the problem more or less difficult to solve. The greater γ is chosen, the more unsymmetrical the system becomes, and the value of β makes the system more or less positive definite. Let's define $\delta = \gamma h/2$.

The basic linear stationary iterative method ($x_{k+1} = Ax_k + b$) on which we extrapolate the solution is either Point-Jacobi or Gauss-Seidel. The corresponding subspace method then solves the related linear system ($Cx = b$ with $C = I - A$). In order to avoid operating in a special subspace, we choose the initial approximation x_0 as a normed random vector.

In our Figures, we show the behaviour of the infinity norm of the *error*-vector over the number of iteration steps or over the elapsed time on our CRAY.

First, we want to show numerically the equivalence between the non-symmetric Lanczos, BiCG and the TEA1. Therefore we choose $\gamma = 96$ $(\delta = 1.5)$ and $\beta = 0$. In Fig 1. we see, that the three methods are

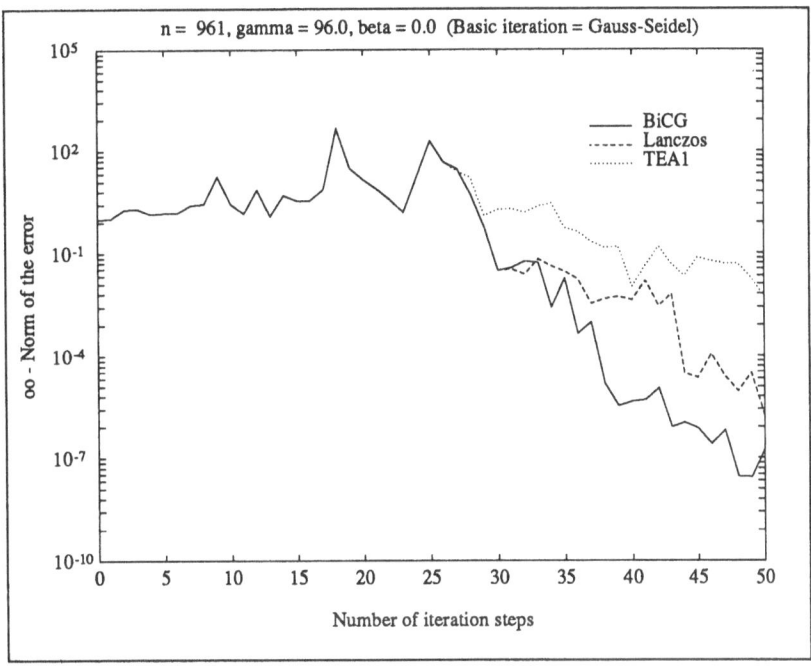

Figure 1: Equivalence of Lanczos and TEA1

equivalent as mentioned in Theorem 3. (Note that one ε-Step uses *two* basic iteration steps.) Lanczos and TEA1 branch off due to rounding errors after 27 steps, Lanczos and BiCG after 30.

However this is not a good way to use the ε-algorithm, because the deeper the ε-scheme is computed the more time is spent, the more memory is used and the more rounding errors influence the results. In finite arithmetic one has to stop the computation of every ε-scheme after some rows and columns. With the best approximate of this scheme, one starts a new ε-scheme. We can describe this procedure in the following algorithm:

1. Choose x_0 and numbers n_k and m_k.

2. for $k = 0, 1, \ldots$ do

$$
\left.
\begin{aligned}
y_0 &= x_k \\
y_{i+1} &= Ay_i, \qquad i = 0, \ldots, n_k - 1
\end{aligned}
\right\} \quad \text{Basic Iteration}
$$

$$
\left.
\begin{aligned}
\varepsilon_0^{(0)} &= y_{n_k} \\
\varepsilon_0^{(i+1)} &= A\varepsilon_0^{(i)}, \qquad i = 0, \ldots, 2m_k - 1 \\
\varepsilon_{2m_k}^{(0)} &= e_{m_k}(y_{n_k})
\end{aligned}
\right\} \quad \text{Extrapolation}
$$

$$
x_{k+1} = \varepsilon_{2m_k}^{(0)} \qquad\qquad\qquad\quad \} \quad \text{best approximate}
$$

In Fig 2, we have applied this truncated extrapolation to the same problem as in Fig 1. We first performed $n_0 = 35$ (low cost) Gauss-Seidel steps and started then the extrapolation of width $m_0 = 16$,

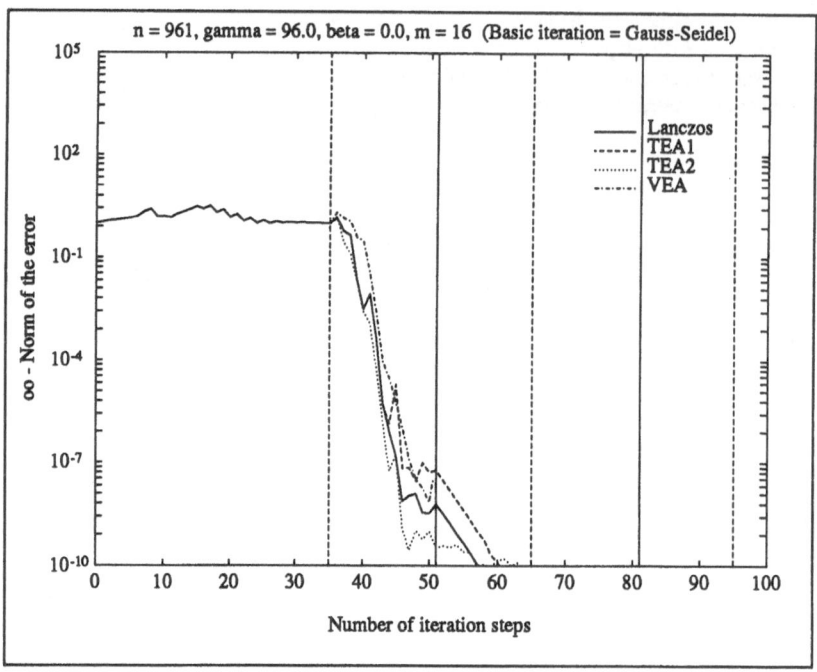

Figure 2: $\delta = 1.5, n_0 = 35$

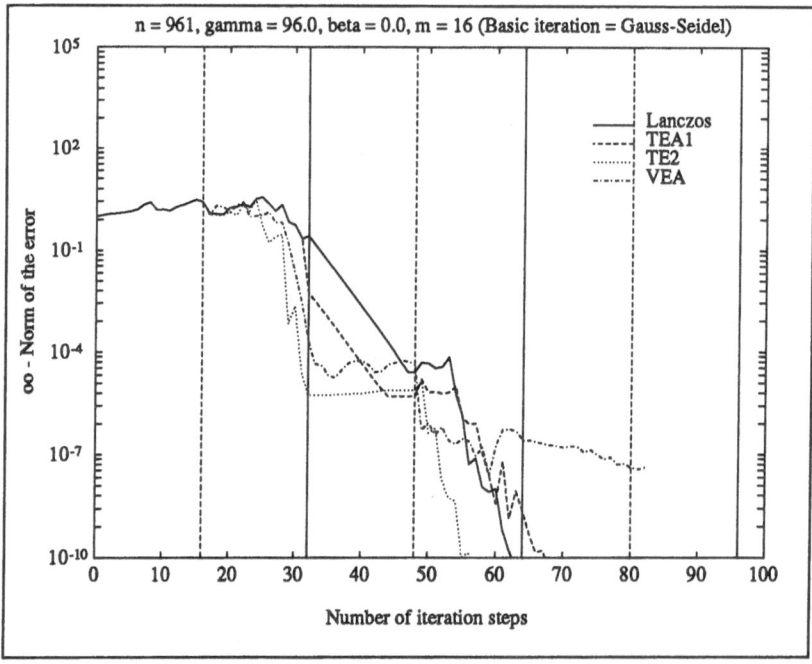

Figure 3: $\delta = 1.5, n_0 = 16$

i.e. we had to compute another 32 Gauss-Seidel steps as basis for the ε-scheme (37). Then we again compute simple basic iteration steps starting with the best approximate ($\varepsilon_{16}^{(0)}$ in our example)., which made the error drop rapidly below 10^{-10}. (In the Figures, these different parts are separated by vertical lines: basic iteration steps between a solid and a dashed line and extrapolation steps between a dashed and a solid line.) We have compared Lanczos, TEA1, TEA2 and VEA, which breaks down after 48 iteration steps.

We see, that we gain up to 10 digits during the extrapolation phase. Looking at TEA1 and Lanczos we see, that after the extrapolation steps, the contributions z_l of the largest eigenvalues of the iteration matrix to the error (cf (3)) must have disappeared in some sense and therefore we get a good convergence rate.

In Fig 2, we see also, that the ε-Algorithm is much more powerful, if we first compute *many* Gauss-Seidel steps. In Fig 3 however, n_0 is set only to 16, and we see, that we use *two* extrapolation phases of width 16 ($m_k = 16, k \geq 0$) to reach the same accuracy as in Fig 2. We can also see, that we gain in each of these two phases only about 5 digits due to the smaller n_0.

In Fig 3 we also notice the difference between the TEA1 and the other two extrapolation methods, TEA2 and VEA. While by TEA1 the contributions of the largest eigenvalues of the iteration matrix disappeared after the extrapolation scheme, we cannot observe this effect by TEA2 and VEA, but the two latter methods gain more accuracy during their extrapolation phase.

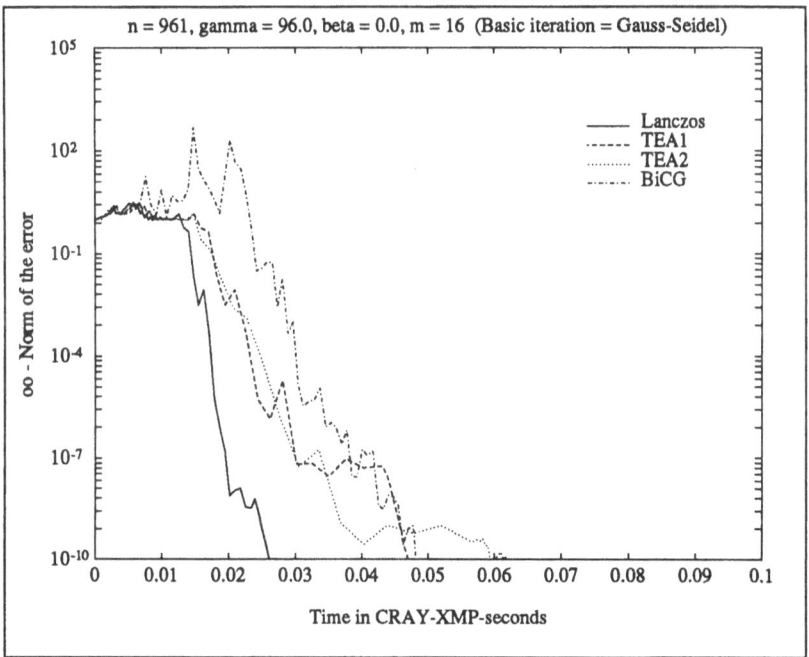

Figure 4: $\delta = 2, n_0 = 35$, related to time

Looking at Fig 4, where the same computation is shown as in Fig 2 but related to time, we see, that all four compared algorithms (Lanczos, TEA1, TEA2 and BiCG) use about the same time (Lanczos is fastest in this example). But the big advantage of the ε-algorithms is, that they don't need the transposed operator of the system!

In the next example (see Fig. 4), we want to show, that the extrapolation works, even if the basic iteration is not convergent. For that, we have chosen with $\gamma = 128$ ($\delta = 2$) and $\beta = 0$ a more unsymmetric example. We are again using Gauss-Seidel as the basic iteration. The width of the extrapolation is always $m = 14$ (based on 28 basic iteration steps) and we always compute $n_k = 23$ intermediate Gauss-Seidel steps. We again see, that after the extrapolation steps of TEA1 or Lanczos, the convergence rate of

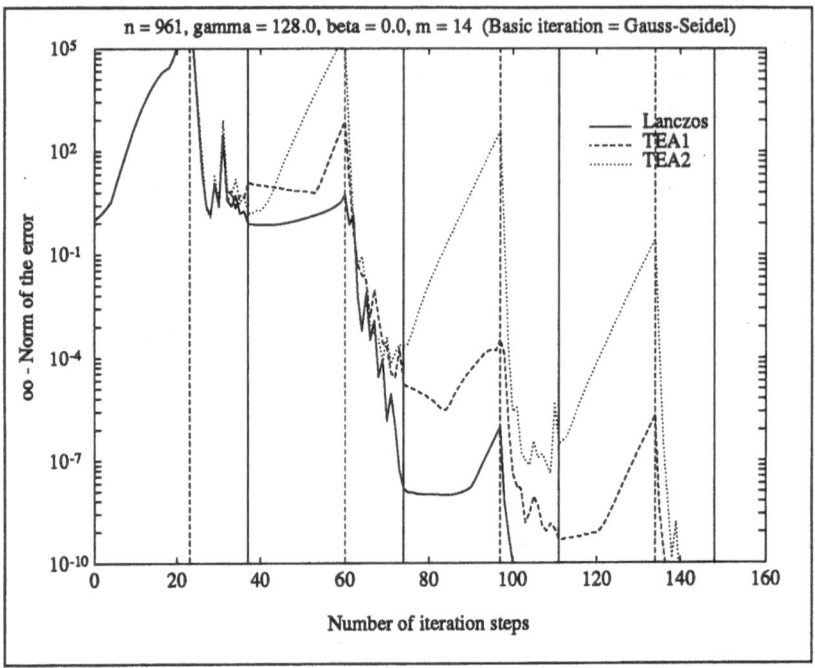

Figure 5: $\delta = 2$

the basic iteration is better for some steps, until the contribution of the largest eigenvalue to the error reappears.

We have observed, that the more Gauss-Seidel steps are performed, the more effective the extrapolation works. Without Gauss-Seidel steps, we would use in this example at least 30 extrapolation steps, that means an extra storage of 60 vectors. This storage is also used when solving the problem with GCR(30).

Conclusion

- We have seen, that the deeper the ε-scheme is computed, the more cancellation occurs. Therefore only some rows and columns of the ε-scheme contain satisfactory approximations. That requires finding stable methods to compute this scheme (like BiCG, which seems to be a good algorithm for TEA2), or one has to transform the formulas in numerically favorable forms. (e.g. by extrapolating the differences u_i of the basic iteration x_i instead of the basic iteration itself.)

- The ε-algorithm is motivation to split the computation into two parts and to use a truncated extrapolation. The BiCG and Lanczos methods don't emphasis this point so this is an interesting new aspect.

- The main problem in the truncated extrapolation is to find the best values for n_k and m_k in every step, preferably in an adaptive way. The values of n_k and m_k can easily be changed during the computation, because the ε-scheme is (favorably) computed row wise. This tuning of n_k and m_k needs more research.

- It seems to us, that the extrapolation works very well even if the system is singular. Such a problem has been discussed in [4] by using the ε-algorithm, but in this area there are still many open questions.

References

[1] C. Brezinski: *Accélération de la Convergence en Analyse Numérique,* Lecture notes in Mathematics 584, Springer 1977

[2] A. Buja, Hastie and Tibshirani: *Linear Smoothers and Additive Models,* Annals of Statistics, Vol. 17, No. 2, June 1989.

[3] J.P. Delahaye: *Sequence Transformations,* Springer 1988

[4] W. Gander and G. Golub: *Discussion of Buja A., Hastie and Tibshirani: Linear Smoothers and Additive Models,* Annals of Statistics, Vol. 17, No. 2, June 1989

[5] W.F. Ford and A. Sidi: *Recursive Algorithms for Vector Extrapolation Methods,* Applied Numerical Mathematics 4, (1988), p. 477-489.

[6] H. Ortloff: *Asymptotic Behaviour and Acceleration of Iterative Sequences,* Numer. Math. 49, (1986), p. 545-559.

[7] Y. Saad: *The Lanczos Biorthogonalization Algorithm and other oblique projection methods for solving large unsymmetric systems,* SIAM J. Numeri. Anal., Vol. 19, No. 3, June 1982.

[8] Y. Saad: *Preconditioning techniques for nonsymmetric and indefinite linear systems,* Journal of Computational and Applied Mathematics 24, (1988), p 89-105.

[9] A. Sidi: *Convergence and Stability Properties of Minimal Polynomial and Reduced Rank Extrapolation Algorithms,* SIAM J. Numeri. Anal., Vol. 23, No.1, February 1986, p. 178-196.

[10] A. Sidi and J. Bridger: *Convergence and stability Analysis for some Vector Extrapolation Methods in the Presence of Defective Iteration Matrices,* Journal of Computational and Applied Mathematics 22, (1988), p. 35-61.

[11] A. Sidi: *Extrapolation vs. Projection methods for linear Systems of Equations,* Journal of Computational and Applied Mathematics 22, (1988), p. 77-88.

[12] A. Sidi: *Application of Vector Extrapolation Methods to Consistent Singular Linear Systems,* Technical Report # 540 Technion, February 1989.

[13] A. Sidi and M.L. Celestina: *Convergence Acceleration for Vector Sequences and Applications to Computational Fluid Dynamics,* NASA Technical Memorandum 101327, ICOMP-88-17, 1988

[14] A. Sidi, W.F. Ford and D.A. Smith, *Acceleration of Convergence of Vector Sequences,* SIAM J. Numeri. Anal., Vol. 23, No.1, February 1986, p. 178-196.

[15] D.A. Smith, W.F. Ford and A. Sidi: *Extrapolation Methods for Vector Sequences,* SIAM Review, Vol. 29, No.2, June 1987, p. 199-233.

[16] D.A. Smith, W.F. Ford and A. Sidi: *Correction to "Extrapolation Methods for Vector Sequences",* SIAM Review, Vol. 30, No. 4, December 1988

[17] R.C.E. Tan: *Implementation of the topological ε-Algorithm,* SIAM J. Sci. Stat. Comput. Vol. 9, No. 5. September 1988

[18] R.S. Varga: *Matrix Iterative Analysis,* Prentice-Hall Inc., New Jersey, 1962

[19] J.H. Wilkinson: *The Algebraic Eigenvalue Problem,* Claredon Press, Oxford, 1988

SUPER PARALLEL ALGORITHMS

D. Parkinson

Active Memory Technology Limited

and Queen Mary College London

ABSTRACT

Serial algorithmic design concentrates on how to solve a single
instance of a given task on a single processor. The natural
extension for parallel processing usually concentrates on how
to solve the given single problem on a set of processors. When
algorithms for massively parallel systems are designed it often
occurs that the algorithm for parallel solution of a given
problem turns out to be more general and automatically solves
multiple instances of the task simultaneously. These algorithms
we call Super Parallel algorithms. This paper discusses some
examples of Super Parallel Algorithms.

INTRODUCTION

Parallelism in computers exploits the fact that we want to be
able to perform multiple operations! Hence at one level the
desire to perform not just one multiply but many multiplications
gives us the opportunity to apply some special architecture to
the task of speeding up the set of multiplications. The two
common routes to parallelism are the vector pipeline and
multiple processors. In both types of system the time to
multiply a single pair of numbers is not a good measure of the
time to create the product of a number of pairs of numbers.

It is well known that a vector pipeline has a start-up time (or
latency) so the time for multiplying N pairs of numbers is
$(s+Nt_v)$, where s is the start-up time and t_v is the 'beat time'

NATO ASI Series, Vol. F 62
Supercomputing
Edited by J. S. Kowalik
© Springer-Verlag Berlin Heidelberg 1990

for the pipeline. Hence a first level approximation to the time for performing N repeats of a single operation is $(s+Nt_v)$. This is however an over-simplification for in practice there is a maximum length of vector that any given vector architecture can perform in a single activation. If this maximum is p_v then we must express N as $M_v p_v + q_v$ with $0 \leq q_v < p_v$. The time to perform the total task is therefore $M_v(s+p_v t_v) + (s+q_v t_v) = (M_v+1)s + Nt_v$.

If we have a multi-computer configuration with p_m processors each with an operation time of t_m then, if $N = M_m p_m + q_m$, the operation time on a multiple processor is

$(M_m+1) t_m$.

Let us define the speed-up due to parallelism (S) as the ratio

S=(time to compute N results singly)/(time to compute the results using the parallel facilities).

For vector pipelines we have

$$S_v = N(s+t_v) / ((M_v+1)s + Nt_v)$$

which is approximately $1+(s/t_v)$ if $p_v \gg 1$. Hence the benefits of the vector pipeline approach are greatest when the start-up time is long compared with the beat time.

For multiple processor systems we have

$$S_m = N_m / (M_m+1)$$

which is approximately p_m. The qualitative difference between S_v and S_m is one of the reasons why vector processing and parallel processing should be considered as separate (and parallel) areas of study.

So far we have only considered using parallel processing for a trivial "single vector" type of problem. The usual parallel algorithm design study is based the use of parallel machines to solve a more complex problem eg. sorting, matrix multiplication, polynomial evaluation, solving sets of linear equations, etc. There is a large and continually growing literature of this type of study.

A major flaw in most studies is that they concentrate on trying to evaluate the maximum number of processors that can be used to implement the given task. For example there is a well known result that matrix multiplication can utilize not more than n^3 processors; therefore, providing one ignores the intercommunication costs, a time proportional to $\log_2 n$ can be expected when multiplying n*n matrices.

These types of result are somewhat pure mathematical existence theorems, often with little practical relevance. For matrix multiplication, in particular, the neglect of inter-processor communications has been criticized by Gentleman[1]. Other problems with these types of result include the failure to give any hint of what to do in the case when the number of processors is not equal to that indicated by the method.

In practice two cases are of interest

a) The number of processors is less than the number suggested ie. we only have n or n^2 processors available for matrix multiplication. This is going to be the usual case when we are interested in large matrix manipulations. The increased parallelism will encourage us to attempt to solve larger applications, and matrices of size 200*200 will not be considered especially large. Theoretically we can use 8 million processors for multiplying such matrices but it is unlikely that we shall have them. It is more likely that we will have access to only 40,000 or 200 processors.

b) The converse of the previous case is also of great interest. The size of matrices that we are dealing with may be only say 4*4 and we have 40,000 processors. The question is now one of an embarrassment of riches, how can we utilize the excess number of processors? This paper is about that problem.

The question of utilizing a larger number of processors is in many ways just the next layer of parallel thinking above the basic question of how we utilize a few processors. The parallelism as discussed initially came from the realization that we had lots of multiplications to perform and the fact that there were lots of multiplies could allow us to use parallel computational facilities. The same argument can be applied now to the problem of our 4*4 matrix multiplications. Many of the applications for which 4*4 multiplication time is important will require us to compute many matrix multiplications, and so we can try and perform many of them simultaneously. In other instances we may which to perform different operations such as polynomial evaluations, FFTs etc. In the terms of programs this may mean that we are examining outer loops as the source of parallelism. This statement appears to me to be rather obvious and banal but seems to be forgotten by some people who artificially seem to limit their consideration of parallelism to the just one algorithmic level above primitive operations, such as add multiply, compare etc.

EMBARRASSINGLY PARALLEL

The most trivial examples of the use of higher level parallelism occurs in the so-called " embarrassingly parallel " problems where one uses the parallel multiple processors to simply perform many repeats of a serial algorithm. Much of the embarrassment seems to be for those people who have not expected their application to be amenable to parallel processing. Embarrassingly parallel problems occur very frequently and should always be searched for. A particular advantage of embarrassingly parallel algorithms is the fact that they operate by performing many repetitions of a serial algorithm simultaneously; for this reason they are often called multi-serial algorithms.

The only snag with embarrassingly parallel problems is sometimes a need for access to large amounts of data, a requirement which might be unsatisfactory, in say real time signal processing, due to an effective increase in the latency. A typical example is in FFT analysis performed on input data streams. If we have to perform FFTs of length say 256 on input signals and we have 10000 processors then, in one sense, the optimal algorithm is to perform one FFT per processor over a set of 10000 samples. Such a strategy implies the collection of 10000*256 measurement points before processing can start. If the data were complex at

4 bytes precision real and imaginary, then 40 Megabytes of data would need to be collected before this " computationally optimal" algorithm can be started. Such a demand might imply a latency time unacceptable for the application.

MULTIPLE PARALLEL ALGORITHMS SUPER PARALLEL ALGORITHMS

For the above reasons it is often the case that we wish to perform simultaneously a number of repeats of a given sub-task using a given processor set. The practical application task confronting somebody implementing an application is therefore of the form:-

" What is the best strategy for solving N repeats of sub-problem Y using P, processors where each of the sub-problems Y has a characteristic size parameter M"?

P is usually a fixed number for a given system at a given time, but the programmer may wish to write a program which will be portable across a range of processor sizes. For small degrees of parallelism it may be reasonable to assume that P can change in steps of unity, but for massively parallel systems (such as the DAP or Connection Machine) P will only change by multiplying factors such as 4.

The problem is, as posed, characterized by a minimum of 3 parameters (N,M,P) and sufficient is known about parallel algorithm design to suggest that different algorithms are optimal in different parts of the parameter space:

If P=1 we are in the serial regime and the best serial algorithm is optimal.

If P=N we are in the embarrassingly parallel regime and again the serial algorithm is optimal.

If N=1 then we are in what we might call the Classical Parallel Algorithm design regime and the literature alluded to above gives many clues about optimal algorithms.

Special cases exist if P divides N exactly or if P has a 'nice' relationship with M.

Currently there seems to a rather naive view prevalent that somehow one should be able to express in a High Level Language this maximally parallel algorithm, and that clever compilers will be able to optimize the implementation onto any particular target set of P processors. The fallacy in this viewpoint is evident when one realizes that such a compiler should, by definition, be able to optimize the use of a single processor and therefore should be able to deduce the optimal serial algorithm from the maximally parallel algorithm! Such a faith in the cleverness of compiler technology is very touching, it is particularly strange to find that faith so prevalent in the high performance FORTRAN community who for years have investigated the tricks which will bluff Fortran compilers into producing efficient code.

The task we are interested in is therefore that of writing algorithms which solve not just one instance of a problem but solve in parallel many instances of the problem. We call these algorithms super parallel. A typical scenario in which the need for super parallel algorithms arises naturally is that of solving sets of tri-diagonal equations. The solution of a set of tri-diagonal linear equations is a common sub-task in many problems derived from partial differential equations arising from 3-point approximations to second derivatives. An efficient tri-diagonal solver is therefore a useful tool for many applications. Parallel algorithms for solving tri-diagonal linear equations have been extensively studied, Hockney and Jesshope[2] is a good source for discussion of the problem. Indeed Hockney and Jesshope do discuss the problem of solving m sets of tri-diagonal systems with n equations, although their study is more closely related to the problem of choosing the optimal algorithm for vector processors. The optimal parallel algorithm for solving one set of tri-diagonal linear equations given n processors is the Cyclic Reduction Algorithm and uses N processors.

Our concern here is as much with programming as with theoretical analysis, so we consider a tri-diagonal set of linear equations as being defined by four one-dimensional arrays A(N), B(N),

$C(N)$, $D(N)$ which define a set of tri-diagonal equations which we write algebraically as

$$b_i x_i = a_i x_{i-1} + c_i x_{i+1} + d_i \qquad i=1\ldots\ldots\ldots n$$

with $a_1 = c_n = 0.0$

The cyclic reduction algorithm transforms these equations into a new set by the operations

$$A_i = a_i/b_i$$
$$C_i = c_i/b_i$$
$$D_i = d_i/b_i$$

then

$$a'_i = \quad A_i A_{i-1}$$
$$c'_i = \qquad\qquad C_i C_{i+1}$$
$$b'_i = 1 \ -A_i C_{i-1} - C_i A_{i+1}$$
$$d'_i = D_i - A_i D_{i-1} - C_i D_{i+1}$$

The equations we now have are

$$b'x_i = a'x_{i-2} + c'_{i+2} + d'_i$$

The algorithm repeats this transformation $\log_2 n$ times to produce the result.

Programming the algorithm in FORTRAN 88 is instructive since the operations all take place on the data structures a,b,c,d. We can write the inner loop as

```
K=1
DO ISTEP=1,LOG2N
      A=A/B
      C=C/B
      D=D/B
      B=1-A*EOSHIFT(C,1,-K)-C*EOSHIFT(A,1,K)
      D=D-A*EOSHIFT(D,1,-K)-C*EOSHIFT(D,1,K)
      A=   A*EOSHIFT(A,1,-K)
      C=                      C*EOSHIFT(C,1,K)
      K=2*K
   END DO
```

The solution is D/B.
An alternative is to use the WHILE construct.

```
K=1
DO WHILE K.LT.N
      A=A/B
      C=C/B
      D=D/B
      B=1 -     A*EOSHIFT(C,1,-K) - C*EOSHIFT(A,1,K)
      D=D -     A*EOSHIFT(D,1,-K) - C*EOSHIFT(D,1,K)
      A=     A*EOSHIFT(A,1,-K)
      C=                      C*EOSHIFT(C,1,K)
      K=2*K
   END DO
```

The benefit of this version is the elimination of the need to compute the variable LOG2N.

One way of describing the functioning of the cyclic reduction algorithm is to consider that each step doubles the number of zeros between the central diagonal and the off-diagonal terms. After $\log_2 n$ steps the off-diagonal terms disappear (become all zero) and we can alter the program to reflect this view to the following:-

```
K=1
DO   WHILE ANY(A.NE.0.OR.C.NE.0)
     A=A/B
     C=C/B
     D=D/B
     B=1-A*EOSHIFT(C,1,-K)-C*EOSHIFT(A,1,K)
     D=D-A*EOSHIFT(D,1,-K)-C*EOSHIFT(D,1,K)
     A=   A*EOSHIFT(A,1,-K)
     C=                    C*EOSHIFT(C,1,K)
     K=2*K
   END DO
```

As Hockney and Jesshope point out one can,,if the equations are diagonally dominant, consider replacing the tests versus zero with a test versus some tolerance factor. The code we have produced has a more surprising property when we realize that there is no explicit reference to the number of equations being solved! At first glance this may seem to be a minor quirk with no great significance, but it hides an observation of much greater significance.

<u>The code that we now have will not only solve one tri-diagonal system it will solve any number of tri-diagonal systems of any combination of lengths, and the number of times that the loop will be traversed is given by the logarithm of the size of largest system in the set!</u>

Although we originally set out to write an algorithm to solve a single instance of our problem we have developed an algorithm which solves multiple instances of the same problem. An algorithm with this property is a SUPER PARALLEL algorithm.

A given tri-diagonal system is defined by the four linear data structures a,b,c and d.

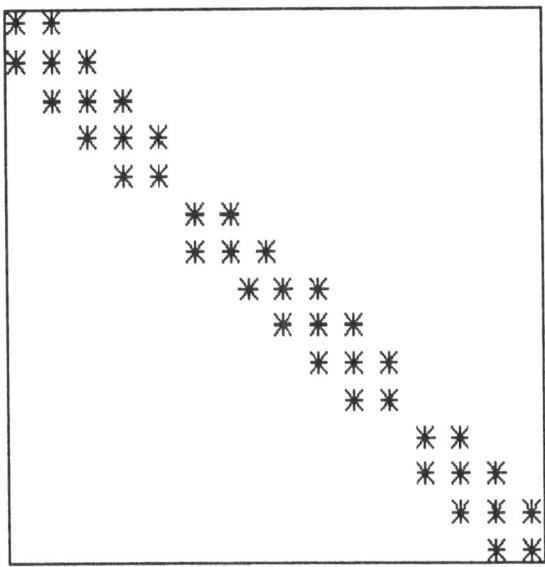

Figure 1 Multiple Tridiagonals

A set of tri-diagonals is usually considered to be defined by sets of these data structures. We can however concatenate all of these small sets into a grand set; typically, in diagrammatic notation, we would have a matrix of the structure shown in Figure 1.

This structure has an overall tri-diagonal structure but does in fact define three distinctly separate tri-diagonal systems with 5,6 and 4 elements respectively.

Applying the general algorithm to the above grand matrix solves the 3 subsets after the loop has been traversed three times.

At first glance the result may seem like magic, but a little thought shows that we should have expected the effect. It is well known that the first step of the cyclic reduction algorithm splits the input problem of solving a system of N equations into two separate systems of about N/2 equations, one set of equations connecting the even numbered unknowns and the other system connecting the odd numbered unknowns. The parallel cyclic reduction algorithm does not exploit this reduction but solves the two distinct sets simultaneously. It is therefore no great surprise that the algorithm handles multiple sets of equations.

One advantage of the above approach comes from the observation that tri-diagonal systems seldom occur in single sets but occur naturally in multiples   Solving multi- dimensional partial differential equations by ADI methods naturally gives rise to sets of tri-diagonal systems.

DATA DEFINING PROBLEMS

The above use of a 'single' solver to solve a multiple set of equations occurs because the data defines the way that the problem is structured. In a very simple way we have used the data to define the 'boundary conditions' and their locations. This technique is a very valuable technique in parallel processing and can be used very effectively to provide parallel algorithms for problems with complex geometries and boundary conditions.
Consider the case of solving a field problem on a grid.
A typical sequential program usually has two parts
a) operate on the interior points and
b) operate on the boundary points.
There are usually far fewer boundary points that interior points and there are often different types of boundary points which the serial programmer treats with different program segments.
The programmer of a parallel system tries to treat all points, interior and boundary simultaneously and usually does this by means of the 'data' rather than the program.
As a simple example we can consider a elliptic equation on a rectangular mesh with a five point difference scheme.
An interior point is related with its neighbors by a relation of the general form.

$$a_{i,j}x_{i-1,j}+b_{i,j}x_{i,j-1}+c_{i,j}x_{i,j}+d_{i,j}x_{i,j+1}+e_{i,j}x_{i+1,j}=f_{i,j}$$

The boundaries may often be included in general form by appropriate manipulation of the local values of the coefficients a,b,c,d,e,f.

For instance fixed points are forced by setting a=b=d=e=0. Derivative conditions are accommodated by carefully setting the appropriate coefficients.

OTHER EXAMPLES OF SUPER PARALLEL ALGORITHMS

It is not difficult to find other examples of Super Parallel examples; most recursive doubling type of operations are ideal candidates. Hence we can use the above methods to solve sets of recurrence relations of the type $x_i = a_i x_{i-1} + b_i$,

$$x_i = (a_i x_{i-1} + b_i) / (c_i x_{i-1} + d_i).$$

The characteristic of these types of operations that we are exploiting is the self- contained nature of the definition, or the possibility of defining the boundary conditions as part of the problem.

Consider the case of first order recurrence relations. One can consider the equations as either a set of relations $x_i = a_i x_{i-1} + b_i$ i=1.....n with x_0 specified, or as a set of relations $x_i = a_i x_{i-1} + b_i$ i=0......n with $a_1 = 0, b_1 = x_0$. This second approach is better for parallel processing.

Those Super Parallel algorithms that we have so far discovered exploit the fact that there is some data structure in the problem which can be used to control the operations. Although there are algorithms which have the recursive doubling structure, they do not have such a control data structure. An example is the FFT algorithm. Although the FFT algorithm functions by splitting the problem of computing an N point FFT into two disjoint problems of size N/2, there does not appear to be a variable which would allow us to easily extend the algorithm to have the Super Parallel property.

References

1]Gentleman, W. M., 'Some complexity results for matrix computations on parallel processors' JACM **25** 112-115(1978)

2]Hockney, R. W. and Jesshope, C. R., Parallel Computers 2 Adam Hilger (1988)

VECTORIZATION AND PARALLELIZATION OF
TRANSPORT MONTE CARLO SIMULATION CODES

Kenichi Miura
Vice President
Computational Research
Fujitsu America, Inc.
3055 Orchard Drive
San Jose CA, U.S.A.

1. INTRODUCTION

In recent years, the demand for solving large scale scientific and engineering problems has grown enormously. Since many programs for solving these problems inherently contain a very high degree of parallelism, they can be processed very efficiently if algorithms employed therein expose the parallelism to the architecture of a supercomputer.

Today's supercomputers such as CRAY X-MP,CRAY Y-MP, CRAY 2, FUJITSU Vector Processor Systems, Hitachi S-820 System and NEC SX System, mainly depend on the vector processing approach to boost their performances, with parallel processing capabilities besides vector processing in some cases[1-4]. One such example is shown in Fig. 1[4].

Supercompulers have successfully exhibited very high performance for applications such as solving the partial differential equations and signal processing. These machines, however, usually give only the scalar performance for the existing Monte Carlo ·simulation codes for solving problems in the neutron and radiation transport in nuclear engineering, phase-space simulation and cascade shower simulation in high energy physics [5-6]. The Ray tracing technique in computer graphics may also be included in this category. In order to fully utilize the architecture of a supercomputer for such applications, development of suitable algorithms is very important.

In this paper, we will discuss the algorithm issues as well as the software engineering issues involved in the high-performance computation of the transport Monte Carlo simulation, particularly how the parallelism in the application programs are to be matched with a given supercomputer architecture. We will also introduce a new law which is suitable for the

NATO ASI Series, Vol. F 62
Supercomputing
Edited by J. S. Kowalik
© Springer-Verlag Berlin Heidelberg 1990

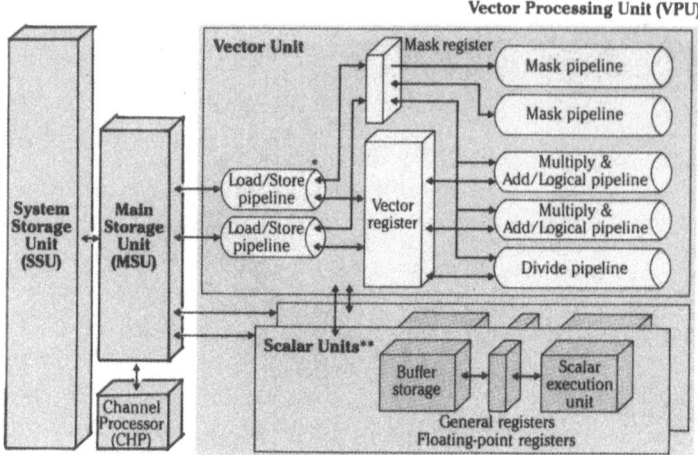

Fig.1 An Example of Vector Architecture (FUJITSU VP2000 Series)

performance characteristics in the <u>asynchronous parallel processing</u>, as an alternative to Amdahl's law in the <u>synchronous parallel processing</u>. The techniques and issues discussed here are not solely for the Monte Carlo applications, but should be applicable to other applications as well.

2   VECTOR PROCESSING OF TRANSPORT MONTE CARLO SIMULATION CODES

2.1 Scalar Nature of Code Structure

In a typical transport Monte Carlo code, each particle is transported through the media, boundary crossing is checked in a given geometrical structure, then it encounters some interactions. This process is continued until the particle escapes from the structure, or until there is no further interest in this particle. Processing of each particle involves many data-dependent IF tests, due to the stochastic nature of the computational model of physical interactions, hence leaving very little parallelism within the particle loop.

It should be noted, however, that a very high degree of parallelism exists at the particle level, since there could be thousands of particles to be simulated, each of which can be treated independently of others.

due to the above-mentioned reasons, the transport Monte Carlo simulation has generally been regarded as a perfect application for the parallel processing rather than for vector processing.

2.2 Basic Strategy for Vector Processing

As the vector supercomputers become widely available, quite a few reports have been recently made regarding the efforts in vectorizing the transport Monte Carlo simulation codes[7-13], most of which strongly indicate that vector approach is indeed worthwhile.

The basic strategy for vectorization is similar in all the reported works, namely to pool particles in a common data structure called stack or queue, and to form vectors with particles possessing identical characteristics by gathering them from the common data structure. In this way, many particles can be processed in one pass in vector mode. In order for this strategy to be successful, it is very important to carefully design the data structure so that the vectorized algorithms can exploit the parallelism contained in the problem; the scalar Monte Carlo codes in many cases adopt inherently sequential data structure (typically a last-in first-out buffer or a push-down stack).

The efficiency of vector processing also depends on the varieties of interaction patterns and/or complexities of geometry, both of which strongly influence the complexity of the codes as well as the effective vector length at each step of simulation. In the actual vectorized codes, most of the loops are heavily populated with the nested IF-THEN-ELSE structures, and the compiler's capability to vectorize such complex loops is essential in obtaining a good vector performance. In practice, code-restructuring involves a deep understanding of the codes and takes considerable amount of effort, since the Monte Carlo codes are usually very large in size.

The reported vector performance improvement over the scalar ranges anywhere from 1.4 to to 85, most of them falling between 5 and 10 [13].

2.3 Random Number Generator

The most commonly used technique for random number generation in the transport Monte Carlo codes is the congruential method due to its simplicity [14-15]. In EGS4 code[6], for example, the following multiplicative congruential method is used:

Loop over i

 Iseed = A * Iseed modulo 2^{32} (Random seed in integer format)

 Ran = Iseed * 2^{-32} (Normalized floating-point random number),

where A = 663608941.

Although this algorithm may seem recursive, it can be easily vectorized if the multiplicative coefficients (A, A^2, A^3, ...A^N) modulo 2^{32} are pre-calculated and stored in an array. At the time of random number generation, each element of this array (up to the desired number of ramdom numbers not exceeding N) is multiplied by the current value of Iseed and the resulting integer random numbers are normalized to obtain the floating-point random numbers, all in the vector mode. Only the last integer random number needs to be stored as the seed for future use. Similar technique is also applicable to the linear congruential method [16].

While the vectorized random number generator generates the identical random number sequence as the original scalar algorithm, the vectorized Monte Carlo code and the scalar Monte Carlo code do not necessarily produce the identical simulation results since the order in which the random numbers are used may be different in two cases.

2.4 Vectorization of Electromagnetic Cascade Shower Simulation Code EGS4

 This section describes EGS4-V, a vector version of the electromagnetic cascade shower Monte Carlo code EGS4, developed by us. The vector supercomputer used for this research is AMDAHL 1200 Vector Processor System with FORTRAN77/VP Vectorizing Compiler [17].

 2.4.1 Overview of EGS4 Code

EGS4 is the latest version of the EGS (Electron-Gamma Shower) Code System which has been developed at the Stanford Linear Accelerator Center by W.R. Nelson, et al.[6]. This code system is a general purpose package for the Monte Carlo simulation of the coupled transport of electrons and photons on an arbitrary geometry. EGS4 is widely used in high energy physics (simulation of electromagnetic cascade showers) and in medical physics.

An electromagnetic cascade shower starts with one particle with very high energy (say, above 1 Gev) which subsequently creates many particles through radiation and collision (Fig. 2). Particles in a shower are transported through the media, and are eventually discarded as they lose their energy below a prescribed threshold through collision and radiation

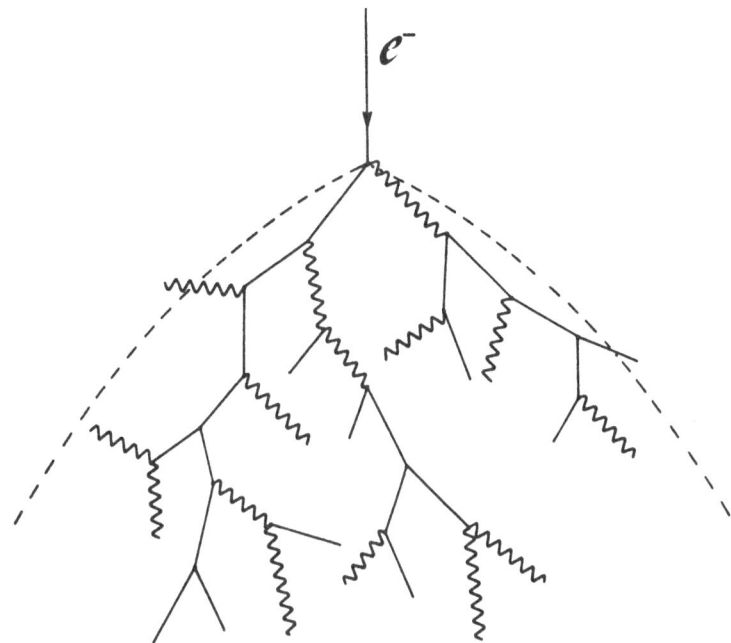

Fig.2 An Electromagnetic Cascade Shower

processes or as they escape from the geometrical structure. The analog
Monte Carlo approach has been adopted in EGS4, and all the multiplicative
processes are simulated.

2.4.2 Timing Measurement of a Sample Problem

A sample problem has been run on AMDAHL 1200 VP System to measure the
vector performance of EGS4-V against the original scalar EGS4. In this
problem, 1 Gev electrons are injected into a lead block of infinite size,
so that no boundary crossing takes place. The number of the cases (that
is, the number of the incident electrons) has been varied from 10 to 200.

The results of the timing measurement are shown in Fig. 3. Since the
publication of the early results [11], the code has beeen improved, and an
asymptotic vector vs. scalar performance ratio of 11.6 has been obtained
for this measurement.

Fig. 4 depicts an interesting dynamic behavior of the active vector
length. The number of cases for this run is 100. The vector lengths have
been averaged over 10 simulation steps in this graph. In Fig. 4, the
vector length varies considerably in spite of the above-mentioned
smoothing process. The vector length starts from 100, reaches the

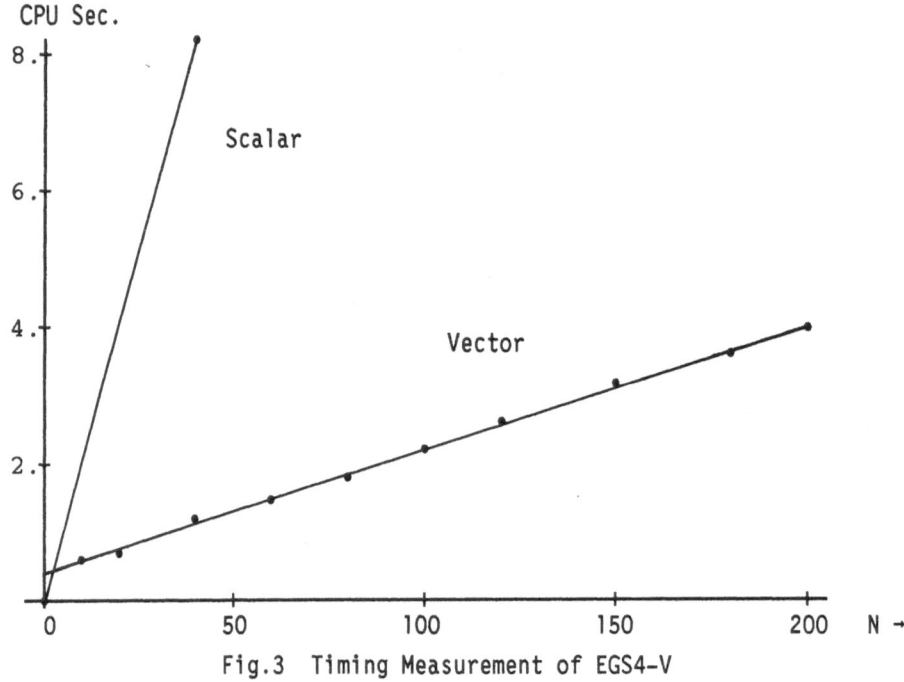

Fig.3 Timing Measurement of EGS4-V

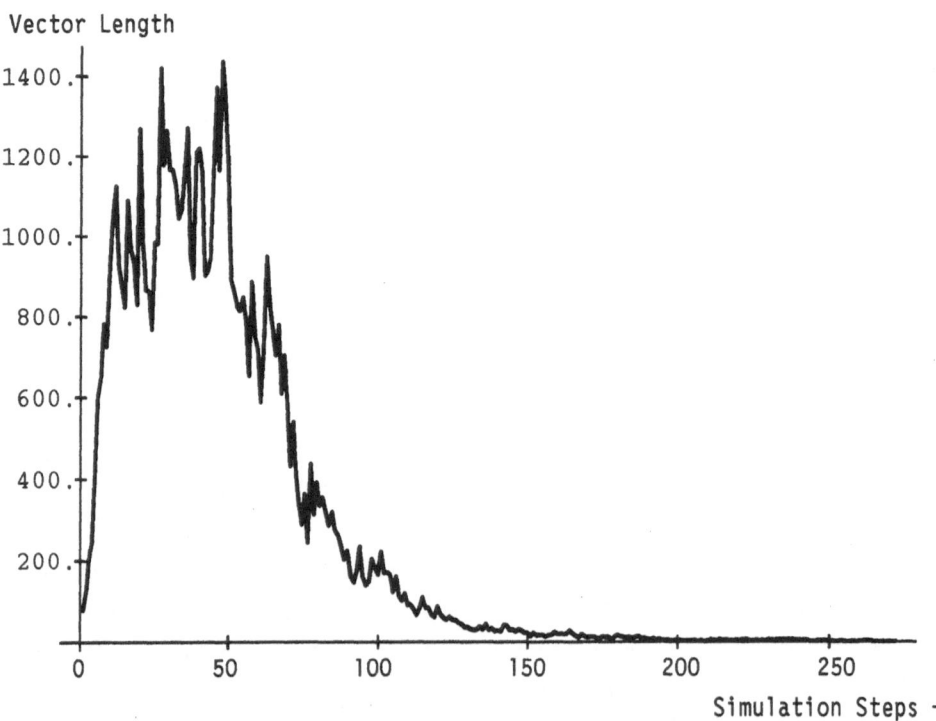

Fig. 4 Typical Dynamic Behavior of Vector Length

instantaneous peak of 2,113 (smoothed out and not shown in Fig. 4), and decreases with frequent spikes.

3   PARALLEL PROCESSING OF TRANSPORT MONTE CARLO SIMULATION CODES

3.1   Issues in Parallelizing Tranport Monte Carlo Codes

In this paper, we will confine our discussions to the parallel processing on the shared-memory architecture. Although the parallelization may seem conceptually more natural and straightforward than the vectorization for the transport Monte Carlo simulations, there are new issues in parallel programming at the same time. These issues are not solely confined to the Monte Carlo simulation, but of a more general nature. Some of them are addressed in the following.

3.1.1   Identification of global and private variables

In a shared-memory architecture, all the variables in the COMMON blocks must be carefully examined whether they should be shared among the processors (global variables), or privately copied for each processor (private variables). Each COMMON block may contain both types of variables, in which case it has to be split into two blocks. This is a very time-consuming task, if done manually. Definitely, good software tools are needed in this area. Another issue to be noted is that the notion of the COMMON block for parallel processing is not well established, and some systems do not support both types of COMMON blocks, hence the portability problem.

3.1.2   Machine-dependent library functions or synchronization primitives for parallel programming

Even within the category of the parallel processing systems with the shared-memory, each system has its own library functions, compiler directives or FORTRAN extensions to describe and/or control parallel processing. There is no standard in this area. This, again makes the porting of the parallel codes very difficult.

3.1.3   Parallel random number generation

Unless great care is taken that each particle uses the same sequence of random numbers in the parallel code as in the scalar code, results are not guaranteed to be the same. Worse yet, it is quite possible to

construct a parallel code which does not produce the same results from run
to run due to the effect of race conditions in obtaining random numbers.
Frederickson et al. proposed the concept called Lehmer-tree, based on two
sets of the linear congruential random number generators [18]. By
adopting this concept and by storing a random seed with each particle, the
same simulation results can be obtained regardless of the order in which
the particles are processed and with any number of processors. This is a
new area, and further research will be needed to establish algorithms for
generating good parallel random numbers.

3.2 Parallelization of EGS4 Code   - Two approaches -

In this section, we describe our experiences in parallelizing EGS4 on
Sequent B21000 Parallel Processing System[19]. There are two basic
approaches in parallelizing the EGS4 code: one is to parallelize the
original scalar code in such a way as to process many independent
particles in parallel (to be called fine-grain approach), and the other
approach is to start with the vectorized version and either to
process each loop in parallel (so called microtasking), or to process the
independently executable vectorized subroutines in parallel (so called
macrotasking, or large-grain approach).

3.3 Fine-grain Parallel Processing

In this approach, each processor fetches a particle from a shared
stack and executes the scalar simulation code. The synchronization is
done by locking and unlocking the stack pointer to the shared particle
stack, thus allowing dynamic load balancing. We have developed the
fine-grain parallel version of EGS4. The Lehmer-tree technique is
incorporated in this code. A sample problem was run with one 50 Gev
electron injected into a lead block of infinite size, and a parallel
speedup factor of more than 25 was obtained with 29 Sequent B21000
processors. (Fig. 5)

The sub-linear characteristics of the speedup curve in Fig. 5 can be
ascribed to the following factors:

(1) Initiation of the parallel tasks

When a new task is spawned, all the private common blocks have to be
replicated. This task replication time is proportional to the number
of tasks, while the total execution time itself is almost inversely
proportional to the number of the tasks. Therefore, this overhead

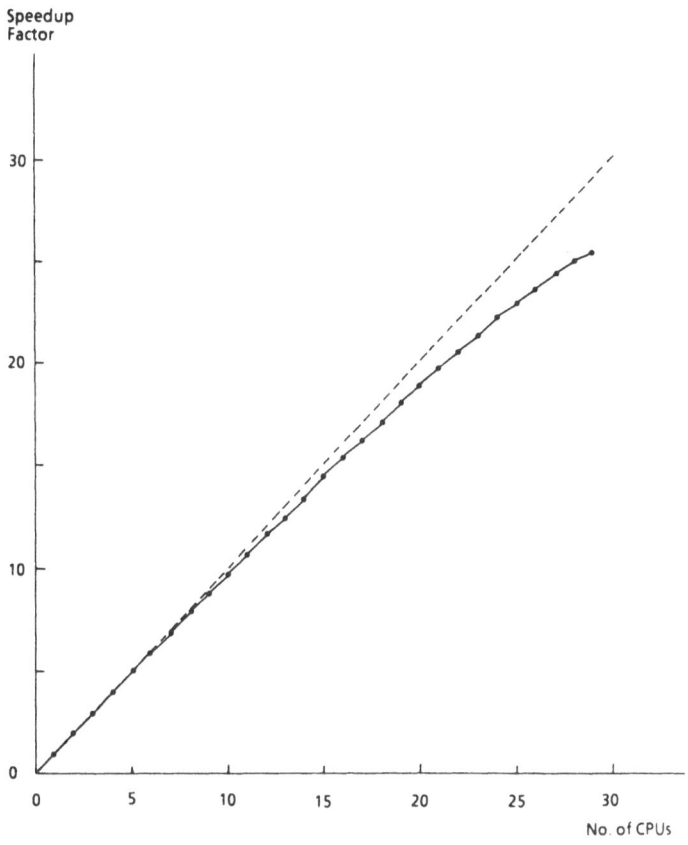

Fig. 5   Performance Measurement of Fine-grain Parallelized EGS4

becomes more significant as the number of the processors increases,
unless the problem size is sufficiently large.

(2) Asymmetric development of a cascade shower

The development of a shower is quitely asymmetric in the initial
stage, so that the degree of parallelism is low until sufficiently
large number of particles have been generated.

(3) Critical section of the code

There are inherently sequential operations in the child tasks due to
the exclusive accesses to the particle stack (the critical section).

(4) Physical resource contentions

In a shared-memory architecture, the common physical resources such
as the memory bus and the memory banks also requires accesses by only
one processor at a time.   The amount of the overhead due to this
factor heavily depends on the system parameters such as the cache

size, the memory bus throughput, the memory configuration etc., and are
usually not known to the users.

While the first two factors may become less significant as the size
of a shower increases, the third and the fourth factors always takes
certain percentage of the CPU time, regardless the size of a shower. In
our numerical experiments on the Sequent Parallel Processing System, the
fraction of the CPU time spent in the critical section is estimated to be
somewhere between 1 and 2 percent. The effects of the physical resource
contentions are not included in this estimate. We further analyze the
effect of the critical section to the speedup factor in Section 4.

3.4 Microtasking and Large-grain Multitasking Approaches

We have investigated the possibilities of parallel processing the
vectorized version of the EGS4 Monte Carlo code. So far, our experiments
revealed that the microtasking approach did not turn out to be attractive
due to the complexity of the DO loop structure and a lack of a software
tool at the time of this study. On the other hand, the large-grain
approach is more promising since the code structure of the vectorized
version of the EGS4 already incorporates independently executable
subroutines[11]. With the advent of the vector multiprocessor systems,
this approach seems to be the right one, and deserves further research.
We are now in the process of developing a large-grain vector-parallel
version of EGS4 code in collaboration with Prof. R.G. Babb of Oregon
Graduate Center.

4   ANALYTICAL PERFORMANCE MODEL FOR FINE-GRAIN PARALLEL MONTE CARLO CODE

In this section, we introduce a simple analytical model which is
suitable for interpreting the sub-linear characteristics of the speed-up
factor which was observed in the fine-grain parallel version of EGS4
code. Since we are in the process of preparing a more detailed report on
this subject[20], only the outline of the derivation and results are given
here.

4.1 Synchronous Model vs. Asynchronous Model for Parallel Processing

It should be emphasized that the commonly known Amdahl's law only applies when the parallelizable part of the code is processed simultaneouly by all the processors before entering the sequential part of the code. We will call this model synchronous parallel processing. Vector processing can be regarded as a special case of the synchronous parallel processing.

The Amdahl's law can be stated as follows[21]:

$$S_p = 1 \; / \; (1 - \alpha + \alpha \; / \; N) \qquad\qquad \ldots \ldots \; (1)$$

where S_p is the speedup factor, α is the fraction of the CPU time to be spent in the parallelizable part of the code, and N is the number of the processors. Fig. 6 illustrates typical speedup curves for the synchronous parallel processing for various α and N.

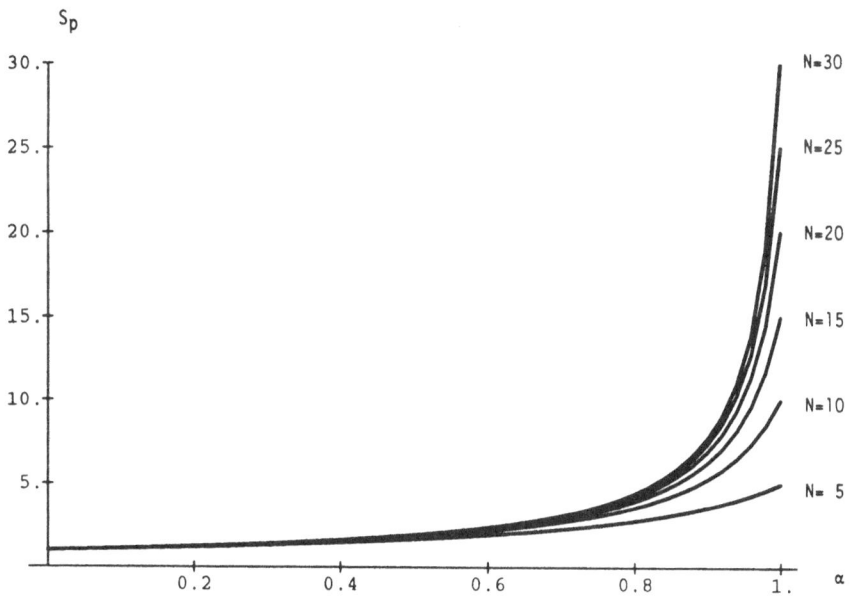

Fig. 6 Speedup Factor for Synchronous Model of Parallel Processing
(Amdahl's Law)

4.2 Analytical Speedup Factor for Asynchronous Model of Parallel Processing

In the fine-grain parallel version of the EGS4 code, on the other hand, processors can run independently of others as long as they do not access the particle stack. The access to the particle stack is exclusive, constituting a critical section, but only those processors which are just about to enter this critical section while it is busy, are kept waiting, not all the processors. In this sense, the parallel processing in this case may be dubbed asynchronous parallel processing.

The analytical formula for the speedup curve which is discussed here is based on the machine service model in the Queuing Theory[25,26], where the parallelizable part of the code (to be called parallel section hereafter) corresponds to the arrival of the events from the N sourses, and the sequential part of the code (to be called critical section hereafter) corresponds to a server. For the sake of the simplicity of the discussion, we assume the N-source/1-server Model with the Poisson arrival and exponential service time with the parameters λ and μ, respectively. Due to the stochastic nature of the execution time of the parallel and critical sections of the code, the parallelization ratio α should be interpreted as the statistical average, and is related to λ and μ as follows:

$$\alpha = \mu /(\lambda + \mu) \qquad\qquad \ldots\ldots (2)$$

Under the above assumptions, the analytical form of the speedup factor for the asynchronous model can be derived by utilizing the readily available results from the corresponding queuing model.

Fig. 7a illustrates the execution profile of the sequential code, from which Fig. 7b is obtained by separating the parallel section (P) from the critical section (C). Fig. 7c illustrates the execution profile of the parallelized version of the code, with the critical section separated from the parallel tasks. Here, W means that the corresponding task is waiting for the critical section to be released. Fig. 7c may be rearranged to obtain Fig. 7d, where the execution profile is sorted according to the number of tasks in the parallel section (P), from N to 0. Here, the k-th section in Fig. 7d (k < N) means that k processors are in P, while one processor is executing the critical section (C) and N-k-1 processors are in W. When k = N, all the processors are in P, and no processor is in the critical section. Since

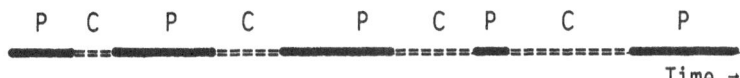

(a) Execution Profile for Sequential Processing (Time Domain)

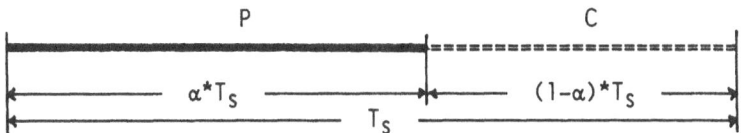

(b) Execution Profile for Sequential Processing (Sorted)

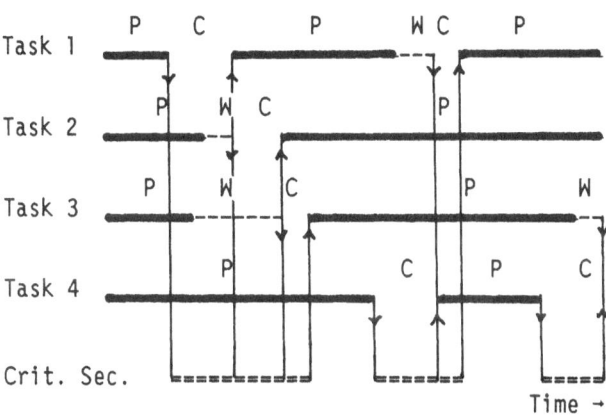

(c) Execution Profile for Parallel Processing (Time Domain)

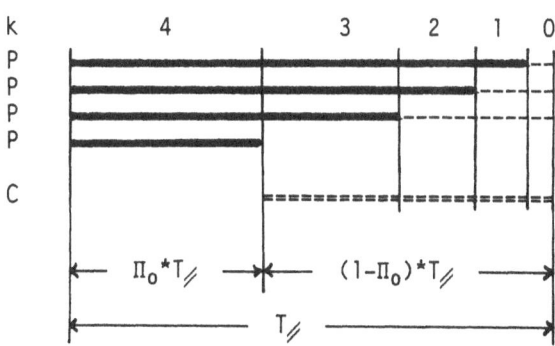

(d) Execution Profile for Parallel Processing (Sorted by k)

Fig. 7 Derivation of Speedup Factor for Parallel Processing (N=4)
(P: Parallel section, C: Critical section, W: Wait state)

the total CPU time to be spent in C should be the same for the sequential and for the N-way parallel processing cases, the speedup factor can be easily obtained with the analytical results in [22,23] as follows:

$$T_{\parallel}*(1-\Pi_0) = T_s*(1-\alpha) \qquad \cdots\cdots (3)$$

$$
\begin{aligned}
S_p = T_s/T_{\parallel} &= (1-\Pi_0)/(1-\alpha) \quad (0 < \alpha < 1) \quad \cdots\cdots (4)\\
&= 1 \qquad\qquad\qquad\ (\alpha = 0)\\
&= N \qquad\qquad\qquad (\alpha = 1)
\end{aligned}
$$

where $T_{\parallel}$ and T_s are the total CPU time for the parallel processing and the sequential processing, respectively, and Π_0 is the probability that no processor is in the critical section, which is a function of α and N as defined in the following:

$$\Pi_0 = 1/(1+N\rho+N(N-1)\rho^2+\ldots+N!\rho^N) \qquad \cdots\cdots (5)$$

and $$\rho = \lambda/\mu = (1-\alpha)/\alpha \qquad \cdots\cdots (6)$$

Fig. 8 illustrates the speedup curves for the asynchronous model of parallel processing.

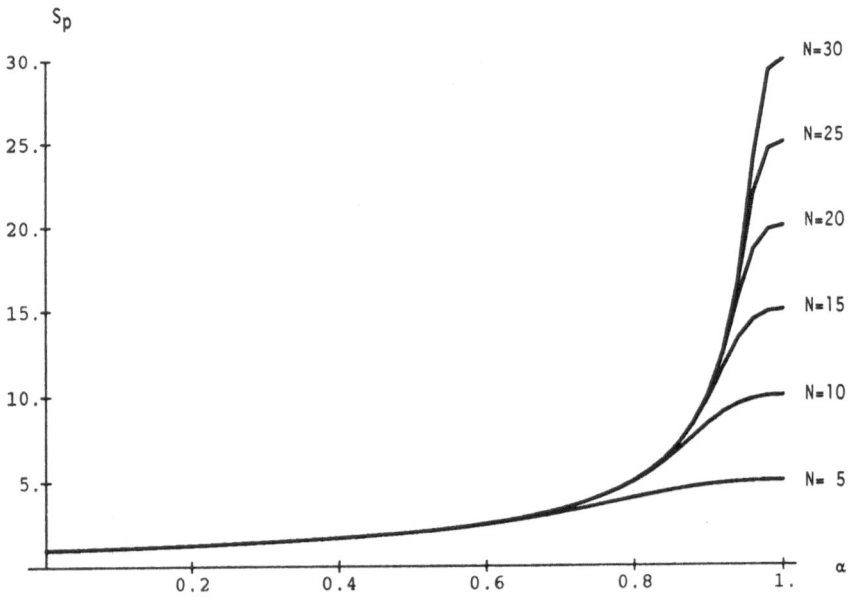

Fig. 8 Speedup Factor for Asynchronous Model of Parallel Processing

4.3 Discussions

By comparing Fig. 6 with Fig. 8, several interesting observations can be made regarding the similarities and the dissimilarities between the synchronous and the asynchronous models of parallel processing.

(1) In both cases, speedup factor is N when $\alpha = 1$.

(2) For a given N and $0 < \alpha < 1$, the asynchronous speedup factor is always greater than the synchronous one, and both are less than $1/(1-\alpha)$.

(3) As α decreases from 1, the asynchronous speedup factor initially gives a slow decay while the synchronous speedup factor gives a very sharp decay.

(4) For a given α and with an increasing N, the asynchronous speedup factor approaches to the asymptotic value $1/(1-\alpha)$ faster than the synchronous one.

(5) As α approaches 0, the asynchronous speedup factor rapidly converges to the synchronous one, both yielding 1 at $\alpha = 0$.

While the asynchronous speedup factor may depend on the actual probalility distribution of the CPU time spent the parallel and critical sections, the qualitative results as described above are believed to hold in general. With regard to our experimental data for the EGS4 code, our analytical model gives $1-\alpha = 3\%$ as the best fit. Since the critical section alone takes from 1 to 2 % of the total CPU time, the bus contention and the memory bank conflict are supposed to take the rest of the 3% according to this model.

5 CONCLUSIONS

This paper discussed vectorization and parallelization techniques for the transport Monte Carlo simulations. The fundamental differences between scalar coding and vector coding, and also between the scalar coding and parallel coding have been addressed, based on our own experiences. It has been pointed out that the transport Monte Carlo codes inherently contain a very high degree of parallelism, and that they can be either vectorized or parallelized efficiently.

As for the vectorization of the transport Monte Carlo codes, typical speedup factors of 5-10 have been reported in the literature. It should also be noted that parallelism at the higher level becomes visible through the vectorization process, which is quite suitable for vector-parallel processing. The vectorization techniques described in this paper are not solely confined to the transport Monte Carlo simulations, but should be applicable to other seemingly unvectorizable problems. The vector data handling capabilities which are accessible from FORTRAN language, are the key factors for implementing vector codes.

Although parallelization of the Monte Carlo codes may be more straight-forward than vectorization, a lot more research and development should be made in the programming environment in general, especially in compiler technology, in debugging tools, and in parallel software development tools which can provide useful information for efficient parallel programming. The necessity for a global scanning capability on the part of the compiler should be emphasized as the architectural trend moves toward various forms of parallel architecture, where the detection of so-called large granularity parallelism is required. Development of a fully automatic compiler may become impractical for such systems. Rather, user-friendly interactive software tools seem to be the right approach[24,25].

We have also introduced an analytical model of the asynchronous parallel processing for the cases where the commonly known Amdahl's law is not applicable. This model is suitable, for example, for describing the performance characteristics of the fine-grain parallel version of the EGS4 Monte Carlo code. With the advent of the asynchronous parallel algorithms in various application areas, the analytical performance models along the line of our approach should be examined for other cases.

Since supercomputing is an application-driven area, very close interactions between researchers in various applications and supercomputer manufacturers will be crucial in order to cope with the ever-increasing demands for large scale scientific and engineering computations.

ACKNOWLEDGEMENT

The author would like to thank Dr. W. R. Nelson of Stanford Linear Accelerator Center for providing EGS4 code and for valuable discussions, Dr. R. G. Babb II for valuable discussions on the fine-grain and the large-grain parallel processing techniques, Amdahl Corporation for providing the computational resources for the development and timing measurements of the vector version of EGS4, and Sequent Computer Systems, Inc. for providing computational resources for the timing measurements of the fine-grain version of EGS4.

REFERENCES

1. S. Fernbach (ed.), Supercomputers, North-Holland (1986).
2. R. Hockney and C. Jesshop, Parallel Computers 2, Adam Hilger (1988).
3. R. Mendez and S. Orszag (eds.), Japanese Supercomputing, Lecture notes in Engineering, 36, Springer-Verlag (1988), 111-127.
4. K. Uchida, VP2000 Series, in this proceedings.
5. R. Alcouffe et al. (Eds.), Monte-Carlo Methods and Applications in Neutronics, Photonics and Statistical Physics, Lecture Notes in Physics, 240, Springer-Verlag (1985).
6. W.R. Nelson, H. Hirayama and D.W.O. Rogers, "The EGS4 Code System", Stanford Linear Accelerator Center Report SLAC-265, December 1985.
7. F. Bobrowicz et al.,"Vectorized Monte Carlo photon transport", Parallel Computing 1 (1984) 295-305.
8. Y. Chauvet, "Multitasking a Vectorized Monte Carlo Algorithm on the Cray X/MP2", Cray Channels 6, No.3 (1984) 6-9.
9. W. Martin and F. Brown, "Status of Vectorized Monte Carlo for Particle Transport Analysis", The International Journal of Supercomputer Applications, 1, No.2 (1987) 11-32.
10. K. Miura: "Vectorization of phase space Monte Carlo code in FACOM Vector Processor VP-200", Computing in High Energy Physics (Eds. L.O. Hertzberger, and W. Hoogland), Elsevier Science publishers B.V. (1986).
11. K. Miura,"EGS4-V: Vectorization of the Monte Carlo cascade shower simulation code EGS4", Computer Physics Communications 45 (1987) 127-136.
12. K. Asai et al, "Vectorization of KENO-IV Code", Nuclear Science and Engineering, 92 (1986) 298.
13. B. Martin, "Particle Transport Monte Carlo on Shared-Memory and Distributed-Memory Parallel Processors", Proc. The Third International Conference on Supercomputing, 2 (1988) 348-353.
14. M. Kalos and P.A. Whitlock, Monte Carlo Methods, Vol.1, Wiley-InterScience, (1986).
15. D. Knuth, The Art of Computer Programming, Vol. 2 (2nd ed.), Addison-Wesley (1981).
16. T. Matsuura, K. Miura and M. Makino, "Supervector Performance without Toil", Computer Physics Communications 37 (1985) 101-107.

17. Vector Processor Overview, MM-142002-005 (July 1986) Amdahl Corp., Sunnyvale California.
18. P. Frederickson et al., "Pseudo-random trees in Monte Carlo", Parallel Computing, $\underline{1}$, No.2 (1984) 175-180.
19. K. Miura and R. Babb II, "Tradeoffs in Granularity and Parallelization for a Monte Carlo Shower Simulation Code", Parallel Computing $\underline{8}$, Nos. 1-3 (1988) 91-100.
20. K. Miura, "An Analytical Speedup Factor for Asynchronous Model of Parallel Processing", in preparation.
21. G.M. Amdahl, "Limits of Expectation", Int. Jour. Supercomputer Applications $\underline{2}$, No.1 (1988) 88-94.
22. W. Feller, An Introduction to Probability Theory and its Applications, Vol. 1, J.H.Wiley and Sons (1968).
23. H. Kobayashi, Modeling and Analysis, Addison Wesley (1978).
24. Pacific Sierra Research Corp., The FORGE EASY REFERENCE GUIDE, Sept. 1989.
25. D. Klappholtz and X. Kong, CFTP: A Tool to Aid in Hand-Parallelizing Sequential Code, Digest COMPCON SPRING '89, IEEE Computer Society Press, (1989) 92-97.

VERY LARGE DATABASE APPLICATIONS
OF THE
CONNECTION MACHINE SYSTEM

DAVID WALTZ and CRAIG STANFILL

Thinking Machines Corporation

245 First Street

Cambridge, Massachusetts

ABSTRACT

The architecture of the Connection Machine system is particularly appropriate for large database applications. The Connection Machine system consists of 65,536 processors, each with its own memory, coupled by a high speed communications network. In large database applications, individual data elements are stored in separate processors and are operated on simultaneously. This paper examines two applications of this technology. The first, which will be examined in the greatest detail, is the use of the Connection Machine System for document retrieval. The second topic is the application of the Connection Machine to associative memory or content addressable memory tasks. This ability has been put to use in a method called "memory-based reasoning" which can produce expert system-like behavior from a database of records of earlier decisions. These classes of applications both scale well; the application of supercomputer power to such problems offers unprecedented functionality and opportunities.

NATO ASI Series, Vol. F 62
Supercomputing
Edited by J. S. Kowalik
© Springer-Verlag Berlin Heidelberg 1990

INTRODUCTION

The rapid growth of on-line databases is a great challenge to information processing technology. First, databases are growing quickly in size: databases with tens of gigabytes are now quite common, and databases with hundreds of gigabytes are by no means unknown. Soon, with advances in optical scanning technology we will have to deal with databases having terabytes or even tens of terabytes of data.

This growth in database size creates two problems. The first is obvious: when a database doubles in size, the amount of storage it requires doubles. The second is perhaps less obvious: as a database grows, it becomes more and more difficult to locate and manipulate the information it contains. For text databases, this second problem manifests itself as deteriorating search quality: as a database grows, finding the right information becomes increasingly difficult, so that the system is likely to miss desired information or to deluge the user with unwanted data. This problem can be (at least partly) overcome by the use of more sophisticated retrieval algorithms such as relevance feedback (see below). These algorithms are, however, computationally more expensive than the algorithms generally used for document retrieval today.

Databases containing numerical and symbolic data also require more intensive processing as they grow in size. In the current paradigm, a database might be abstracted to a much smaller set of statistical characters (e.g. "The average salary of the CEOs of Fortune 500 coporations"). However, in abstracting a database to a few statistical parameters information is inevitably lost. We believe a good solution to this problem is to search the database for precedents, a technique called "Memory Based Reasoning" (see below). In this method, decisions are made by searching a database for episodes similar to the problem facing the user, then basing conclusions on the data so located. Again, problems associated with large databases may be attacked, by computationlly intensive algorithms.

Thus, it is desirable that computational power increase as fast as or faster than the size of the database. Until recently, just the opposite had been happening, as the performance of serial computers reached a plateau while databases continue to get bigger and bigger. A solution to this problem is the use of new generation parallel computers, such as the Connection Machine System, which have much greater computational power than conventional (serial) machines. The remainder of this paper will consider some specific applications of this new machine to problems associated with large databases.

DOCUMENT RETRIEVAL

Document retrieval has traditionaly been implemented as Boolean search on an inverted file. The main difficulties of Boolean search are that: 1) users require considerable training in the use of a query language, and 2) users generally alternate between being overwhelmed by too many documents if one uses a too general search pattern, or too few documents if one is more restrictive.[1]

We have built an easy-to-use document retrieval system that allows simultaneous searches of very large databases by a large number of users.[2] The system mixes AI ideas with methods from information science. Its basis is a weighted associative memory algorithm. In contrast to a Boolean search system, a naive user can be trained to use our system in a few minutes. The system operates very rapidly, and has high precision and recall. ("Precision" refers to the fraction of retrieved documents that are actually relevant to the user's query; "recall" is the fraction of all relevant documents retrieved. Both assume a fixed number of retrieved documents.)

Using the algorithm described here, a single Connection Machine system allows a 6 GByte free text database to be searched and browsed rapidly (subsecond response time) and conveniently by several hundred simultaneous users. Large memory chips (4 Mbit) will allow 25 GByte databases to be searched at $1/4$ the speed by $1/4$ the number of users, given the same number of processors. Adding more processors can speed the search linearly in proportion to the number of processors. Other algorithms that are dependent on disk arrays, a high speed multiple disk mass storage unit, will allow much larger databases to be searched.

Relevance Feedback

From the user's point of view, the search process on the Connection Machine document retrieval system has two distinct phases. In the first phase, the user types a list of a few keywords, for example, "Promising New Drugs". The retrieval system returns a list of documents, ordered according to how many of the keywords they contain, and how important each keyword is (the rarer the word, the more important). In the second phase, the user browses through these documents and finds one or more that bear on the topic of interest. As relevant documents are located, the user may command the system to search for related documents by marking the documents (e.g. with a mouse), and then performing a full-text to full-text comparison between the documents he has already found and every document in the database. This is done by automatically extracting the words from the text of the known relevant documents, and rating the documents in the database according to how many of those words they contain, weighted according to their frequency of occurrence in the database (rare words are given large weights, common words low or zero weights, since they do not distinguish between documents).

This method, termed "relevance feedback", has been known since the 1960s to yeild high quality searches, but to the best of our knowledge, it has not been used commercially because of its high computational requirements. We have implemented a commercial system that has been in continuous use since January 1989 by Dow-Jones as the basis for a news retrieval service called "DowQuest". The description below is based on the DowQuest system.

Implementation

Relevance feedback may generate queries containing hundreds or thousands of terms, where each term consists of a word and the weight assigned to that word. Each processor in the system is assigned a 30 word segment from a single document. To execute a query, a serial front end processor braodcasts the word-weight pairs to the Connection Machine System. As this is done, each processor tests its segment for the presence of each word. When a word is found, the processor accumulates the word's weight to form a score for that document. After the complete query has been broadcast, the results are sorted in order of decreasing score, and the pointers to the documents with the highest scores are sent to the user (*n* is typically between 10 and 40, though any number could be used).

Documents are represented using a method of *surrogate coding* described by Stanfill and Kahle.[2] In this method, the groups of 30 words are collected and used to set bits in a 1024 bit vector. Bits are set by applying 10 hash code functions to each term. Thus, to represent a document with 90 terms, 300 bits in each of three vectors of 1024 bits each would be set. Occasionally, more than one hash function may set the same bit. This is handled by superimposing the results. The vectors for each document are stored in a contiguous group of processors.

In order to search for a document, the same hash codes that were used to construct the bit vectors are applied to each of the terms in the search pattern, one at a time, and the positions of the hash code bits are broadcast to all the processors. Each processor checks to see whether it has all 10 bits set for a given term and if so, it adds a score for that term to the total score for the document stored in its document mailbox. This algorithm is probabilistic: there is a small chance that a given term will hash into 10 locations which were set by other terms, causing the system to interpret it as a term occuring in a document when in fact it has not. The probability of this happening is dependent on the number of hash functions applied to each term, the size of the vector, and the number of terms in the overall database. The probability of a false hit can be made arbitrarily small by applying more hash functions or increasing the length of a bit vector. With the parameters we have been using, the probability of a false hit is about 1/1,000,000. Additionally, since each query in revelance search contains an average of 75 terms, one or two false hits cannot make much of a difference in the overall relevance calculation.

Performance

The performance figures contained in this section assume the use of a Connection Machine system with 65,536 processors and 2 GBytes of fast memory, unless otherwise noted. Furthermore, we assume the data preparation algorithms <ref. Smith's tech. rep.> have been applied, yielding an overall data compression rate of 3.3:1. Each search returns the top 16 documents, and search patterns are truncated at 100 terms.

The bit vectors are much smaller than the memories of the individual processing elements. This allows us to use the Connection Machine's "virtual processor" mechanism to increase the amount of data stored in the system. To do this, each processor's memory is segmented, and the processors sequentially perform any computation on each memory segment. To the programmer or user, it appears that there are n times as many processors, each with $1/n$ as much memory as a real processor, each operating at $1/n$ the speed.[3, 4]

The DowQuest System uses a 32,768 processor CM-2 with 256 MBytes of memory to store about 800 MBytes of text. Total time for processing a query ranges from about 130 msec. for a 5-term query to about 870 msec. for a 100-term query. These times can be reduced to about 65 msec. and 430 msec. respectively by switching to a DEC VAX 8800 or Sun 4 front end, rather than the VAX 8350 actually used by Dow-Jones. (The VAX 8350 is not fast enough to drive the CM-2 at its full rate). These times include serial preparation of queries on the front end. Assuming that half of all queries have 5 terms (initial queries) and that half have 100 terms (relevance feedback), and also assuming that users take two minutes (browsing, examining headlines, making documents, reading text, etc.) between queries, the current system can support 180 simultaneous users. With the largest memories currently offered, and assuming a 64K processor CM-2, with a fast front-end, we can support about 100 users with about 900 msec per search average response time on about 6 GBytes of text. With 4 mbit chips installed (projected 1990), a CM could support 25 users with 3.6 sec. response time on 25 GBytes of text.

A PARALLEL INDEXED ALGORITHM FOR IR

With its parallel disk array option allowing fast I/O, the CM-2 can also be used to support interactive access to much larger databases, ranging up to 1000 GBytes (1-TByte).[5] It would be much too slow to load the signature files from the method described above from secondary storage. Instead, a variant of the standard inverted file algorithm can be adapted for the CM-2 to yield fast performance on databases far larger than can be handled by current serial machines.

Our method, called the "mailbox algorithm", stores inverted files on fast DataVault secondary storage devices. Inverted files exist for each searchable word or term in the database; each inverted file consists of a list of pairs: (<document ID>, <word weight>), called "postings". Each occurrence of a term corresponds to a posting in the inverted file. To search for documents containing several terms, the inverted lists of postings are loaded from one or more DataVaults into a reserved portion of memory in the CM-2. The rest of memory is devoted to a large set of "mailboxes", one for each document in the database. Each mailbox has enough memory space to store the largest possible score a document could receive (if we truncate any search to no more than n terms, and the largest possible term score is k, we need enough memory space in a mailbox to store the number nk.) The "document ID" portion of each posting consists of a pointer to the mailbox corresponding to the document in which it occurs; we design functions, so that given a mailbox location (processor number + memory location) we can specify the location of its document (disk track + sector) and vice-versa. The actual selection of the best m documents then requires the following steps: 1) load postings for all search items; 2) send the term weights in each posting to the mailbox address (document ID) specified in the posting; 3) add the term weight to the mailbox score; 4) when all terms have been processed, find the *n* mailboxes with the largest scores, and return their addresses. These can then be used to retrieve the text of the documents (or headlines) for the user.

Using this method, and engineering considerations described in Stanfill, Thau & Waltz[5], we estimate a 3.5 second response time for 168 simultaneous users on a 100 GByte database, using a system with an 8K CM-2 and a single DataVault. A 1000 GByte database could be searched with similar numbers of users and similar response times, using a 64K CM-2 with 10 DataVaults.

MEMORY-BASED REASONING

Memory-Based Reasoning (MBR) is a paradigm for AI in which an associative memory using a best-match algorithm takes the place of rules. It is particularly well-suited to massively parallel computers such as the Connection Machine System. Memory-Based Reasoning places memory at the foundation of intelligence, rather than at the periphery. Memories of specific events are used directly to make decisions, rather than indirectly (as in systems which use experience to infer rules). In its purest form, memory-based reasoning uses the global nearest match computation to find the items in memory most similar to a current situation, and then uses the actions associated with these items to deal with the current situation. In essence, reasoning is reduced to perception; the current situation is observed, it reminds the system of something it has seen before, and an immediate reaction is forthcoming without further analysis.

We can contrast memory-based reasoning in this extreme form with models based on heuristic search. In heuristic search, solutions are generated rather than looked up. To give a concrete example, in solving a medical reasoning problem, a memory-based reasoning system finds the patient or patients most similar to the current patient by a global nearest match operation, and uses the diagnosis, treatment, and outcome to find a diagnosis and treatment, and to predict an outcome for the current patient. A rule-based forward chaining system takes the patient's symptoms and applies rules one after another until it arrives at a diagnosis.

The advantages of memory-based reasoning are: 1) it is much easier to generate examples than to generate rules, so the knowledge acquistion for a memory-based reasoning system is much simpler; 2) memory-based reasoning systems inherently have a mechanism for judging confidence in an answer - if there is a very close match in memory to the current situation, one can be quite confident of the outcome. If the nearest match is far away, the system can note that its results are uncertain, "it can know that it doesn't know"; 3) memory-based reasoning systems can scale well to very large problems, and a single system can handle simultaneously a number of different kinds of problems.

Significantly, parallel hardware reverses the relative efficiency of memory-based reasoning and rule-based reasoning. On serial hardware, the best-match operation is very expensive because every data item must be considered in turn, while rule-triggering is relatively cheap due to existence of algorithms (e.g. Rete networks) that allow a database of rules to be efficiently searched. On parallel hardware, the best-match algorithm takes constant time, while rule-invocation takes time proportional to the number of rules which must be chained to obtain an answer.

The remainder of this section will discuss work to date on memory-based reasoning, including experiments with pure MBR and MBR augmented with a generalization mechanism. We will then present our plans for further development of the paradigm.

Work to Date

In this section we will discuss memory-based reasoning applied to the classification problem. Given a database of objects, each object belonging to one of a set of mutually exclusive classes, we classify new objects by finding the best match in the existing database and looking at its class.

As reported in Stanfill and Waltz's "Toward Memory-Based Reasoning,"[6] we implemented a memory-based reasoning "shell". This shell computes the nearest match to an input pattern (which may be incomplete) using a set of weighting and distance measures, the net effect of which is to find the distance from every individual example in memory to the current example to be classified.

The shell assumes a relational database-like format for examples. More formally, a database is a set of records. Each record has a fixed set of field, and the other fields are predictor fields. Novel records which are to be classified are targeted records.

The computation of similarity is fairly complex. Field weights are computed by judging how tightly a particular predictor field constrains values of the goal field. The distance between two records is then computed by summing the values. For example, if a patient reports having a sore throat, this constrains the range of diseases he/she might be suffering from to a relatively small range (e.g. a viral infection, a strep infection, smoking). Thus, if a patient reported a sore throat, all records in the database which did not include a sore throat would receive a large distance measure. On the other hand, having a low fever places relatively few constraints on the possible maladies, so it would receive a small weight.

More recently, we have added a generalization algorithm to the memory-based reasoning shell. This algorithm searches for patterns in the database (e.g. "a high fever accompanied by a sore throat indicates a strep infection"), and remembers which records obey them. These stored patterns are then used to augment the best-match process: if a target record matches a stored pattern, the data records used to generate the pattern will have their similarity-measures boosted. The primary benefit of this generalization mechanism is to significantly reduce the system's sensitivity to noise. In all cases, it produces a significant improvement in the quality of MBR's decisions.

An Experiment

Memory-Based Reasoning was applied to the problem of pronouncing English words. The formulation of the task is deliberately similar to Sejnowski and Rosenberg's NETtalk system.[7] In this case, the database is a dictionary.[8] Each record in the database consists of a seven-letter window in a word (a letter, the three previous letters, and the next three letters); a three-phoneme window in the pronunciation (the phoneme plus the two preceding phoneme); and the stress of the letter (primary stress, secondary stress). For example, the word "file" which has pronunciation "fAL-" and stress pattern "l--", would yield the following four records:

*	*	*	f	i	l	e	*	*	f	+
	*	f	i	l	e	*	*	f	A	1
		f	i	l	e	*	*	f	A	l	-
f	i	l	e	*	*	*	A	l	-	-

The 20,000 words in the dictionary yield 146,951 records.

It must be noted that perfect performance on the pronunciation task, as outlined above, is fundamentally impossible. First, many English words are borrowed from other languages, often retaining their original pronunciations. Thus, any system which pronounced "montage" correctly would almost certainly mispronounce "frontage". Second, the stress patterns of English words often depend on their part of speech. In some cases, a word will even have two acceptable pronunciations, depending on whether it is used as a noun or a verb ("to object" versus "an object").

In spite of the difficulty of the pronunciation task, Memory-Based Reasoning does quite well. Given the three preceding letters, the three succeeding letters, the two preceding phonemes, and the stress, MBR produces the correct phoneme 92% of the time. If we omit the previous two phonemes and the stress, MBR gets the correct phoneme 87% of the time. With a database of 128K records running on a 32K processor Connection Machine System, each classification is accomplished in 30 milliseconds. Recently, Woplert[9] has reported 93% correct generalization on the smaller NETtalk database, using the weighed averaged outputs of the four nearest neighbors. This result uses a method like that in Sejnowski and Rosenberg's original experiment[8]: instead of using phonemes as outputs, Wolpert uses a two stage process. The first stage outputs vectors of phonetic features (e.g. voiced/unvoiced, fricative, labial, etc.); in the second stage these features are combined for the four nearest neighbors, and compared to all "legal" phonemes; the closest legal phoneme is the overall output of the system.

Evaluation

The MBR algorithm described above was evaluated according to sensitivity to database size, distraction, and noise. The results of these experiments are discussed more fully in Stanfill's "Memory-Based Reasoning Applied to English Pronunciation".[9] In these experiments, two different measures of similarity were evaluated.

Database size.
Both similarity measures exhibit graceful degradation as the size of the database shrinks from 128K down to 4K. With 4K records (approximately 700 words), 78 percent of phonemes were correct.

Distraction.
In an effort to distract the algorithms, between one and seven fields containing random values were added to each record in the database. These had no effect on either similarity measure.

Noise.

Two types of noise were considered. First, between 10% and 100% noise was added to the predictor fields.* Neither similarity measure was significantly affected until noise exceeded 90%, at which point performance collapsed. Second, between 10% and 100% noise was added to the goal fields. For one measure, performance fell about linearly with added noise; for the other, performance degraded less than linearly until the noise level exceeded 60%.

Prospects for Memory-Based Reasoning

Work, so far, has concentrated on the application of pure memory-based reasoning to "flat" relational databases, with the representation fixed by the system builder.
The next stages will be to relax some of these restrictions. Part of this work has already been started, with the addition of generalization to MBR. We also plan to allow MBR to modify its representations, as well as to allow for a greater flexibility in their form (e.g. allowing networks and heirarchies).

In the long run, we believe that memory-based reasoning will provide a unifying paradigm for Artificial Intelligence. Most aspects of intelligence, we believe, can be expressed as operations on or augmentations to memory. This includes perception, attention, generalization, learning, and deduction. Indeed, aspects of some of these phenomena appear as emergent behavior of the simple MBR model presented above.

---

*For 10% noise 10% of the predictor-fields in the database would receive a random value.

SUMMARY

Massively parallel computers present new horizons for applications to large database problems, both for text and numerical databases. Massively parallel solutions scale extraordinarily well: if we compute cost-performance as a function of database size, massively parallel systems are constant in cost-performance whereas the cost-performance for serial systems falls off as the reciprocal of the database size. To see this, note that the response time in a massively parallel database remains constant since we can add more processors to match database size. The cost of a system grows linearly with database size. Thus the product of the response time and the system cost is a linear function of the database size. Thus the overall cost-performance (in dollars per GByte-sec) remains constant as the database grows larger. In contrast, in a serial system, the cost-performance falls off linearly as the database size grows larger. The response time is linear in the database size, and the cost of the system is a constant for the processor plus a linear factor for the cost of disc storage. The cost-performance thus grows as the square of the database size, so that the overall cost performance is O (1/ [database size]).

Cost effective solutions to text problems have already been fielded commerically and have operated successfully. Applications to other database problems are currently being investigated. Memory-based reasoning methods are currently being applied to protein structure prediction[10], learning for robot arm control[11], medical diagnosis using the Framingham Heart Database, and other problems in trouble-shooting, diagnosis, and signal processing.

In addition, while it is outside of the scope of current paper, massively parallel systems such as the Connection Machine, are eminently well matched to other methods for handling large databases, such as statistical analysis, and artificial neural network methods.

A number of artificial neural net models have also been implemented on the Connection Machine. A very fast method was devised by Zhang[12], and is capable of 80 million weight updates per second, at the time this system was written the fastest artifical neural net system in the world. More recently, Singer[13] has implemented a system which is capable of speeds at least four times as fast. Massively parallel system are also capable of representing networks with tens of thousands of nodes and millions of weights.

In summary, massively parallel systems are uniquely well suited for dealing with large database problems, and we expect that such systems will come to dominate the standard large database applications, and expand the horizons of problems that can be solved by computers.

References

1.   Blair, D.C., and M.E. Maron. "An Evaluation of Retrieval Effectiveness for a Full-Text Document-Retrieval System". *Communications of the ACM,* 28 (1985) 3, pp. 285-299.

2.   Stanfill, C., and B. Kahle. "Parallel Free Text Search on the Connection Machine System." *Communications of the ACM,* 29 (1986) 12, PP. 1229-1239.

3.   Thinking Machines Corporation, *Introduction to Data Level Parallelism.* Cambridge, MA, April, 1986.

4.   Hillis, D. *The Connection Machine.* Cambridge, MA. MIT Press, 1985.

5.   C. Stanfill, R. Thau, and D.L. Waltz, "A Parallel Indexed Algorithm for Information Retrieval", *Proceedings of the 12th International Conference on Research and Development in Information Retrieval* (SIGIR-89), Cambridge, MA, June 26-28, 1986.

6.   Stanfill, C. and D.L. Waltz. "Toward Memory-Based Reasoning", *Communications of the ACM,* 29 (1986) 12, pp. 1213-1228.

7. Sejnowski, T.J. and C.R. Rosenberg. "NETtalk: A Parallel Network that Learns to Read Aloud". *The John Hopkins University Electrical Engineering and Computer Science Tecnical Report JHU/EECS-86.*

8. *Merriam Webster's Pocket Dictionary*, 1974.

9. Stanfill, C. "Memory-Based Reasoning Applied to English Pronunciation". Porceedings, *Sixth National Conference on Artificial Intelligence (AAAI -87)*, Seattle, Washington, July 13-17, 1987.

10. Zhang, X., D.L. Waltz, and J.P. Mesirov, "Protein Structure Prediction by a Data-level Parallel Algorithm", to appear in Proc. of Supercomputing-89, Nov. 1989, Reno, NV.

11. Atkeson, C., "Roles of Knowledge in Motor Learning", *Massachusetts Institute of Technology Technical Report 942*, Sept. 1986.

12. Zhang, X., M. McKenna, J.P. Mesirov, and D.L. Waltz, "An Efficient Implementation of the Backpropagation Algorithm on the CM-2", to appear in Proc. of NIPS Conf., Nov. 1989, Butler, Co.

13. Singer, A., paper forthcoming.

DATA STRUCTURES FOR PARALLEL COMPUTATION ON SHARED-MEMORY MACHINES*

Narsingh Deo
Department of Computer Science
University of Central Florida
Orlando, Florida 32816

Abstract: One of the major bottlenecks in developing parallel algorithms is the lack of "parallel data structures," particularly for nonnumerical and seminumerical problems. Many of the traditional data structures that have served so well in sequential computing do not easily lend themselves to parallel processing. In this paper we develop three practical paradigms for designing parallel data structures in a shared-memory environment and illustrate them through parallelization of a queue, a stack, and a heap. Other related structures are also discussed.

1.0 Introduction

One of the major bottlenecks in development of parallel algorithms is the lack of "parallel data structures." There are, of course, some problems where the data have no particular structure at input, output or intermediate stages. A simple example would be computing the average of, say, a million numbers using 32 processors. It is straightforward to partition the numbers among the processors and then compute the average of the averages. In this case individual pieces of data are like grains to be ground into flour and the order in which they arrive at the grinding stone does not matter.

There are, however, many problems, particularly in nonnumerical and seminumerical applications, where the information is contained not only in the individual data elements but in the structure in which they are organized through various steps of the computation. In parallel-processing environments, sharing or partitioning of such structures, with a minimum of contention, while the structures themselves change dynamically as the computation progresses, is often at the heart of the algorithm.

---

* This work was supported by a research contract (#N61339-88-G-0002) from the U.S. Army's PM-TRADE, and by a research grant from Florida High Technology and Industry Council.

Many of classic elementary data structures — e.g., stacks, queues, linked lists, heaps, and so forth — that have served us so well in sequential computation, do not lend themselves to parallel manipulation. They seem to be inherently sequential.

Before going further into the subject, we must specify the model of the parallel computer being employed. Parallel computers mean different things to different people. Our model is the tightly-coupled, shared-memory, MIMD machine — such as, IBM's rp-3; NYU's Ultracomputer; or the HEP. The corresponding theoretical model would be the PRAM. Of the various PRAMs we will use either EREW model or the CREW model, but not the CRCW.

Obviously, the type of the parallel machine available will have a strong bearing on the design and implementation of the parallel data structure. In a shared-memory machine, one would like to have a single shared-structure with multiple access paths for different processors, while avoiding contentions, collisions, and deadlocks during accessing data items. In a distributed-memory MIMD machine, where the processors have only local memories, such as, Intel's iPSC, we would either have to partition or replicate the data structure, and use message-passing for updating. In the shared-memory MIMD model that we consider here, sharing a data structure is more appropriate than partitioning or replicating.

The rest of the paper is organized as follows: In Section 2, we discuss the parallelization of a queue, and through it highlight one paradigm of parallel data structures design, which we call "private plus public." That is, for some concurrent operations the data structure is treated as partitioned into private substructures — one for each processor; and for other operations the same structure is treated as one public structure accessible to all. In Section 3, we deal with parallelizing a stack, and through it, we illustrate our second paradigm, called "transformation." The seemingly sequential data structure (stack) is transformed into a parallel data structure (binary search tree) while preserving the properties essential for the efficiency of the algorithm. In Sections 4 and 5, parallel operations on a heap are discussed, and the paradigm of "partial locking" is discussed. That is, rather than locking the entire data structure, a processor should lock as small a portion as possible, in order that different processors can be scheduled to access different parts of the same data structure, without conflicts. In Section 6, parallel operations on other data structures are mentioned briefly, and some concluding remarks are made.

2.0 Parallelization of a Queue

A queue (a first-in-first-out list) is used in implementation of many sequential algorithms. Consider, for instance, the Bellman-Ford-Moore (BFM) Algorithm for solving the one-to-all shortest distance problem in a given weighted directed graph $G = (V, E, len)$, with vertex set V, edge set E, and lengths of edges, len. The distance $dist(v)$ is assigned value ∞ to every vertex except the specified source vertex s, for which $dist(s)$, the distance from s to itself, is 0. A queue, Q, contains vertices which are to be explored further; initially Q contains just s. In each iteration, a vertex u from the head of Q is deleted and all edges (u, v) emanating from u are examined to see if a path (from s to v) shorter than the existing one has been found. If so, $dist(v)$ is reduced and vertex v is added to the tail of the queue, if it is not already in the queue. The exploration continues till the queue is empty.

```
procedure BMF (s);
initialize Q to contain only s;
dist(s) := 0;
for all v ∈ V-{s} do dist(v) := ∞;
while Q ≠ ∅ do
begin
     delete head node u from Q;
     for each edge (u,v) emanating from u do
          if dist(v) > dist(u) + len(u,v) then
          begin
               dist(v) := dist(u) + len(u,v);
               insert v at the tail of Q if v not in Q;
          end
end
```

In 1979, Deo, Pang, and Lord [7] parallelized this algorithm and implemented it on the HEP computer with a maximum of eight processors. They found that with an 8-fold parallelism the maximum speedup was only 3, and they observed that the primary bottleneck was that "every process demands private use of Q."

Assigning each processor a separate and independent queue may lead not only to load imbalance and coherence problems but may even produce incorrect results. As a compromise, a new data structure called "linked array" was suggested by Quinn and Yoo [15]. It allows each processor (in fact, a process) to insert elements in its own private space. But these private lists are linked together so that for deletion purposes, each of the p processors examines every pth element in this linked array, by traversing the entire data structure along the pointers (or links) in the linked array.

The iterations are carried out synchronously in insert-delete cycles in parallel. In other words, in delete phase the p processors remove (one vertex each) from the linked array, as if it were a shared public structure. Then each processor (after traversing all the edges emanating from its current vertex) inserts the vertices it must insert into its own preassigned portion of the linked array, as if it were its private structure. Further details of the implementation and performance can be found in Quinn [16], Quinn and Yoo [15].

This approach of a "public plus private" data structure appears to be a useful and interesting paradigm for designing parallel data structures for shared-memory machines.

3.0 Parallelization of a Stack

A stack (a last-in-first-out list) is another ubiquitous data structure, which appears to be inherently sequential. At a time only one element can be inserted into or deleted from a stack. In this section we will show how to get around this bottleneck by employing an alternate data structure, which allows parallel insertions and deletions.

Perhaps the best-known use of a stack is in matching parentheses, where we are given a legal sequence of length n of left and right parentheses and are asked to find the mate for each. The standard sequential algorithm for this problem requires $O(n)$ time and employs a stack. Parallel algorithms for matching parentheses have been given by Dekel and Sahni [9], Bar-On and Vishkin [1], and Sarkar and Deo [20]. Bar-On and Vishkin's algorithm requires $O(\log n)$ time and $O(n/\log n)$-processor CREW-PRAM machine, which was a significant improvement over Dekel and Sahni's. Sarkar and Deo, achieve the same complexity as Bar-On and Vishkin (both being fast and optimal), but the algorithm of Sarkar and Deo is much simpler. Using their approach, we will discuss the second paradigm for designing parallel data structures.

3.1 Parenthesis-Matching Algorithm

Let S be a legal sequence of balanced parentheses stored in a linear array. Let s_1 and s_2 be two arbitrary consecutive subsequences in S. A right parenthesis in s_2 without a matching left parenthesis in s_2 must have its matching parenthesis in the substring to the left of the substring s_2. Similarly, a left parenthesis in s_1 without a matching right parenthesis in

s_1 must have its matching parenthesis in the substring to the right of substring s_1. Let r_1 (respectively r_2) be the number of right parentheses in s_1 (s_2) that do not have their matching parentheses in s_1 (s_2) and l_1 (l_2) be the number of left parentheses in s_1 (s_2) that do not have their matching parentheses in s_1 (s_2). Therefore the concatenated string $s = s_1 s_2$ must have $r_1 + r_2 - \min(l_1, r_2)$ right parentheses whose matching parentheses are in the substring to the left of s. Similarly, s must have $l_1 + l_2 - \min(l_1, r_2)$ left parentheses whose matching parentheses are in the substring to the right of s.

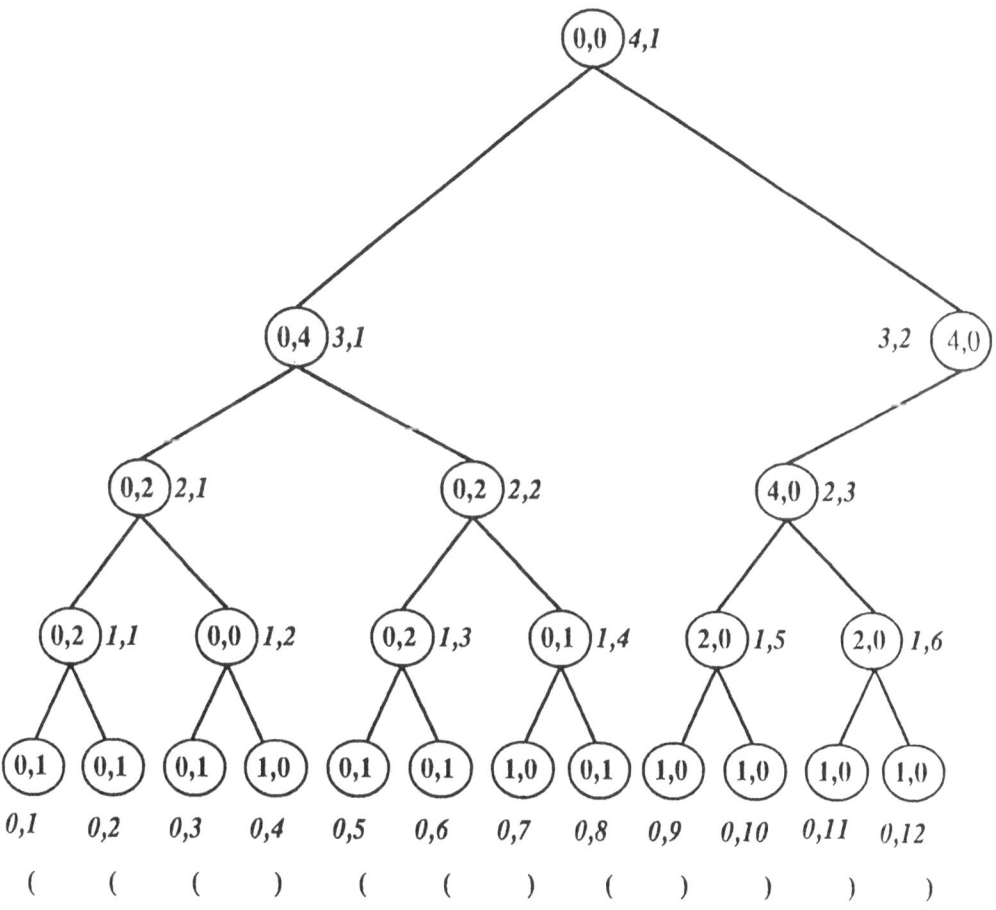

Fig. 1 A Sequence of parentheses and its search tree. Node labels are italicized and shown next to the nodes; the (r, l) values are shown within the ellipses.

3.2 Construction of the Binary Search Tree

The preceding observation is utilized for constructing the ith level of a balanced binary search tree in a bottom-up fashion using values r_x and l_x at every node x at $(i - 1)$th level of the tree (the leaf nodes are at the zeroth level of the tree). To find the mate of a parenthesis, a search is conducted on this binary tree. Each parenthesis in the given sequence S is a leaf-node in the tree. We label a node of the binary tree by an ordered pair $<i, j>$ of nonnegative integers, where the first integer represents the level of the node (from the bottom) and the second represents its position from the left in that level. Nodes $<i - 1, 2j - 1>$ and $<i - 1, 2j>$ are then the left and the right children of the node $<i, j>$. We also use two arrays, $r[i, j]$ and $l[i, j]$, to store the values at nodes $<i, j>$, $0 \le i < \lceil \log_2 n \rceil$ and $1 \le j \le 2^i$. If the parenthesis at position j (in string S) is a right (respectively left) parenthesis then $l[0, j] = 0$ ($l[0, j] = 1$) and $r[0, j] = 1$ ($r[0, j] = 0$). The search tree for a given balanced sequence of parentheses is shown in Fig. 1. The algorithm for construction of the binary search tree follows.

Input: A legal sequence of n parentheses stored in a linear array.

Output: A binary search tree of height $\lceil \log_2 n \rceil$ stored in two arrays.

Step 1: {Initialization} **for all** j **do in parallel**
$\qquad$ **if** there is a right parenthesis at position j **then**
$\qquad\qquad$ **begin** $r[0, j] := 1$; $l[0, j] := 0$; **end**;
$\qquad$ **else**
$\qquad\qquad$ **begin** $r[0, j] := 0$; $l[0, j] := 1$; **end**;

Step 2:
$\qquad$ **for** $i := 1$ **to** $\lceil \log_2 n \rceil$ **do**

$\qquad\qquad$ **for** $j := 1$ **to** $2^{\lceil \log_2 n \rceil - i}$ **do in parallel**
$\qquad\qquad$ **begin**
$\qquad\qquad\qquad$ $r[i, j] := r[i - 1, 2j - 1] + r[i - 1, 2j] -$
$\qquad\qquad\qquad\qquad$ $\min (l[i - 1, 2j - 1], r[i - 1, 2j])$;
$\qquad\qquad\qquad$ $l[i, j] := l[i - 1, 2j - 1] + l[i - 1, 2j] -$
$\qquad\qquad\qquad\qquad$ $\min (l[i - 1, 2j - 1], r[i - 1, 2j])$;
$\qquad\qquad$ **end**;

In the following discussion, two descendants of a node in the search tree are identified as its left child and right child; left (respectively right) child is called left (right) brother of the right (left) child. Let us consider the search procedure to find the matching right parenthesis of a left parenthesis at position x of the input string (for convenience we shall call parenthesis at position x as parenthesis x or simply x). We search on the search tree constructed by the previous procedure. Obviously, the matching right parenthesis of a left parenthesis x is in the substring to the right of the x. Suppose in the searching process we have arrived at a node of the search tree such that there are c_1 unmatched left parentheses to right of x in the substring corresponding to this node. If the present node is a right child, its string concatenates with a substring on the left side of its substring, and hence, no left parentheses comes to the right of x in the concatenated string, and value of c_1 does not change. If the present node is a left child, we determine how many right parentheses in the substring corresponding to its right brother do not have a match i.e., find the value of r corresponding to its right brother. If $r \leq c_1$ then the match for x is not in the string corresponding to the right brother and we climb up to the father of the present node. The number of left parentheses to the right of x in the concatenated string is given by $c_1 - r + l$. Thus, c_1 is assigned $c_1 - r + l$ (l is the number of unmatched left parentheses of its right brother). We continue climbing towards the root by these rules until we reach a node whose right brother has r unbalanced right parentheses such that $r > c_1$. At this point we know that the mate of x is in the substring corresponding to the right brother, and we move to the right brother and start to climb down towards the leaves until we reach a leaf-node which has the mate of x. While we are climbing down towards the leaf-nodes, we test the number of unbalanced right parentheses in the string corresponding to the left child. If its r-value is greater than c_1 then we follow the left subtree, otherwise c_1 is assigned the value $c_1 - r$ and we follow the right child. We continue this process until a leaf-node is reached. A formal description of the search procedure is as follows.

3.3 Searching for the Right Matching Parenthesis

Input: Search tree and the position of a left parenthesis in the
            sequence of parentheses.
Output: Position of the matching right parenthesis is j.

```
{for the mate of a left parenthesis at position x}
begin
    count := 0;  i := 0;  j := x;
    {climbing-up phase}
    while ((count ≥ r[i, j]) or (j is even)) do
        if j is odd then {present node <i, j> is a left child}
            begin {update count and climb up}
                count := count + l[i, j + 1] - r[i, j + 1];
                i := i + 1;  j := (j + 1)/2;
            end
        else {present node is a right child — climb up}
            begin
                i := i + 1;  j := j/2;
            end;
    {end of climbing up — climb down to the mate}
    j := j + 1;  {move to the right brother}
    while (i ≠ 0) do
        if (count ≥ r[i - 1, 2j - 1]) then
            begin {move down to the right child}
                count := count - r[i - 1, 2j - 1];
                i := i - 1;  j := 2j;
            end
        else
            begin {move down to the left child}
                i := i - 1;  j := 2j - 1;
            end;
    {now, right parenthesis at location j is the mate}
end
```

A similar procedure for searching the mate of a right parenthesis can be constructed. It is not difficult to implement these two procedures in $O(\log n)$ time with $(n/\log n)$ processors using Brent's scheduling principle [5].

What we have done here is a transformation of one data structure (stack) into another (binary search tree) in order to achieve concurrent operations. As a result a fast (polylog-time) algorithm was achieved without sacrificing the original optimality of the sequential algorithm. The time-processor product of the parallel algorithm is $O(n)$, the same order as that of the best sequential algorithm.

However, this does not imply that a sequential stack can blindly be replaced by a binary search tree for parallelization in every algorithm without any loss in efficiency. Such an assertion would immediately imply (incorrectly) that a depth-first search on a graph can be conducted in $O(\log|E|)$ time with $O(\frac{|E|}{\log|E|})$ processors, where $|E|$ is the number of edges in the given graph.

4.0 Parallelization of a Heap

A heap is a balanced binary tree in which the value (or the key) of the parent never exceeds the value of its children. Although heapsort is probably its best-known application, a heap is used in many other algorithms. Consider, for example, the problem of finding a minimum spanning tree (MST) in a weighted, undirected, connected graph. In Kruskal's (smallest-edge-first) algorithm the edges are examined in nondecreasing order of their weights, one at a time. An edge which does not form a cycle, with the edges already selected, is included in the solution. The examination continues till $(n - 1)$ edges have been selected to form a minimum spanning tree of the given n-vertex graph [19].

In sequential version of this MST algorithm, it is best to maintain a heap of the edges, rather than sorting them. (We may be wasting a lot of effort on sorting edges that may never be examined.) The next smallest edge to be examined is removed from the top of the heap (root) and examined for its inclusion in the solution. Then the vacancy at the root is filled in $\lfloor \log m \rfloor$ sequential steps by updating the heap level by level going from top to bottom, m being the number of elements in the heap at this time.

The following log m-fold parallelization of this serial-access heap was achieved by using a simple scheme called "software pipelining" proposed by Yoo [23], which in turn was based on the concept of a tree machine implied in Muller and Preparata [14]: log $m = p$ processors are assigned to the heap — one to each level. A linear array of size m is used to implement the heap in the usual fashion. That is, the root occupies location 1, and the children of node at location i are in locations $2i$ and $(2i + 1)$.

A node is "empty" if its value has been transmitted to its parent and no replacement has been received from either of its children. An array "status" (of length log m) indicates which levels have an empty node; $status[i]$ = empty if level i has an empty node, otherwise $status[i]$ = full.

If status[i] = empty, then the value empty-node [i] indicates which node at level i is actually empty. Note that at a given level in the heap, at most one node can be empty. In software pipelining of the heap, processor i is assigned the job of keeping the nodes at level i - 1 full. If status[i - 1] = empty and status[i] = full, then processor compares the keys of the two children of empty-node[i - 1], puts the smaller of the two in empty-node[i - 1] position, sets status[i - 1] := full and status[i] := empty. In order to ensure that no child of an empty node is also empty, this concurrent upward shifting of elements is performed in two phases. In Phase 1 all processors at odd levels (except level 1) perform the updating operation (just described) and in Phase 2 all processors at even levels perform the updating. Thus, every two units of time, one element is pumped out of the heap. The pumping is continued as long as necessary or till the heap becomes empty. Thus, a speedup of $O(\log n)$ with $O(\log n)$ processors is achieved. Figure 2 shows such a heap and the corresponding arrays.

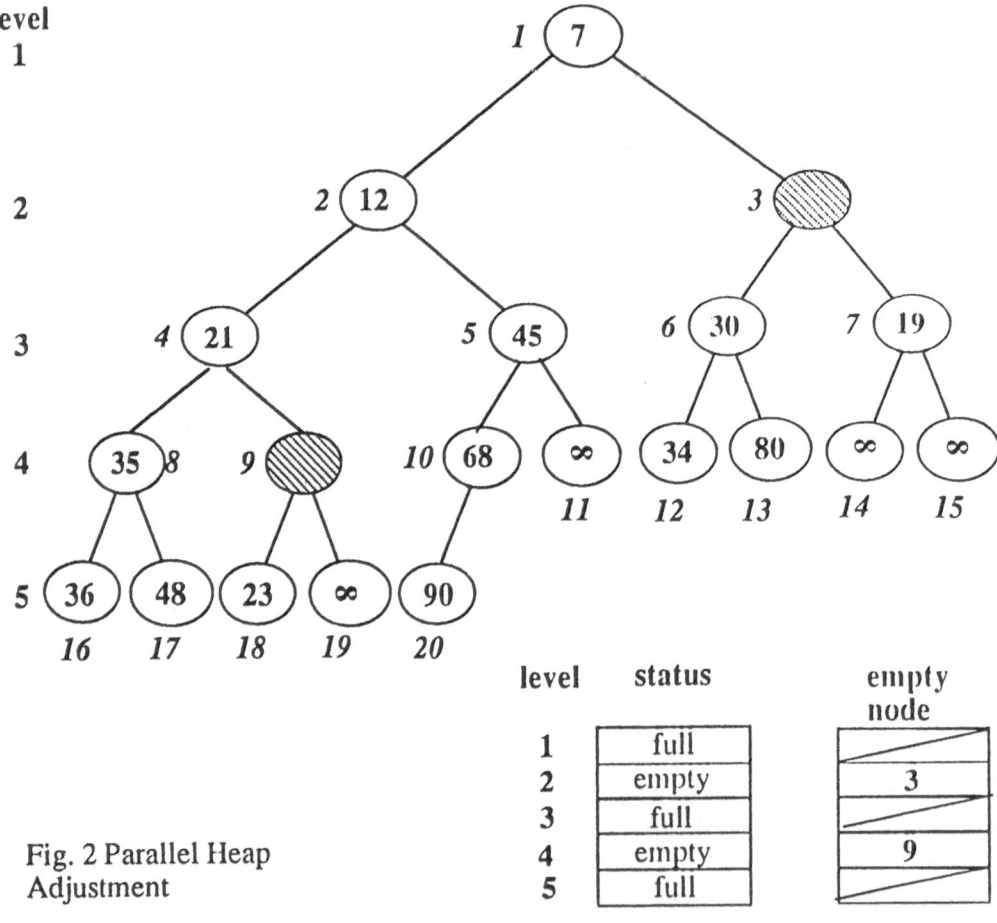

Fig. 2 Parallel Heap Adjustment

level	status	empty node
1	full	
2	empty	3
3	full	
4	empty	9
5	full	

Initial building of the heap can also be done parallel as is done in the sequential case [19]. The only difference would be that in the parallel case, all internal nodes at a given level (whose subtrees are heaps) are restored in parallel. Thus, $m/4$ processors can create an initial heap in $\lceil \log m \rceil$ iterations, in which each iteration takes at most $O(\log m)$ time. Thus, initial building of the heap can be done in $O(\log^2 m)$ time with $O(m)$ processors.

4.1 Heaps with Insertion and Deletion

The use of the heap, just described, in Kruskal's MST algorithm was only for deleting elements. There are applications where elements are removed from the heap as well as inserted into it. Concurrent deletion and insertion operations on a heap have been considered by Biswas and Browne [4], Rao and Kumar [18], and Jones [11]. Typically, these heap-based applications consist of the "delete-think-insert" cycles, in that the item with the smallest value (or the highest priority) is deleted from the top of the heap, then some processing is done on that item, which produces some more items to be inserted into the heap.

Deletion is performed from the top as described in the previous section. Insertion could be performed at the bottom or at the top. In the former case, the item is inserted at the node following the last node in the heap, which is then followed by a concurrent heap adjustment phase, employing $\log n$ processors. In bottom insertion, there is a possibility of a deadlock, because deletion and insertion are moving simultaneously in opposite directions each performing a window locking (locking a sliding window of three nodes — a node and its two children — for adjusting the heap). Biswas and Browne [4] have proposed a scheme for handling this problem, but its performance is poor because of the heavy overhead. This problem is elegantly solved by Rao and Kumar, by performing both insertion and deletion at the top. The details are given in [18].

5.0 Generalized Parallel Priority Queue

The standard serial priority queue is almost always implemented as a heap (and in fact the two terms are often used interchangeably). Since there is only one processor, one needs only one item (of highest priority), at a time, from the queue (which is taken off from the top of the heap). In a parallel environment, on the other hand, at a given instant, one may need to delete k items (of k highest priorities). As an example, consider a parallel version of the usual branch-and-bound algorithm for solving the traveling salesman problem [13, 17]. In each cycle, k processors delete the k least-cost nodes from the branch-and-bound tree, generate two successor subproblems for each, compute their costs and insert them (or ignore them if their partial-costs exceed the current bound). In such an application, we need a true priority queue, where not just one but k highest-priority items are identified, for any specified $k > 1$.

The other serious problem with the sequential-access parallel heap, discussed in the previous section, is that we could not utilize more than $O(\log n)$ processors, and hence every parallel algorithm for a problem of polynomial (sequential-time) complexity will remain polynomial, no matter how many processors become available. In order to obtain a polylog parallel algorithm, we must be able to use effectively a polynomial number of processors in handling the priority queue.

Both these problems have been solved recently by Deo and Prasad and reported elsewhere [8]. The outline will be presented here; for the details with an illustrative example, [8] should be referred to.

Instead of having just one element per node of the parallel heap, we allow up to r elements per node; where r is a fixed parameter, a function of n (the maximum number of elements in the heap) and p (the number of processors available). To each level of the heap, $\dfrac{r}{\log r}$ processors are assigned, which operate in a pipelined fashion to maintain the heap property [19]. The *parallel heap property*, of course, means that no item at any node has a value greater than that of any item at either of its children. In addition to these $\dfrac{r}{\log r}$ maintenance processors at each level of the heap, there are r general processors, which perform "delete-process-insert" cycle, synchronously. That is, these r general processors delete the r smallest-valued items from the root, process these items producing up to $2r$ new items, and insert them into the heap.

The $2r$ newly-produced items are sorted, using Cole's parallel sort [6] employing r general processors, and then merged with the r smallest items of the heap at the root using the Bilardi-Nicolau [3] bitonic merging and employing $\dfrac{r}{\log r}$ maintenance processors at the root. Out of the resulting $3r$ sorted items, the first r are kept for the next deletion phase. The next r are kept at the root for the heap adjustment and deletion phase. The remaining are sent down along an insertion path to the next empty node (called the target node) at the bottom level of the heap. In case the (current) last node in the heap is only partially filled (i.e. has fewer than r elements) some of the elements would go to fill that node and the rest to the target node.

6.0 Conclusion

We have discussed three useful paradigms for designing parallel data structures. They are: (1) Private plus public: During the insertion operation a queue was regarded as collection of private subqueues, one for each processor, but for the deletion operation, the same structure was regarded as a single common queue. (2) Transformation: The stack, as used in the parentheses-matching algorithm, was transformed into a binary search tree (which is parallel by nature) without sacrificing speedup or efficiency. (3) The software pipeline or partial locking: By only partially locking the heap, $O(\log n)$ processors could concurrently adjust a heap. This approach was further extended and refined in Section 5.

There are several important issues in parallel data structures that were not touched upon here. Not discussed was Wyllie's [22] "recursive doubling" for concurrent operations on linked-list, which is a powerful and widely-used technique. Likewise, we omitted discussion of various concurrent implementations of search trees which have been investigated, since mid-1970's, primarily in context of data base systems [11, 12]. Nor did we touch upon the "distributed data structures," for DRAM type parallel machines. In this paper we were concerned with finding multiple access paths to and through a single large common data structure, and not with its partitioning across processors.

7.0 References

[1] Bar-On, I., and Vishkin, U., "Optimal Parallel Generation of a Computation Tree Form", *ACM Trans. on Prog. Lang. and Systems,* 7, pp. 348-357, 1985.

[2] Berkman, O., Breslauer, D., Galil, Z., Schieber, B., and Vishkin, U., "Highly Parallelizable Problems", *Proc. 21st ACM Symp. on Theory of Computing,* 1989.

[3] Bilardi, G., and Nicolau, A., "Adaptive Bitonic Sorting: an Optimal Algorithm for Shared-Memory Machines", *SIAM J. Comput.,* 18, pp. 216-228, 1989.

[4] Biswas, J., and Browne, J. C., "Simultaneous Update of Priority Structures", *Proc. Int. Conf. Parallel Process.,* pp. 124-131, 1987.

[5] Brent, R. P., "The Parallel Evaluation of General Arithmetic Expressions", *J. ACM,* 21, pp. 201-206, 1974.

[6] Cole, R., "Parallel Merge Sort", *SIAM J. Comput.,* 17, pp. 770-785, 1988.

[7] Deo, N., Pang, C. Y., and Lord, R. E., "Two Parallel Algorithms for Shortest Path Problems", *Proc. Int. Conf. on Parallel Process.,* pp. 244-253, 1980.

[8] Deo, N., and Prasad, S., "Parallel Heap", *Technical Report No. CS-TR-89-10,* Dept. of Computer Science, University of Central Florida, Orlando, FL, 1989.

[9] Dekel, E, and Sahni, S., "Parallel Generation of Postfix and Tree Forms", *ACM Trans. on Prog. Languages and Systems 5,* pp. 300-317, 1983.

[10] Ellis, C. S., "Distributed Data Structures", *IEEE Trans. Comput.,* 34, pp. 1178-1185, 1985.

[11] Jones, D. W., "Concurrent Operations on Priority Queues", *Comm. ACM,* 32, pp. 132-137, 1989.

[12] Manber, U., "Concurrent Maintenance of Binary Search Trees", *IEEE Trans. Softw. Engg.,* SE-10, pp. 777-784, 1984.

[13] Mohan, J., "Experience With Two Parallel Programs Solving the Traveling Salesman Problem", *Proc. Int. Conf. Parallel Process.,* pp. 191-193, 1983.

[14] Muller, D. E., and Preparata, F. P., "Bounds to Complexities of Networks for Sorting and for Switching", *J. ACM,* 22 (2), pp. 195-201, 1975.

[15] Quinn, M. J., and Yoo, Y. B., "Data Structure for the Efficient Solution of Graph Theoretic Problems on Tightly-Coupled MIMD

Computers", *Proc. Int. Conf. Parallel Process.*, pp. 431-438, 1984.

[16] Quinn, M. J., "The Design and Analysis of Algorithms and Data Structures for the Efficient Solution of Graph Theoretic Problems on MIMD Computers", Ph.D. Dissertation, Computer Science Dept., Washington State University, Pullman, WA, 1983.

[17] Quinn, M. J., and Deo, N., "Parallel Graph Algorithms", *ACM Computing Surveys*, 16, pp. 319-348, 1984.

[18] Rao, V. N., and Kumar, V., "Concurrent Access of Priority Queues", *IEEE Trans. Comput.*, 37, pp. 1657-1665, 1988.

[19] Reingold, E. M., Nievergelt, J., and Deo, N., *Combinatorial Algorithms*, Prentice-Hall, Englewood Cliffs, NJ, 1977.

[20] Sarkar, D., and Deo, N., "Parallel Algorithms for Parenthesis Matching and Generation of Random Balanced Sequences of Parentheses", *Proc. Int. Conf. on Supercomputing*, Springer-Verlag LNCS, Vol. 297, pp. 970-984, 1987.

[21] Syslo, M. M., Deo, N., and Kowalik, J. S., *Discrete Optimization Algorithms with Pascal Programs*, Prentice-Hall, Englewood Cliffs, NJ, 1983.

[22] Wyllie, J. C., "The Complexity of Parallel Computation", Ph. D. Dissertation, TR 79-387, Dept. of Computer Science, Cornell University, Ithaca, NY, 1979.

[23] Yoo, Y. B., "Parallel Processing for Some Network Optimization Problems", Ph.D. Dissertation, Computer Science Dept., Washington State University, Pullman, WA, 1983.

PARALLEL EVALUATION OF SOME RECURRENCE

RELATIONS BY RECURSIVE DOUBLING

*

A. Kiper & D.J. Evans
Parallel Algorithm Research Centre
Department of Computer Studies
Loughborough University of Technology
Loughborough, Leicestershire,
U.K.

ABSTRACT

The second order linear recurrence formulae which result from the Fourier series coefficients of the Jacobian elliptic functions $sn^m(u,k)$, $cn^m(u,k)$ and $dn^m(u,k)$ with $m \geq 1$, are evaluated by the method of recursive doubling on a parallel computer and performance results are discussed.

1.   INTRODUCTION

In the evaluation of functions using series approximations, the accuracy increases with the increasing number of terms involved in the expansion which results in an increase in computer time. In the last decade, advances in the area of "Parallel Computing" have led the numerical analysts to develop numerical algorithms for the solution of various problems, including the recurrence relations.

The Fourier series expansion for twelve Jacobian elliptic

---

*
  On leave: Department of Computer Engineering, Middle East
            Technical University, Ankara, Turkey.

NATO ASI Series, Vol. F 62
Supercomputing
Edited by J. S. Kowalik
© Springer-Verlag Berlin Heidelberg 1990

functions (JEFs) have been studied and given by several authors (Abramowitz and Stegun [1], Byrd and Friedman [2], Du Val [3], Whittaker and Watson [8]). Two-term recurrence formulae have been obtained for the coefficients of these series corresponding to powers of the JEFs (Kiper [5]). The resulting recurrence formulae are of the second order and linear. A parallel evaluation of these recurrences using the nested recurrent product form algorithm has been considered by Kiper [6]. In this paper, another parallel evaluation of these coefficients is formulated using the method of recursive doubling.

2. GENERALISATION OF RECURRENCE FORMULAE OF THE FOURIER COEFFICIENTS FOR POWERS OF THE JEFs

The analysis of the relations for the Fourier series coefficients for powers of the JEFs (Kiper [5]) shows that for a prescribed k (k is the modulus of the elliptic functions) these recurrences may be represented by the common expression,

$$
\left.
\begin{aligned}
&\Omega_n^{(0)} = \alpha \\
&\Omega_n^{(1)} = \beta \\
&\Omega_n^{(r)} = a(n,r)\,\Omega_n^{(r-1)} + b(r)\,\Omega_n^{(r-2)} \quad , \quad r=2,3,\ldots,\ell
\end{aligned}
\right\} n=0,1,\ldots,m
\tag{1}
$$

where ℓ is the required power and m is the required number of terms in the expansion. Equation (1) is a second order linear recurrence relation $R<\ell,2>$.

3. PREVIOUS EVALUATIONS

The coefficients for the Fourier expansion of the JEFs

$sn^m(u,k)$, $cn^m(u,k)$ and $dn^m(u,k)$ with $m \geq 1$ were evaluated sequentially and the numerical values were listed for various values of k ($0.1 \leq k^2 \leq 0.9$) (Kiper [4], [5]). It is seen that the rate of convergence decreases as the value of k and the power of the JEFs increase.

Kiper [6] also considered the parallel evaluation of these coefficients using the nested recurrent product from algorithm in which the relation (1) has been expressed as the solution of a matrix system,

$$\underline{\Omega} = A\underline{\Omega} + \underline{b} , \tag{2}$$

where,

$$
A = \begin{bmatrix}
0 & & & & & & 0 \\
a(n,3) & 0 & & & & & \\
b(4) & a(n,4) & 0 & & & & \\
0 & & & & & & \\
& & & & & & \\
& & & & & & \\
0 & & & b(\ell) & & a(n,\ell) & 0
\end{bmatrix}
$$

$$
\underline{\Omega} = \begin{bmatrix}
\Omega_n^{(2)} \\
\Omega_n^{(3)} \\
\Omega_n^{(4)} \\
\vdots \\
\Omega_n^{(\ell)}
\end{bmatrix}
, \quad
\underline{b} = \begin{bmatrix}
a(n,0)\Omega_n^{(1)} + b(0)\Omega_n^{(0)} \\
b(1)\Omega_n^{(0)} \\
0 \\
\vdots \\
0
\end{bmatrix}
$$

If the first m terms of the expansion for a power ℓ is required, then a matrix system of size ($\ell-1$) must be solved (m+1) times with n=0,1,2,...,m. The size of the system increases with the increasing required power, and the number of solutions of the system increases with the increasing number of terms.

4. **PARALLEL DETERMINATION OF THE COEFFICIENCY BY THE METHOD OF RECURSIVE DOUBLING**

The generalisation of the recurrence relations for the expansion coefficients of the JEFs $sn^m (u,k)$, $cn^m (u,k)$ and $dn^m (u,k)$ with $m \geqslant 1$ results in a linear second order form (1) which yields a form of linear recurrent equation which generates a vector result in 2 dimensions. The method of recursive doubling is one of the parallel processes that can be used in the computations [7].

Let

$$V_n^{(r)} = \begin{bmatrix} \Omega_n^{(r)} \\ \Omega_n^{(r-1)} \end{bmatrix} \quad \text{and} \quad A_n^{(r)} = \begin{bmatrix} a(n,r) & b(r) \\ 1 & 0 \end{bmatrix} \tag{3}$$

$r=2,3,\ldots, ; \quad n=0,1,2,\ldots,m.$

If the first m terms of the expansion for the prescribed power is required, then (1) can be written as,

$$V_n^{(\ell)} = A_n^{(\ell)} \cdot A_n^{(\ell-1)} ,\ldots, A_n^{(2)} \cdot V_n^{(1)} , \quad n=0,1,2,\ldots,m. \tag{4}$$

The associative property of matrix-matrix multiplication leads us to use the recursive doubling process in $0(\log_2 \ell)$ steps for each n $(n=0,1,\ldots,m)$. (In sequential mode, computations need $0(\ell)$ steps for each n).

It must be noted that the determination of the coefficients for a prescribed power can be done in any order (or parallel) since they are independent for $n=0,1,2,\ldots,m$.

5.   RESULTS

The   recurrence   relation (4) can be written as   the   finite product form,

$$V_n^{(\ell)} = \prod_{j=2}^{\ell} A_n^{(j)} V_n^{(1)} \qquad , \quad n=0,1,2,\ldots,m$$

and by exploiting the use of the associative law its numerical evaluation can be completed in parallel by the use of the log sum (fan-in) algorithm illustrated in Figure 1.

The   main   differences between the sequential   and   parallel evaluation of the recurrence relation (4) is that the former is based on (ℓ -1) matrix-vector operations whilst the latter employs $\log_2$ (ℓ) matrix-matrix operations followed by a final matrix-vector operation. So the speed-up for parallel evaluation is proportional to (ℓ /$\log_2$ ℓ) if sufficient processors are available.

For second order recurrence relations such as (1), (2x2) matrix multiplication occurs simultaneously at each of the leaf nodes in Figure 1.   Then the sequential evaluation is proportional   to 2ℓ and the parallel evaluation takes $2 \log_2 \ell$ which in general further reduces the speed-up by a factor of 2^3 .

The   fan-in algorithm was programmed to perform the sequence of operations given in (5) and described in Figure 1 on the Sequent Balance 8000 multiprocessor at Loughborough University with 5 processors and a series of numerical experiments carried out   to determine the efficieny of the proposed   parallel algorithms for the evaluation of functions represented by the given second order recurrence relations.

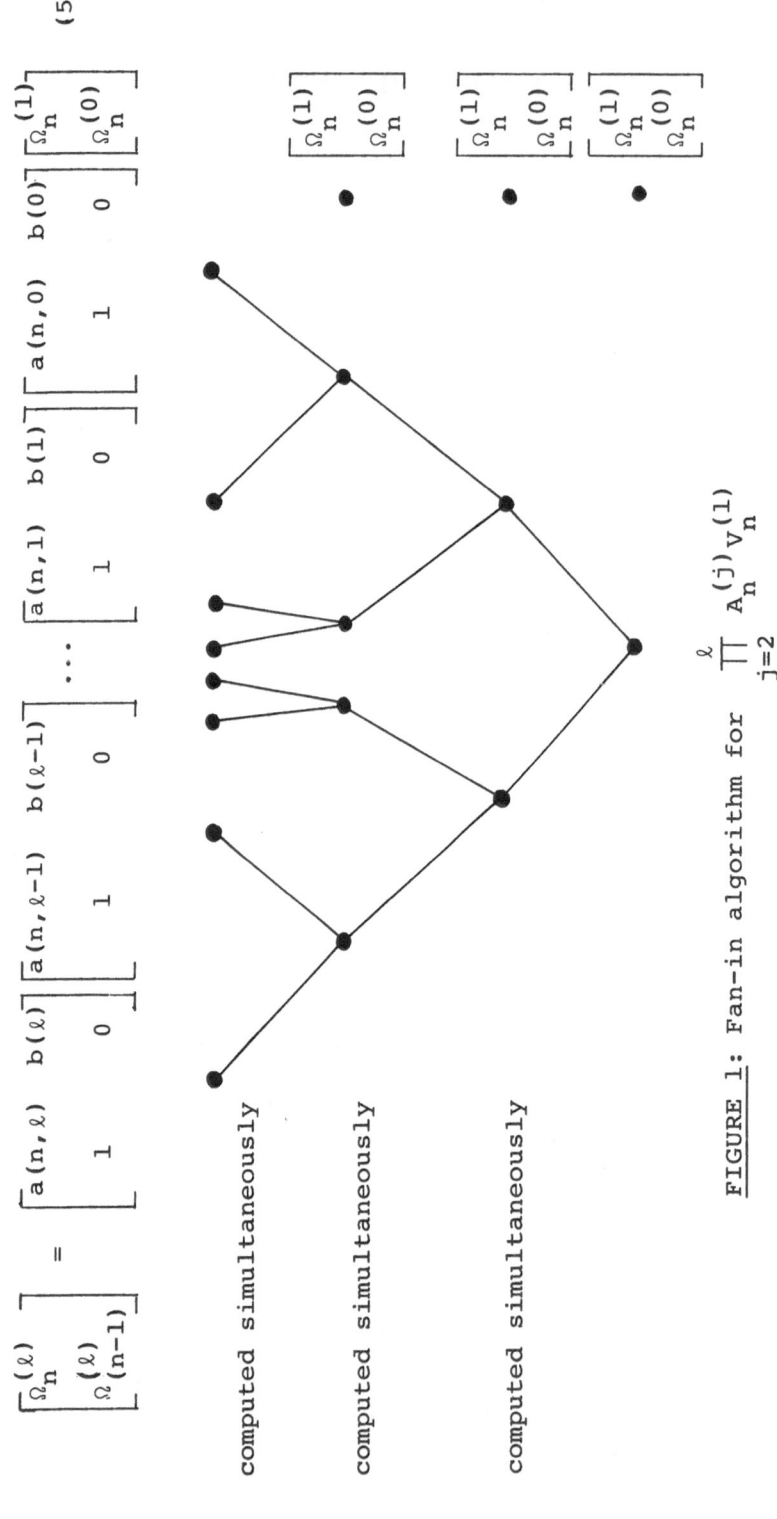

FIGURE 1: Fan-in algorithm for $\prod\limits_{j=2}^{\ell} A_n^{(j)} V_n^{(1)}$

Firstly, the numerical evaluation of the Jacobian elliptic functions (JEFs) using the Fourier series expansions were obtained by the method of recursive doubling and the results compared favourably with previous published results [5] in accuracy and numerical stability against round-off error. These will be published in a separate report.

However of more relevance to the nature of this workshop is the performance of the parallel algorithms and their comparison with their sequential counterparts. In Table 1 the results of the elapsed times using 1 processor for the sequential evaluation and parallel evaluation by the fan-in algorithm of the recurrence relation (4) are given. The $0(\ell/\log_2 \ell)$ order of improvement in the performance of the algorithms are reflected in these figures but are obscured by the parallel software overheads for the smaller values of ℓ. It can be noticed that the efficiency increases with the number of terms considered in the recurrence relation.

ℓ	Sequential Alg. Elapsed Time in secs.	Parallel Alg. (Fan-in) Elapsed Time in secs.	Speed-up
256	0.075	0.06	1.25
512	2.66	0.19	14
1024	9.47	0.37	25.6

TABLE 1: Comparison of sequential and parallel (fan-in) algorithms for recurrence relation evaluation

Further, when the tasks at each node of the fan-in algorithm are computed in parallel and are scheduled asynchronously through p=5 processors the elapsed times become:

ℓ	1 Processor	5 Processors	Speed-up
	Elapsed Time (in secs.)	Elapsed Time (in secs.)	
10,000	3.74	1.35	2.77
50,000	18.59	4.58	4.059
100,000	37.18	8.64	4.303

TABLE 2: Elapsed times of the fan-in algorithm on the Balance multipprocessor

To avoid the difficulties of accurate clock timing here we have increased the number of terms to eliminate the cost of software overheads and demonstrate the parallelism gains.

Finally, the inefficiencies caused by the parallel software overheads i.e. the cost of forking processes on the fan-in algorithm involving both shared and private data and the cost of page table make up when each of the child processes are allocated can be substantial. When these overheads are eliminated, the actual parallel computation times of the recurrent relations are revealed in Table 3.

ℓ	Execution Time (in secs.)	Speed-up	p
10,000	0.76	4.921	5
50,000	3.82	4.866	5
100,000	7.51	4.951	5

TABLE 3: Execution times with parallel software overheads eliminated

REFERENCES

1. Abramowitz, M. and Stegun, I.A.: Handbook of Mathematical
   Functions With Formulas, Graphs and Mathematical Tables,
   Nat. Bur.Standards. Appl. Math. Series No.55, December 1954
   (Also: Dover, New York, 1968).
2. Byrd, P. F. and Friedman, M. D.: Handbook of Elliptic
   Integrals for Engineers and Scientist, 2nd ed., Springer-
   Verlag, Berlin, 1971.
3. Du Val, P.: Elliptic Functions and Elliptic Curves, London
   Mathematical Society Lecture Note Series 9, Cambridge Univ.
   Press, Cambridge, 1973.
4. Kiper, A.: Computational Aspects of Series Expansions for
   the Jacobian Elliptic Functions, Technical Report, Dept. of
   Comp. Eng., METU, 1983.
5. Kiper, A.: Fourier series coefficients for powers of the
   Jacobian elliptic functions, Math. Comp. 43 (1984) 247-259.
6. Kiper, A.: Some recurrence relations and their parallel
   evaluation using nested recurrent product form algorithm,
   (To appear in J.Comp. Appl. Math. 27 (1989), presented at
   the CAM Conference, August, 1988, Leuven).
7. Schendel, U.: Introduction to Numerical Methods for
   Parallel Computers, Ellis Harwood, Chichester, 1984.
8. Whittaker, E. T. and Watson, G.N.: Modern Analysis,
   Cambridge Univ. Press, Cambridge, 1962.

AN ASYNCHRONOUS NEWTON-RAPHSON METHOD

Freerk A. Lootsma
Faculty of Technical Mathematics and Informatics
Delft University of Technology
P.O. Box 356, 2600 AJ Delft, The Netherlands

Abstract

We consider a parallel variant of the Newton-Raphson method for unconstrained optimization, which uses as many finite differences of gradients as possible to approximate rows and columns of the inverse Hessian matrix. The method is based on the Gauss-Seidel type of updating for quasi-Newton methods originally proposed by Straeter (1973). It incorporates the finite-difference approximations via the Barnes-Rosen corrections analysed by Van Laarhoven (1985). At the end of the paper we discuss the potential of the method for on-line, real-time optimization.

1. Introduction

In the turbulent development of parallel computers, three types of architectures emerged, with their own particular advantages and disadvantages:

a) vector computers which apply arithmetic operations, subdivided and in assembly-line style, on vectors of numbers;

b) single-instruction, multiple-data (SIMD) computers where each processor carries out the same instruction at the same time (fully synchronized) on its own data set;

c) multiple-instruction, multiple-data (MIMD) computers, where each processor works independently from the other ones (asynchronously), usually on shared memory.

As a result of the hardware development, the concept of parallel computing has also been subject to various alterations, accordingly as the views on coordination and synchronization of computational processes came to maturity. Roughly speaking, we can now distinguish two categories of algorithms:

a) synchronous algorithms designed in such a way that the processors carry out predetermined, independent tasks; moreover, the processors signalize completion so that the next set of tasks can be distributed and carried out on predetermined data;

b) asynchronous algorithms where the processors, working on local algorithms, do not wait for predetermined data to become available; on the contrary, they keep on computing, trying to solve the given problem with the data that happen to be available.

Research in parallel non-linear optimization has initially been concerned with synchronous algorithms. Only recently, the possible benefits and the convergence properties of asynchronous algorithms came under investigation. Note that the boundary between synchronous and asynchronous algorithms does not coincide with the line separating the SIMD from the MIMD computers. The parallel algorithms proposed so far are mostly synchronous, even if they have been designed for MIMD computers: they consist of well-defined tasks to be carried out completely between successive synchronization steps.

In this paper we consider a particular type of parallellism for unconstrained optimization: simultaneous evaluations of the objective function, and possibly its gradient, at various points. These are independent tasks, in general suitable for execution on MIMD computers. If there are no branching instructions in the respective routine, the tasks are even identical so that they may be carried out on SIMD computers. Obviously, we ignore the effective, but highly problem-dependent introduction of parallellism into the function-evaluation routine itself. We also disregard parallellism in the matrix and vector operations of a method, on the assumption that in practical problems the function and gradient evaluations are much more expensive than the overheads of the algorithm. The reasons why we concentrate on Newton and quasi-Newton methods are simple to explain. The first and second derivatives of the function to be minimized can easily be approximated by finite differences of function values and gradients, particularly on parallel computers. Moreover, we can readily use the powerful updating mechanism of quasi-Newton methods to incorporate second-order information in the approximate inverse Hessian. We shall particularly be dealing with a Gauss-Seidel type of updating, and with an asynchronous Newton method using the second derivatives that happen to be available.

We recommend the textbook of Bertsekas and Tsitsiklis (1989) for further reading on Gauss-Seidel type substitution methods and on asynchronous algorithms. Although the book does not really discuss the computational implementation of parallel optimization methods, so that it does not contain a performance evaluation either, it provides a theoretical basis for the analysis of parallel algorithms.

2. The Newton-Raphson Method

We consider first the classical Newton-Raphson method for minimizing a non-linear objective function f over the n-dimensional vector space E_n. The method starts from an arbitrary point $x^0 \in E_n$. At the beginning of the k-th iteration we calculate the gradient $\nabla f(x^k)$ and the Hessian matrix $\nabla^2 f(x^k)$ at the current iteration point x^k. Thereafter, the k-th iteration proceeds as follows.

a) Generation of a search direction. The search direction s^k is solved from the system

$$\nabla^2 f(x^k)s = -\nabla f(x^k). \tag{1}$$

b) Exploration of the search direction (one-dimensional search, linear search). We seek an approximation λ_k to a local minimum of the function $\psi_k(\lambda) = f(x^k + \lambda s^k)$.

c) Optimality test. We set $x^{k+1} = x^k + \lambda_k s^k$. If x^{k+1} and x^k are close enough, according to the user-supplied tolerances, the procedure terminates.

This method has the property of quadratic termination: it minimizes a quadratic function with positive-definite Hessian matrix in a finite number of exact linear searches (step b, exploration of the search direction, with λ_k exactly equal to the minimum of ψ_k). Unconstrained minimization methods with this property are usually fast and accurate, at least when they operate in the vicinity of a local minimum. The sequential Newton-Raphson method is excellent for small problems ($n \leq 30$), and workable for medium-size problems ($30 < n \leq 100$).

Supplying the first as well as the second derivatives of f is a major stumbling block in actual applications of the Newton-Raphson method. Several optimization routines therefore provide facilities for numerical differentiation using the following finite-difference approximations:

$$\frac{\partial f(x)}{\partial x_j} \sim \frac{f(x + h_j e^j) - f(x - h_j e^j)}{2h_j},$$

$$\frac{\partial^2 f(x)}{\partial x_j^2} \sim \frac{f(x + h_j e^j) - 2f(x) + f(x - h_j e^j)}{h_j^2},$$

$$\frac{\partial^2 f(x)}{\partial x_i \partial x_j} \sim \frac{f(x + h_i e^i + h_j e^j) - f(x + h_i e^i) - f(x + h_j e^j) + f(x)}{h_i h_j},$$

where e^j denotes the j-th coordinate vector, and h_j a suitable step size in that direction. Calculating the expressions on the right-hand sides by simultaneous function evaluations is an ideal job for parallel computers, but the total number of processors that would be required to do it simultaneously is of the order n^2, and this may be prohibitive for several computers. The alternative is to calculate the user-supplied gradients in parallel. The vector

$$\frac{\nabla f(x + h_j e^j) - \nabla f(x)}{h_j}$$

approximates column j and row j of the Hessian matrix; hence, $n + 1$ processors would be required to approximate the second derivatives simultaneously. This mode of operation has the advantage that the structure of the Hessian matrix is maintained: if $\partial f / \partial x_i$ does not depend on x_j, then both $\partial^2 f / \partial x_i \partial x_j$ as well as its approximation are exactly zero.

The linear search can be carried out in two steps (regardless of the dimension of the problem) when parallellism is introduced. In the bracketing step, designed to locate an interval which indeed contains a local minimum of ψ_k, we calculate the function f simultaneously at the points of a sufficiently large grid; this can readily be generated since $\lambda_k = 1$ is usually a good approximation to a line minimum. In the subsequent interpolation step, we take a grid of points in the so-called embracing interval, we calculate the objective function simultaneously at the gridpoints, and we use the minimum of a high-order interpolating polynomial or rational function to approximate a line minimum. It is important to guarantee stability: we must have $f(x^k + \lambda_k s^k) < f(x^k)$ in order to prevent divergence of the method. The number of processors will not be prohibitive here. We have the impression from the comparative studies in sequential optimization (see Lootsma (1985)) that some ten will be sufficient, even for the variants of Newton's method that we do not discuss in the present paper. A further improvement of Newton's method is deferred updating: then we keep the approximation to the Hessian matrix unchanged during a number of iterations. This idea is to be found in Mukai (1981), who considered deferred updating of the Jacobian matrix of a system of non-linear equations. Computational experience with the above parallel Newton-Raphson method is rare. Much attention has been given, however, to the variants described in the remainder of this chapter.

3. Quasi-Newton Methods

No class of unconstrained-minimization methods attracted so much theoretical and computational attention as the class of quasi-Newton methods. They are excellent for small and medium-size problems. Saving the user the trouble to supply second derivatives, they nevertheless provide many advantages of Newton's method.

In designing methods for unconstrained minimization, we usually study their behaviour on quadratic models of the form

$$q(x) = \frac{1}{2}x^T A x - b^T x, \tag{2}$$

with symmetric and positive-definite Hessian matrix A. We follow the same mode of operation in the present paper.

The quasi-Newton methods employ the difference of two successive iteration points and the difference of the corresponding gradients to approximate the inverse Hessian matrix. The iterative procedure starts at an arbitrary point x^0 with an initial approximation H_0, usually the identity matrix. Let H_k denote the current approximation to A^{-1} at x^k, and $s^k = -H_k \nabla q(x^k)$ the search direction, let $x^{k+1} = x^k + \lambda_k s^k$ be the next iteration point conceptually obtained by an exact linear search along s^k, then the vectors

$$\begin{aligned} \sigma^k &= x^{k+1} - x^k, \\ y^k &= \nabla q(x^{k+1}) - \nabla q(x^k), \end{aligned}$$

are connected by the relation

$$\sigma^k = A^{-1} y^k, \tag{3}$$

but it is generally true that

$$\sigma^k \neq H_k y^k.$$

Hence, we update H_k to obtain a new approximation H_{k+1} such that

$$\sigma^k = H_{k+1} y^k. \tag{4}$$

This is tentatively carried out with an additive correction term D_k. Thus,

$$H_{k+1} = H_k + D_k, \tag{5}$$

and we try to find an expression for D_k satisfying the quasi-Newton property

$$D_k y^k = \sigma^k - H_k y^k. \tag{6}$$

Moreover, we impose the hereditary property which implies that the information gathered in previous iterations remains valid. So if

$$\sigma^i = H_k y^i \quad \text{for } 0 \leq i \leq k - 1,$$

then the correction term D_k must be chosen such that

$$\sigma^i = H_{k+1} y^i \quad \text{for } 0 \leq i \leq k - 1,$$

whence

$$D_k y^i = 0 \quad \text{for } 0 \leq i \leq k - 1.$$

On the basis of formula (6) it is possible to find several expressions for the correction term D_k. The resulting updating formulas are well-known, both in the literature and in practical applications. The Barnes-Rosen formula is given by

$$H_{k+1} = H_k + \frac{(\sigma^k - H_k y^k)(\sigma^k - H_k y^k)^T}{(\sigma^k - H_k y^k)^T y^k}, \tag{7}$$

a so-called rank-one formula with the interesting feature that the hereditary property is satisfied when quadratic functions are minimized, even if the linear search is not exact. Because this property plays an important role in the sections 4 and 5 devoted to synchronous and asynchronous parallel variants of the quasi-Newton methods, we shall briefly establish it here. We consider two vectors y and σ connected by the relations

$$H_k y = \sigma \quad \text{and} \quad A^{-1} y = \sigma. \tag{8}$$

Then

$$(\sigma^k - H_k y^k)^T y = (\sigma^k)^T y - (y^k)^T H_k y =$$
$$= (\sigma^k)^T A\sigma - (y^k)^T \sigma = (\sigma^k)^T A\sigma - (A\sigma^k)^T \sigma = 0.$$

For any two vectors y^i and σ^i with the properties (8), it follows immediately from (7) that

$$H_{k+1} y^i = H_k y^i = \sigma^i,$$

which proves the assertion.

The original DFP formula of Davidon, Fletcher, and Powell is given by

$$H_{k+1} = H_k + \frac{(\sigma^k)(\sigma^k)^T}{(\sigma^k)^T(y^k)} - \frac{H_k(y^k)(y^k)^T H_k}{(y^k)^T H_k(y^k)}, \tag{9}$$

a rank-two formula provided that the vectors σ^k and $H_k y^k$ are linearly independent. Huang (1970) studied the generalized class of updating formulas written as

$$H_{k+1} = H_k + a_{11}^k(\sigma^k)(\sigma^k)^T + a_{12}^k(\sigma^k)(y^k)^T H_k +$$
$$+ a_{21}^k H_k(y^k)(\sigma^k)^T + a_{22}^k H_k(y^k)(y^k)^T H_k. \tag{10}$$

In order to satisfy the relation (6), the coefficients $a_{11}^k, \cdots, a_{22}^k$ are subject to the restrictions

$$\left. \begin{array}{l} a_{11}^k(\sigma^k)^T(y^k) + a_{12}^k(y^k)^T H_k(y^k) = 1, \\[2mm] a_{21}^k(\sigma^k)^T(y^k) + a_{22}^k(y^k)^T H_k(y^k) = -1. \end{array} \right\} \tag{11}$$

An interesting member of Huang's class is the BFGS formula of Broyden, Fletcher, Goldfarb, and Shanno. It has a widespread popularity, but because we do not need it in the present paper we omit the explicit formulation.

In order to establish the hereditary property for the updating formulas (9) and (10), we have to use the condition that the successive iteration points are obtained by exact linear searches. In practice, this condition does not impose unacceptably high requirements on the quality of the search, but nevertheless one always has to be aware of these limitations, which are not present when the Barnes-Rosen correction (7) is employed.

4. A Synchronous Parallel Quasi-Newton Method

The method which we discuss in this section has originally been proposed by Straeter (1973). The theoretical analysis is due to Van Laarhoven (1985). Computational experience is mainly reported by Dayde (1986, 1989).

For ease of exposition we rewrite the updating formula (7) as

$$H_{k+1} = H_k + \tau_k r^k (r^k)^T, \tag{12}$$

using the residual vector

$$r^k = \sigma^k - H_k y^k \tag{13}$$

so that

$$\tau_k = ((r^k)^T y^k)^{-1}. \tag{14}$$

At the beginning of the k-th iteration we find ourselves at the iteration point x^k with the approximate inverse Hessian matrix H_k. Let $\sigma^{k1}, \cdots, \sigma^{kn}$ denote n linearly independent displacements (usually multiples of the unit vectors), then we evaluate the function f and its gradient ∇f in parallel at x^k and at the points

$$x^{kj} = x^k + \sigma^{kj}, \quad j = 1, \cdots, n.$$

In a tentative approach to update H_k via a generalization of (12) - (14) we define

$$y^{kj} = \nabla f(x^{kj}) - \nabla f(x^k),$$

and we write

$$H_{k+1} = H_k + \sum_{j=1}^{n} \tau_{kj} r^{kj} (r^{kj})^T, \tag{15}$$

$$r^{kj} = \sigma^{kj} - H_k y^{kj}, \tag{16}$$

$$\tau_{kj} = ((r^{kj})^T y^{kj})^{-1}. \tag{17}$$

It will be obvious that formula (15) updates H_k via n additive, independent corrections. In that sense, the procedure is similar to a Jacobi algorithm (see Bertsekas and Tsitsiklis (1989)) which generates the next iterate by processing current pieces of information independently. Straeter (1973), however, launched a more sophisticated procedure, similar to a Gauss-Seidel algorithm which uses the updated pieces of information immediately to process the remaining ones. He introduced the so-called partial updates whereby the information collected via the displacements $\sigma^{k1}, \cdots, \sigma^{k,j-1}$ is used to make an update in the direction σ^{kj}. This leads to the following procedure:

$$V_{ko} = H_k, \tag{18}$$

$$r^{kj} = \sigma^{kj} - V_{k,j-1} y^{kj}; \quad j = 1, \cdots, n, \tag{19}$$

$$V_{kj} = V_{k,j-1} + \tau_{kj} (r^{kj})(r^{kj})^T; \quad j = 1, \cdots, n, \tag{20}$$

$$H_{k+1} = V_{kn}. \tag{21}$$

The search direction s^k is generated by

$$s^k = -H_{k+1}\nabla f(x^k),$$

whereafter a linear search along s^k is performed to find x^{k+1}. This completes the k-th iteration of Straeter's (1973) method.

It is easy to demonstrate that the method has the property of quadratic termination. When a quadratic model is minimized, it must be true that $H_1 = A^{-1}$. Hence, it will take one exact linear search to find the minimum, regardless of the starting point and the dimension of the model.

Van Laarhoven (1985) studied the class of parallel quasi-Newton methods obtained by a generalization of Huang's (1970) updating formula (10). He replaced (18) - (21) by

$$V_{ko} = H_k, \tag{22}$$

$$B_{kj} = a_{11}^{kj}\sigma^{kj}(\sigma^{kj})^T + a_{12}^{kj}\sigma^{kj}(y^{kj})^T V_{k,j-1} +$$
$$+ a_{21}^{kj} V_{k,j-1} y^{kj}(\sigma^{kj})^T + a_{22}^{kj} V_{k,j-1} y^{kj}(y^{kj})^T V_{k,j-1}, \tag{23}$$

$$V_{kj} = V_{k,j-1} + B_{kj}; \quad j = 1, \cdots, n, \tag{24}$$

$$H_{k+1} = V_{kn}, \tag{25}$$

and established sufficient conditions for quadratic termination of the method as follows. It is sufficient to show that

$$V_{kj} y^{kl} = \sigma^{kl}, \quad l = 1, \cdots, j, \tag{26}$$

holds for any $j = 1, \cdots, n$, provided that the displacements $\sigma^{kl}, l = 1, \cdots, n$ are taken to be linearly independent. On the basis of (24) and (26), the parameters $a_{11}^{kj}, \cdots, a_{22}^{kj}$ must be chosen to satisfy

$$B_{kj} y^{kl} = 0; \quad l = 1, \cdots, j - 1, \tag{27}$$

$$B_{kj} y^{kj} = \sigma^{kj} - V_{k,j-1} y^{kj} = r^{kj}. \tag{28}$$

So, for $j = 1$ we must have

$$B_{k1} y^{k1} = \sigma^{k1} - H_k y^{k1}, \tag{29}$$

an expression which clearly resembles the requirement (6). Turning to the case $j > 1$ we observe, prior to working out expression (27), and using (26), that

$$(y^{kj})^T V_{k,j-1} y^{kl} = (y^{kj})^T \sigma^{kl} =$$
$$= (y^{kj})^T A^{-1} y^{kl} = (\sigma^{kj})^T y^{kl}, l = 1, \cdots, j - 1. \tag{30}$$

Moreover, we assume that the terms appearing in (30) are generally non-zero, and that the two vectors σ^{kj} and $V_{k,j-1}y^{kj}$ are independent. With these results it follows easily from (23) and (27) that

$$a_{11}^{kj} = -a_{12}^{kj},$$
$$a_{21}^{kj} = -a_{22}^{kj},$$

which reduces (23) to

$$B_{kj} = a_{11}^{kj}\sigma^{kj}(\sigma^{kj} - V_{k,j-1}^T y^{kj})^T +$$
$$+ a_{22}^{kj}V_{k,j-1}y^{kj}(\sigma^{kj} - V_{k,j-1}^T y^{kj})^T. \tag{31}$$

Finally, the requirement (28) yields

$$a_{11}^{kj} = -a_{22}^{kj} = ((r^{kj})^T y^{kj})^{-1},$$

whence

$$B_{kj} = \frac{(r^{kj})(\sigma^{kj} - V_{k,j-1}^T y^{kj})^T}{(r^{kj})^T y^{kj}}. \tag{32}$$

This result yields Straeter's formula (20), provided that the matrices $V_{k,j-1}$ are symmetric. And they are, if H_k and the first partial update B_{k1} are symmetric matrices, which can be guaranteed by a proper utilization of (29). Hence, in the class of quasi-Newton methods with the generalized updating formulas (22) - (25), only Straeter's method (1973) has the property of quadratic termination.

Computational experience is limited here. Straeter (1973) and Van Laarhoven (1985) used small traditional test problems only. Dayde (1986, 1989) employed medium-size problems and observed that the method is competitive with other quasi-Newton methods, even on a sequential computer. In a parallel environment, however, it has to compete with Newton's method using finite differences of gradients to approximate second derivatives (see sec. 2). Newton's method, however, preserves sparsity and structure.

5. An Asynchronous Newton Method

As we have seen in sec. 3, the current approximation of the inverse Hessian matrix is updated on the basis of simple pieces of information, namely the difference y between two gradients and the difference σ between the corresponding arguments. When we concern ourselves with the quadratic model (2) we know that y and σ are connected by $\sigma = A^{-1}y$. Thus, y and σ demonstrate how the inverse Hessian matrix operates. In doing so, they provide the valuable second-order information which is successfully employed in the updating mechanism of quasi-Newton methods.

In the previous section, we only used the hypothesis that the displacements $\sigma^{kj}, j = 1,\ldots,n$, are linearly independent. A particular situation arises when the displacements are multiples of the unit vectors so that they can be written as

$$\sigma^{kj} = h_{kj}e^j \tag{33}$$

where e^j represents the j-th unit vector and h_{kj} a suitable step size. In the quadratic case, it is obvious that

$$y^{kj} = h_{kj} A e^j = h_{kj} a^j.$$

with a^j standing for the j-th row and column of the Hessian matrix A. In the non-quadratic case, the method of Straeter and Van Laarhoven generates finite-difference approximations

$$d^{kj} = y^{kj}/h_{kj} \tag{34}$$

to the second derivatives in row j and column j of the Hessian matrix evaluated at the current iteration point x^k, but the method has a remarkable feature. The finite differences are not used to replace the corresponding rows and columns of the current approximate Hessian matrix itself, but they are immediately employed to update the current approximation H_k of the *inverse* Hessian matrix. Moreover, the intermediate partial updates

$$V_{kj} = V_{k,j-1} + \frac{(e^j - V_{k,j-1} d^{kj})(e^j - V_{k,j-1} d^{kj})^T}{(e^j - V_{k,j-1} d^{kj})^T d^{kj}} \tag{35}$$

are symmetric and positive -definite. This is an attractive property, as we will see later in the present section. Note that intermediate updates of the approximate Hessian matrix itself, obtained when we replace only *some* rows and columns by finite-difference approximations to the second derivatives, are generally indefinite.

The powerful updating mechanism of Straeter's and Van Laarhoven's method is particularly useful when we do not wait for a complete "scan" along n independent displacements. In fact, it may be sufficient to process only a limited number of displacements per iteration, provided that each displacement is regularly scanned during the computations, possibly in a cyclic order. In general, the updating mechanism uses as many displacements per iteration as possible, given the number of available processors. It does not have to wait until a prescribed number of rows and columns becomes available, as Byrd, Schnabel, and Shultz (1988) propose. Their experiments show, not unexpectedly, the gradual transition between quasi-Newton and Newton methods. Fischer and Ritter (1988) use indeed the finite differences obtained so far to replace some rows and columns in the current approximation to the Hessian matrix itself. They accept that the approximate Hessian so obtained is not necessarily positive-definite. Thus, they loose the attractive property of Newton's method, that a stepsize of 1 in the search direction is a good guess at the line minimum. We need the property, not because accurate line minimization would be required, but in order to maintain stability of the method, in the sense that the function f decreases at the subsequent iteration points. There are no numerical experiments, however, to show whether it is really desirable to guarantee that the approximate Hessian remains positive-definite throughout the computations.

Final Remarks

The potential of the asynchronous Newton method is still unexplored, but it could be surprisingly high, particularly if there is a large number of processors available for the continual computation of rows and columns of the Hessian matrix. Moreover, the possible benefits of automatic differentiation are still under study (see Grandinetti and Conforti (1988)); such a technique could also contribute to the power of parallel optimization. In summary, there is a promising area of research and development here, with a variety of applications in on-line and real-time optimization.

Lastly, a few words about asynchronous computing. We cannot deny that many parallel algorithms are still sequential by nature. True, they contain independent tasks which are allocated to various processors, but the host computer still has overall control in a sequential way: it does not allocate new tasks before the previous ones have been carried out properly. Asynchronous computing, using the information that happens to be available, presents a new viewpoint on parallel computing, and it requires a new type of analysis.

REFERENCES

[1] D.P. Bertsekas and J.N. Tsitsiklis, Parallel and Distributed Computation. Prentice Hall, Englewood Cliffs, New Jersey, 1989.

[2] R.H. Byrd, R.B. Schnabel, and G.A. Shultz, Parallel Quasi-Newton Methods for Unconstrained Optimization. Mathematical Programming 42, 273-306, 1988.

[3] M. Dayde, Parallélisation d'Algorithmes d'Optimisation pour des Problèmes d'Optimum Design. Thèse, Institut National Polytechnique de Toulouse, France, 1986.

[4] M. Dayde, Parallel Algorithms for Nonlinear Programming Problems. To appear in the Journal of Optimization Theory and Applications, 1989.

[5] H. Fischer and K. Ritter, An Asynchronous Parallel Newton Method. Mathematical Programming 42, 363-374, 1988.

[6] L. Grandinetti and D. Conforti, Numerical Comparisons of Nonlinear Programming Algorithms on Serial and Vector Processors using Automatic Differentiation. Mathematical Programming 42, 375-389, 1988.

[7] H.Y. Huang, A Unified Approach to Quadratically Convergent Algorithms for Function Minimization. Journal of Optimization Theory and Applications 5, 405-423, 1970.

[8] P. van Laarhoven, Parallel Variable Metric Algorithms for Unconstrained Optimization. Mathematical Programming 33, 68-81, 1985.

[9] F.A. Lootsma, Comparative Performance Evaluation, Experimental Design, and Generation of Test Problems in Nonlinear Optimization. In K. Schittkowski (ed.), Computational Mathematical Programming.
Springer, Berlin, 249-260, 1985.

[10] F.A. Lootsma and K.M. Ragsdell, State-of-the-Art in Parallel Nonlinear Optimization. Parallel Computing 6, 133-155, 1988.

[11] H. Mukai, Parallel Algorithms for Solving Systems of Nonlinear Equations. Computation and Mathematics with Applications 7, 235-250, 1981.

[12] T.A. Straeter, A Parallel Variable Metric Optimization Algorithm.
NASA Technical Note D-7329, Langley Research Center, Hampton, Virginia 23665, 1973.

PART 8

NETWORKING

INTERCONNECTION NETWORKS FOR HIGHLY PARALLEL SUPERCOMPUTING ARCHITECTURES

A. Antola   R. Negrini   M.G. Sami   R. Stefanelli

Dipartimento di Elettronica, Politecnico di Milano,
Piazza Leonardo da Vinci 32, I-20133 Milano, Italy

ABSTRACT

This paper treats (Multiple) Interconnection Networks (MINs), dealing in particular with their adoption in supercomputing architectures. The main properties and characteristics of MINs are recalled and some typical parallel computer architectures adopting MINs are summarized. Two main application classes of MINs are considered: parallel computer systems implemented by connecting together powerful processors and large shared memories, and dedicated supercomputing structures directly implementing highly parallel algorithms. For both application classes, the adoption of fault tolerance methods is discussed. Fault tolerance can be usefully adopted both to overcome production defects and faults arising during systems working life. Classic approaches to fault tolerance in MINs for parallel computer systems and some recent results in the less known field of fault tolerance in dedicated supercomputing structures are surveyed.

1. INTERCONNECTION NETWORKS AND SUPERCOMPUTING

A very promising approach to supercomputer architectures is based upon massive parallelism, i.e., adoption of hundreds or thousands of identical computing units. Compared with approaches based upon extremely powerful electronic technologies and processes, this approach exploits in a cheaper way the best known characteristics of off-the shelf technologies, i.e., the capability of producing highly integrated, complex, identical circuit modules.

The reachable processing capabilities of massively parallel systems heavily depend on the throughput of the adopted interconnection networks, since — especially in supercomputer environments — the exploitation of parallelism implies the use of many cooperating processes (executed by different processors) exchanging data in heavy and complex ways.

General purpose supercomputers have been designed exploiting parallelism in many ways, ranging from SIMD arrays of bit-serial processors (each with simple instruction decoders) to MIMD structures based upon powerful CPUs. This last case, for example, has been afforded by means of *multibus* structures. Many of these structures are tightly connected, meaning that data transfers between interconnected modules are executed by means of bus cycles within instruction cycles. Totally different examples of tightly coupled powerful systems can adopt the *cross-bar switch* interconnection, i.e., an interconnection matrix of switches allowing for direct connection of all the members of an homogeneous class of elements to all the elements of a second homogeneous class of elements (e.g., processing units and memory banks, respectively).

This paper will consider Interconnection Networks (INs) based upon more complex but more flexible structures. In particular, Multiple Interconnection Networks (MINs) will be treated;

NATO ASI Series, Vol. F 62
Supercomputing
Edited by J. S. Kowalik
© Springer-Verlag Berlin Heidelberg 1990

MINs can be defined as interconnection structures obtained by cascading either identical or not identical interconnection networks.

A complementary approach to supercomputing by means of general purpose highly parallel structures is the one that considers dedicated circuits. In this case, that can be classified as the *hardwired algorithms* approach, the goal consists in the implementation of structures directly dedicated to the computation of a particular algorithm or class of strictly similar algorithms. Of course, this approach is mainly valid for the cases when inherently parallel algorithms must be applied to very large data sets, with real-time requirements. This happens, for example, in the field of digital signal or image processing. Typically, these structures directly map the flow graph of the algorithm onto the circuit structure: only slight changes are allowed for in the structure (e.g., for the change of numeric values of constants of the algorithm, or for the change of the size of the data set, or for switching from an algorithm to another, very similar one). These structures can be used as powerful stand alone embedded systems, or as modules connected to standard workstations or systems, specializing them as supercomputers for signal or image processing. In recent years this approach has given rise to new interesting concepts and structures as, e.g., the systolic arrays and the wavefront arrays.

As it will be seen in the sequel, many important algorithms of the fields of signal or image processing show flow graphs that are amenable to the graphs of well known MINs. This fact suggests that some basic principles and methods can be shared between these distant approaches. As an example of this possibility, this paper is dedicated to the analysis of some important approaches to fault- tolerance both in MINs and in hardwired, dedicated arrays.

2. ANALYSIS OF INTERCONNECTION NETWORKS

Interconnection networks can be classified by their communication methods (see, e.g., [LIP87]). In *circuit-switched* networks, each node contains at most digital switches that connect input wires to output wires. Data move through a node from an input wire to an output wire through an enabled amplifier, being delayed by gate delays only. Links can of course be constituted by single wires (bit-serial networks) or by many wires in parallel (up to a computer bus). Many paths can be contemporarily active between different couples of input and output nodes of the interconnection networks.

In *packet-switched* networks, a node contains a data register, and bits of data are grouped into packets as large as a data register. During a communication *cycle*, a packet is sent from one node — through a link — to another node, where it is stored waiting for the next cycle. Here also, links can be either bit-serial or parallel up to the dimensions of the data registers. In the network, many packets can be routed at the same time. If nodes contain more than one data register, each register can sustain a communication path, and more paths can be active simultaneously at the same node: if these paths do not require the same links, this method can be implemented quite simply. Furthermore, a *packet train* constituted by several packets (in a fixed number) can be sent in sequence through the network along the same path. A full *message-switching* communication mode is supported when the packet train can be of arbitrary length. Hardware protocols can be defined so that even asynchronous communication between adjacent nodes is allowed for.

In both cases, information addressing and controlling the routing of data can be added to the true data and used by the control algorithms implemented at the nodes, in order to set up a communication path. These mechanisms allow for the dynamic creation of communication paths through the network.

A network can be characterized by various parameters and figures of merit; it is often important to determine how these parameters change with the *size* of the network defined by the total number of nodes, or by the sum of the number of input nodes s_i and output nodes s_j.

The most important parameters and characteristics are the following ones:

(1) the number of input nodes s_i, where paths can start. Input nodes appear as source nodes of the graph of the network, having outgoing arcs only;

(2) the number of output nodes s_j, where information paths can end; these nodes are sink nodes of the graph;

(3) the total numbers of nodes and of arcs (i.e., the cost of the network);

(4) the maximum number of arcs entering and of arcs leaving a node (i.e., the part of the costs of a node that depends from interconnections);

(5) the maximum number of arcs along a path from a source to a sink (*diameter*). This parameter influences the total communication (worst case) delays;

(6) parallelism, i.e., how many communication paths the network can support simultaneously, stating whether it is independent from *what* paths are created, or not;

(7) address mechanism of the paths (absolute, or relative to the input node). In some networks, an absolute value (a binary number) labels each output node, and the control algorithm in each node uses this number to decide how to prolong the path towards the desired output. It is convenient that the control algorithm is independent from the position of the nodes;

(8) connection capabilities of nodes (i.e., the number of possible connections that a node can contemporarily hold between its input and output arcs, and constraints upon the choice of contemporary connections).

Well known examples of interconnection networks are *rings, meshes, trees, crossbar switches, n-cubes*. In this paper, we will treat in particular the interconnection networks whose structure is represented by graphs called *banyans*: a Banyan is a directed graph such that for all s_i source nodes and for all s_j sink nodes there is exactly one single path from s_i to s_j (no directed loops are allowed).

Looking at their definition, banyan networks appear particularly interesting since they (1) do not contain redundant links and (2) allow for interconnection of each input node with each output node. In other words, a banyan is a fully connected mapping from a set of sources to a set of sinks, with intermediate nodes allowing to share connection links. Various kinds of banyans have been defined. Some of them show *regularity* and *rectangularity*, as defined below.

Definition: A Banyan is *regular* if all the nodes (except inputs) have the same number of incoming arcs (*in-degree*), and all the nodes (except outputs) have the same number of outgoing arcs (*out-degree*). Of course, banyans that are *not* regular are called *irregular*.

Definition: A Banyan is *l-level* if every path from inputs to outputs is of length l. For these banyans, it is possible to group intermediate nodes into intermediate *levels*: i.e., level k is the set of nodes at the end of all the paths of length k starting from input nodes. As an example, a binary tree is an *l*-level regular banyan.

Definition: A Banyan is *rectangular* if there is the same number of nodes at each level.

A regular *l*-level banyan is often classified by the triple (i, o, l), where i is the in-degree, o the out-degree and l the number of levels. Regular *l*-level banyans can be built by means of replicated

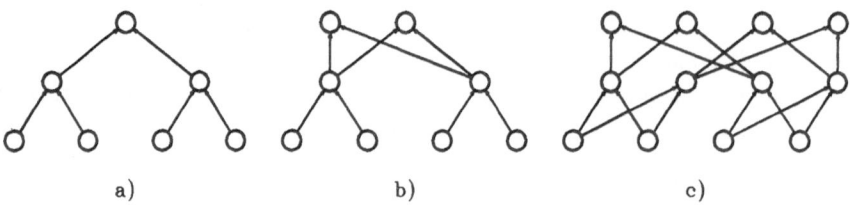

a)   b)   c)

Fig. 1 Construction of a (2,2,2) SW-banjan: (a) the original tree; (b) duplication at level 1; (c) duplication at level 2.

identical modules. In a l-level regular rectangular banyan, the number of levels is $log(n)$ (n
being the number of source nodes, and log being base i, the in-degree). Every level has n nodes
and is linked to the following one by $n \times i$ arcs. It is straightforward that the in-degree and the
out-degree are identical.

An important class of regular banyans is that of *SW-banyans* (SWitching banyans) that
can be built starting from a tree, by means of iterated duplications as shown in Figure 1 for the
case of (2,2,2) SW-banyan [LIP87], starting from the root. The duplication at level k consists
in copying the subgraph from level 0 to level $k-1$ and then reproducing from level $k-1$ of the
copy to level k of the complete graph all the links from level $k-1$ to k in the complete graph.
At each duplication step the number of sources is duplicated: by construction, the graph is a
regular banyan.

Some useful interconnection networks, while not banyans, are strictly related to them. The
perfect shuffle, for example, is an homomorphism of a banyan. Other well known networks
are compositions of banyans, or can be obtained by augmenting banyans (i.e., they contain
banyans as partial graphs). (See [REE87] and [KUM87].) These augmented networks are often
introduced in order to add fault-tolerance capabilities to banyans, as we will see in Sections 3
and 4.

In Section 4, this paper will show how banyans and related networks can be used in order
to build dedicated processing arrays allowing for very high computing power.

Many highly parallel machines based upon interconnection networks have been proposed,
or built. We only recall few well- known examples, following [LIP87], which gives more detailed
comments and long references lists.

The first version of *TRAC* (the "Texas Reconfigurable Array Computer") is an experimental
prototype of a large scientific computer, being built at the University of Texas at Austin. The
machine was designed to be a "number cruncher", exploiting parallelism in order to build a
scientific supercomputer out of commercial, powerful microprocessors. TRAC has a regular
structure, constructed from an array of identical cells interconnected in a regular, non rectangular
banyan. Both circuit and packet switching can be used in the same network. The banyan is a
regular (2,3,2) SW-banyan.

PASM is a partitionable SIMD/MIMD multicomputer, result of a long research activity
at Purdue University. It can be structured as one or more independent SIMD and/or MIMD
machines of various size. Its intended application fields are image processing and pattern recognition. It can also be applied to speech understanding and biomedical signal processing. PASM's
multistage network is a generalized cube (a (2,2,2) SW-banyan) quite similar to that of TRAC.

RP9 is the IBM Research Parallel Prototype, introduced in cooperation by the IBM T.J.
Watson Research Center and the Courant Institute of New York University. It is a research

and prototyping effort to investigate both hardware and software aspects of highly parallel computation. The machine will consist of 512 state-of-the art microprocessors, and two multistage networks. The networks are based on expanded Omega networks (equivalent to SW-banyans), with an extra stage for improved performance and fault-tolerance.

The *ULTRACOMPUTER* is a multi-purpose, MIMD machine designed at Courant Institute of New York University. The full design consists of 4096 processors sharing memory through an Omega network (equivalent to a (2,2,12) SW-banyan) using message switching.

3. FAULT-TOLERANT INTERCONNECTION NETWORKS FOR PARALLEL PROCESSORS

Fault-tolerance in highly parallel supercomputers can be dealt with from a number of different points of view. A possible aspect deals with faults located in the *terminal (i.e., input and output) nodes*: in that case, the interconnection network must route communications with the faulty-free terminals so as to support the reconfiguration/reassignment procedure foreseen for the terminals themselves. We do not consider here this problem, that relates more in general to reconfiguration of parallel systems; rather, the faults taken into account are those affecting *the interconnection network* proper, and the policies examined aim at keeping as far as possible the communication capacities of the network itself.

The networks discussed in the present section are *circuit-switched networks* providing a complete communication within a set of N *processors* constituting the *input nodes* of the network and a set of N *memories* constituting its *output nodes*. The primary goal of such a network is to grant that any output node be *reachable* from any input node; the secondary goals are to achieve the highest parallelism of communication (at least, in statistical terms) while keeping the cost and complexity of the interconnection network within acceptable bounds. Given the nature of the terminal nodes and of the operation performed by input-output routing, whenever a connection is activated the information exchange thus initiated is terminated prior to de-activating the connection and no memory of previous routings is required.

Referring to the elementary 2×2 switches considered in most cases, accepted switch settings range from the cross-through pair of positions for the so-called basic Omega network (extended to $K \times K$ crossbar switches for generalized Omega networks) to the augmented set comprising also lower broadcast and upper broadcast for the generalized cube (see Figure 2). Correspondingly, a variety of fault models has been devised. In the simplest instances [AGR82], *stuck-at* faults have been assumed for both control and signal lines, so that a switch is incapable of either:

(a) assuming one of its positions (stuck-at fault on control line), or

(b) transmitting correct data on one of its data lines (stuck-at fault on an input line).

A classical solution to the diagnosis of a MIN utilizing 2×2 switches and adopting the above fault model was presented in 1981 by Feng and Wu [FEN81], who proved that any such MIN, whatever its dimensions, could be fully diagnosed for any single fault by a test sequence of exactly *four* tests. To the input s_i a 0 is applied if i is odd (or a 1 if i is even) during tests number 1 and 3 (inputs are reversed in tests 2 and 4). All switches are set in direct connection in tests 1 and 2, and in cross-connection in tests 3 and 4. If the MIN is fault-free, the outputs are identical to the inputs; in case of a single fault, the output values allow for detection and localization of the fault.

A more comprehensive fault model was adopted by Cherkassky, Opper and Malek [CHE84],

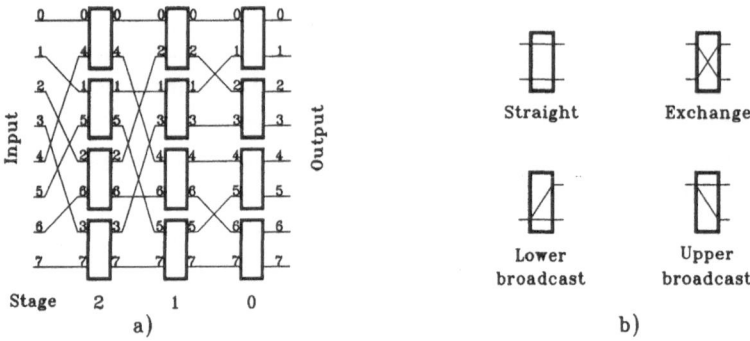

Fig. 2 An 8 × 8 MIN and states of its 2 × 2 switches.

where switches of type $f \times f$ are considered for creation of a *fault-tolerant* banyan network (including therefore redundant stages); faults allowed involve:

(a) stuck-at faults on data and control (*tag routing*) lines;

(b) bridge faults, i.e., AND (OR) shorted pairs of adjacent physical lines.

Results achieved by Cherkassky, Opper and Malek lead to state that at most $2f$ tests are required to detect a single fault in a banyan network with switches of type $f \times f$ and l stages, whatever the value of l, and that at most $2f + |log(l + 1)|$ tests are required to locate any single fault.

A broader, *functional* definition can be accepted when *fault-tolerance* proper, i.e., capacity of overcoming a fault through reconfiguration of the network, is envisioned. The fault assumption adopted states simply that *a faulty switch is incapable of correct operation,* so that reconfiguration actions must be undertaken to isolate and circumvent it; no residual functionality is taken into account.

A further point obviously concerns faults of connections between switches. Assumptions vary widely depending on whether a discrete-element network is envisioned or a VLSI structure is taken into account. While in the first case probability of fault of a link is quite high (some authors even stated it to be higher than for logic elements), in the second case statistics show greater probability of faults in logic rather than in inter-logic connections, so that authors dealing with VLSI devices often assume interconnections to be fault-free. Usually, a fault on an interconnection link can be assimilated to a suitable fault of one of its terminal switches; still, taking into account such faults obviously leads to considering larger distributions of faults.

An interesting analysis was performed by Gazit and Malek [GAZ88], with the aim of identifying the *inherent fault-tolerance* of banyan networks. The basic (non-redundant) network provides exactly one path from any input node to any output node, so that any single fault (of a switch or of a link) will make at least one input-output connection unavailable; if a *degraded* performance is accepted, by allowing the system to continue operation on a subset of its input and output nodes (either terminals and processors, respectively, or processors and memories), faults can be tolerated. To this end, degraded capabilities can be defined, such as:

(1) $QPRT$: the capability of at least Q operational processors (input nodes) to communicate with at least R operational terminals (output nodes);

(2) $SPST$: the capability of a single operational processor to communicate with a single operational terminal;

(3) $SPAT$: the capability of a single operational processor to communicate with all the terminals (assumed to be operational);

(4) *APST*: the capability of all processors (assumed to be operational) to communicate with a single terminal;

(5) *QPAT*: *Q* processors to all terminals;

(6) *APRT*: all processors to *R* terminals;

(7) *APAT*: all processors to all terminals (this, obviously, is the fault-free instance).

The influence of any single fault in the interconnection network can be determined by extracting the subtrees leading to and originating from the faulty element: thus a fault in a single *internal* switch of the network will result in a decrease of both the processors and the terminals capable to communicate with each other. The main results obtained by Gazit and Malek are that:

(a) any single fault (unless it involves an output terminal) allows *SPAT* capability;

(b) any single fault (unless it involves an input terminal) allows *APST* capability;

(c) an (f,f,l) regular rectangular banyan allows *SPAT* capability for up to $(f-1)$ faults, provided no output terminals are faulty; the minimum number of faults excluding such capability is f;

(d) an (f,f,l) regular rectangular banyan allows *APST* capability for up to $(f-1)$ faults, provided no input terminals are faulty; the minimum number of faults excluding such capability is f.

Many authors have dealt with the opposite problem, i.e., that of adding some form of redundancy to the basic interconnection network to grant complete functionality even after any fault in a pre-defined pattern has occurred. Probably the best-known solutions in this context derive from the *Extra-Stage Cube* network proposed by Adams and Siegel [ADA82, ADA84] with reference to the generalized cube network. A Generalized Cube Network is a MIN with N inputs and N outputs, where N is a power of 2, with a number of stages equal to $log_2 N$ and in which each stage consists of N links connected to $N/2$ switches. 2×2 switches are used, and each switch (individually controlled) is characterized by *four* acceptable states, namely *cross*, *through*, *lower broadcast* and *upper broadcast* (see Figure 2). For routing purposes, input and output ports of all switches are marked by ordered numerals; upper (lower) input and output port of any switch have the same ordering numeral, and the labels are such that upper and lower port for a switch in the $i-th$ stage differ only in the $i-th$ bit.

The Extra Stage Cube is obtained from this network by adding an extra stage at the network's inputs and by introducing, respectively, two-way multiplexers on each output port of the added stage and two-way demultiplexers on the input ports of the switches of the output stage. The control signals to all multiplexers of stage 0 and all demultiplexers of stage n are sent in parallel; thus, stage 0 or stage n can be *enabled* (information is forwarded to the input ports of its switches) or *disabled*. If all switches are fault-free (or if a switch in stage n is faulty), stage n is disabled and stages 0 to $n-1$ operate normally; if a switch in stage 0 is faulty, stage 0 is disabled and stage n is enabled. If a faulty switch located in any other stage, or a faulty link, are diagnosed, both stages 0 and n are enabled, thus providing *two* alternative paths between any input and any output and therefore granting survival to any single fault. (The underlying assumption is that multiplexers and demultiplexers should be fault-free: this is an acceptable assumption, limiting the *hard-core* section to a restricted portion of the extended system). The Extra-stage cube is also *robust* to multiple faults, meaning that a number of patterns of multiple faults can be survived to: more precisely, all multiple faults whose distribution is such as not to affect *both* paths connecting a given input-output pair can be overcome. It should be noticed that while connectivity is kept unchanged in the presence of single faults, *parallelism* (i.e., number of *simultaneous connections activated*) may decrease.

In general, the fault-tolerance criterion for multistage interconnection networks can be summarized as providing multiple paths between input and output terminals so that total connectivity can be kept even in the presence of suitable fault distributions. The *F-network* structure, [CIM86], for instance, reaches this goal by introducing additional links, rather than an additional stages (and actually restricting the number of allowable faults, since input and output switches constitute now the hard-core of the structure). The architecture in Figure 3, has thus 4×4 switches; again, the network is tolerant to single faults (with the exceptions already underscored) and robust to multiple faults. Redundancy is actually much more extended than could be apparent from the abstract scheme: switches are far more complex than the basic 2×2 ones, their control requires more lines and the more intricate structure of the links adds to the area overhead in a relevant fashion.

An alternative approach consists in adding a *row* (or rows), rather than a stage, to the basic structure; this solution is presented by the *Dynamic Redundancy (DR) Network* [JEN86], derived from the flow graph of the generalized cube. Figure 4 shows the scheme of a DR network (derived from the cube in Figure 2). Here also, any component may fail (faults are assumed to be independent) and fault tolerance to any single fault is achieved: Figure 4.a shows how a generalized cube can be embedded in a DR network with two spare rows, whereas Figure 4.b shows how a single fault (or more faults hitting two adjacent rows) can be overcome, by simply reordering inputs and outputs. (Thick lines represent active connections.) Thus, the system retains possibility of performing any set of *simultaneous* connections allowed by the fault-free cube, even in the presence of such faults. Apparently, the level of redundancy is lower than for the extra-stage system (only two rows of switches have been added) even though higher performances are granted: still, it should be noticed that the far greater complexity of the system of links adds considerable area redundancy when a VLSI implementation is envisioned.

The approaches examined in this section have exemplified the four main philosophies concerning fault-tolerance of MINs: apart from graceful degradation (the first solution), redundancy has taken the form of added *stages*, added *rows* of switches or added *links* (and, implicitly, added complexity of the switches and of the related control). While more complex solutions (involving, e.g., two different approaches at the same time) are possible, two main figures of merit - related to each other - must be taken into account. First of all, switches in the basic MIN are relatively simple, and added complexity - in particular, with respect to their control - becomes easily relevant; secondly (and consequently) the increase in silicon area leads to sharp reliability drops above a *balance point* that has been well exemplified by such authors as Mangir and Franzon [MAN82, FRA86].

4. FAULT-TOLERANT INTERCONNECTION NETWORKS FOR DEDICATED SUPER-COMPUTING

Dedicated supercomputing is here referred as high performance processing implemented by computing architectures derived from the translation and the direct mapping of an algorithm (or, more generally, a functionality) onto an hardwired structure. Translation and mapping may be done, in a formal way, maintaining the inherent parallelism of an algorithm, using its *data dependency graph* and *signal flow graph* (see, e.g., [RAO88]).

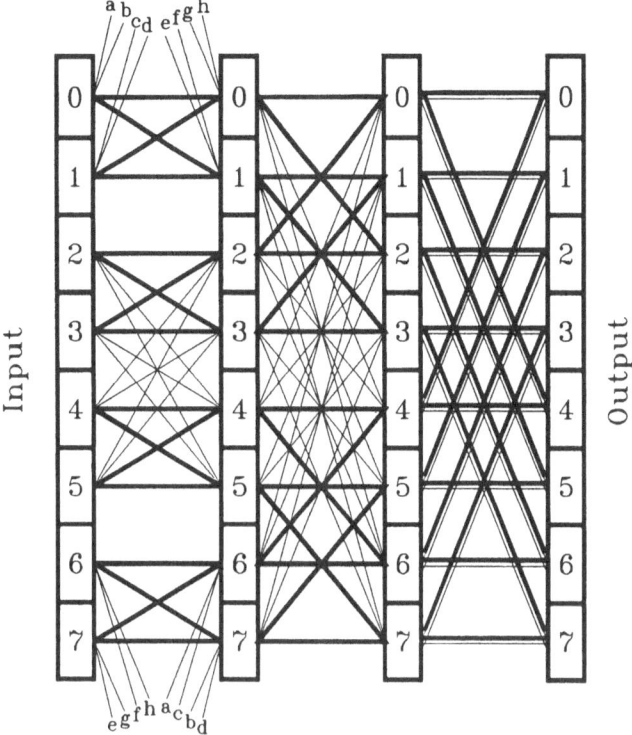

Fig. 3 An F-network structure; thin lines represent the redundant links.

In a more intuitive way, this translation and mapping is obtained by identifying the iterative solution of the algorithm itself. The single basic iteration represents then the operation performed by the processing elements of the dedicated structure, for this reason the graph represents not only the algorithm, but also the hardwired architecture performing it. The computational graph is obtained by interconnecting the basic iterations and exploiting both the temporal and the spatial parallelism. In most cases, this graph is characterized by a number of *stages* identifying the iterations of a given-size problem. The number of nodes (i.e., processing elements) associated with each stage defines the spatial parallelism of the algorithm. Stages are connected through a network whose topology defines the dependencies between the partial results produced in an iteration and the inputs necessary to the following iteration.

Algorithms (operations) that can be mapped onto these structures are characterized by a fine-grain parallelism. This is the case for example of operations concerning digital (low-level) processing of images and signals. These application fields are in fact characterized by highly structured input and output data and by repetitive operations, and require a very high throughput in order to grant real-time operations.

The best known architectural solutions based upon the described approach are *array processor architectures*, characterized by a number of identical processing elements interconnected in a regular way, usually in a near- neighbour scheme. Array processor architectures here considered have a computational graph whose topology may be ranked in one of the classes of MIN graphs previously examined.

Dedicated supercomputing may then take advantage of the concepts previously described

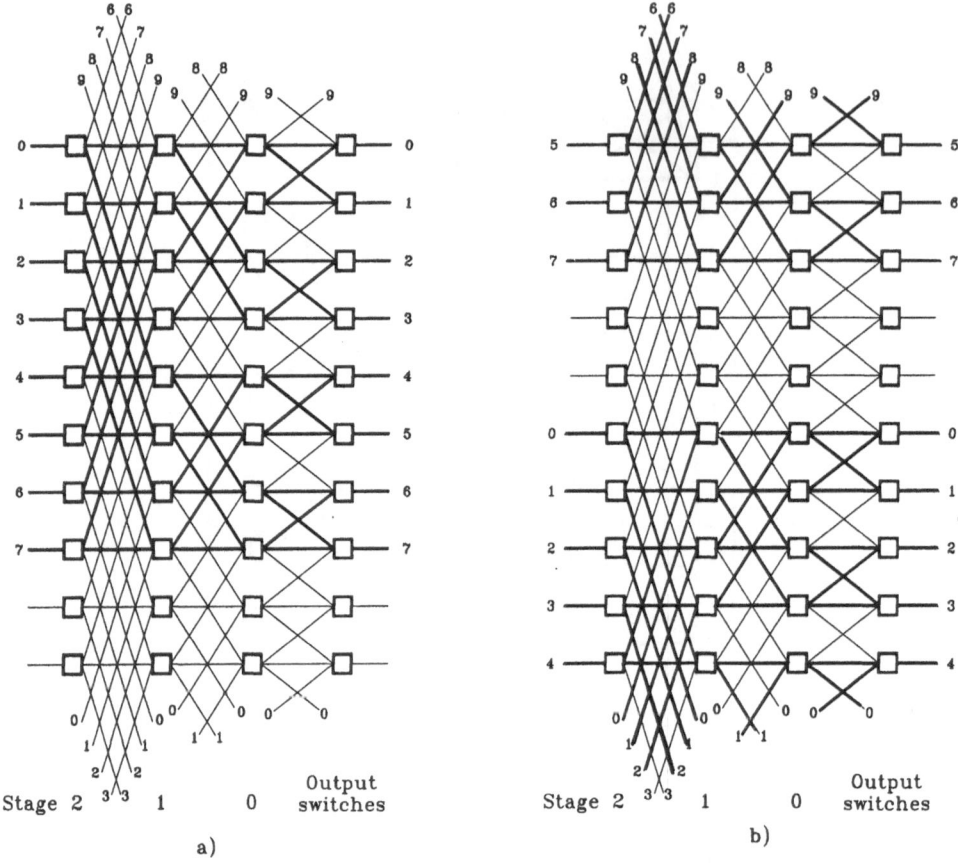

Fig. 4 DR-network for N=8 and S=2. The generalized cube embedded into
a fault-free network (a) and into a reconfigured network (b).

for the MINs: as an example, consider the case of the N-point Fast Fourier Transform algorithm
and its associated radix-r computational graph. This graph, for $r = 2$, is constituted by $log_2 N$
stages of $N/2$ processing elements, each performing the basic iteration, i.e., the *butterfly*. (The
same approach as for the Fast Fourier Transform may be in fact adopted for all algorithms
allowing for a fast version.) In Figure 5, two homomorphic graphs are shown for $N = 16$: all
these graphs are homomorphisms of regular rectangular banyans. A third homomorphism can
be obtained from the graph of figure 5.b by vertically mirroring the graph.

In order to reduce the analysis of dedicated supercomputing networks to that of MINs, it is
necessary to consider in higher detail the operating capabilities of the processing elements. The
following case are particularly significant:

a. processing elements are simple arithmetic circuits (as for the Fast Fourier Transform);

b. processing elements are CPUs or computers;

c. processing elements are packet- switch processors.

While cases a. and b. represent typical examples of dedicated supercomputing networks, case
c. refers to proper packet- switched MINs. Also cases a. and b. can be described as packet-
switched networks: the communication protocol between linked processors is organized making

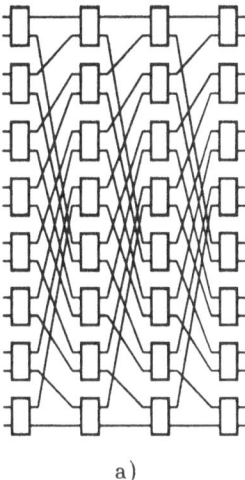 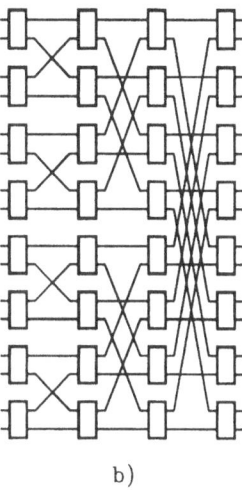

a)   b)

Fig. 5 Radix-2 homomorphic computational flow graphs: (a) homogeneous
stages and (b) incremental distance interconnection network

use of data packets. From the operational point of view, all these cases are characterized by processing elements that alternate a *computational* phase with a *communication* phase, operating globally in a synchronous way.

When considering definition of fault- tolerance policies for these networks, basic differences arise, related to the fault models and to the reconfiguration criteria, in comparison with the approaches described in the previous section. It is now necessary to rebuild a *complete* computational graph, maintaining its complete topology. Moreover, as regards the fault model, the following assumptions can be done: (1) multiple faults are allowed in processors, links, and redundant circuits and links required by the reconfiguration algorithms; (2) all the faults are considered at the *functional* level; this assumption is suggested by the increased complexity of the elements with respect to the ones (switches) considered in the previous section; and (3) a faulty element, whether processor or link, is completely unusable. In general, the concept of *inherent fault- tolerance* of the network cannot be applied, since in a computing network the global functionality is always required, as the computing network is tailored on the size of the problem to be solved. A common goal of the reconfiguration algorithms is *tolerance* to single faults and *robustness* to *k* multiple faults. Tolerance to multiple faults usually depends on the particular fault distributions.

In our approach, redundancies required by the reconfiguration algorithms foresee spare interconnection links, physically added to the network, and spare operational (i.e., computation and communication) phases that are granted by spare processing elements (*structure redundancy*) or by fault - free nominal processing elements performing also execution of phases nominally attributed to faulty processing elements (*time redundancy*). Typical figures of merit for these fault- tolerance algorithms are *survivability*, i.e., the capability of tolerating up to *k* faults, and *reliability* taking into account circuit redundancy required by the algorithm.

General criteria for definition of reconfiguration algorithms for array based architectures exploit the modularity, regularity and locality of the architectures themselves: these parameters have to be considered in order to trade- off efficiency of the algorithm against an increase of redundant circuitry leading ultimately to a decrease in the reliability of the fault- tolerant structure. Extensive literature has been developed in this field especially referring to locally interconnected structures (see, e.g., [NEG88]). Computational graphs and MINs here considered show the characteristic of a reduced locality, still maintaining regularity: the lack of locality has to be taken into account in the definition of the reconfiguration algorithm, and - at an higher decision level - when considering the applicability of structure or time redundancy. As a major consequence of non-locality in the interconnections, the wide class of algorithms for both end-of-production and life-time reconfiguration is not straightforward applicable: recent examples mainly consider radix-r flow graphs (see [CHO88], [JOU88], [ANT88a], [ANT88b], [ANT89a]).

Besides problems related to definition of reconfiguration algorithms, two other instances have to be considered in the definition of fault- tolerant structure: the first one - testing and diagnosing - is instrumental to the reconfiguration proper, while the second one - decreased performances - is a consequence of the reconfiguration algorithm and can be considered as an evaluation figure together with survivability and reliability.

As regards testing, general techniques such as *recomputation by alternate path* or *redundant in time recomputation* may be used. Specific testing techniques referring to the topology of the computational graphs have been developed for radix-r graph by, e.g., [ANT89b].

As far as *performances degradation* is concerned, the adoption of reconfiguration policies based on structure redundancy grants to maintain the nominal performances also in the presence of faults, whatever the number and the actual distribution. In case of reconfiguration policies based on time redundancy, introduction of added operational phases to rebuild the computing network imply an intrinsic decrease of the throughput: this appears as soon as the first (single) fault has to be tolerated. In this case, an interesting goal of the reconfiguration algorithm is the granting of a throughput decrease independent from the actual number and distribution of faults.

Different reconfiguration policies adopting time redundancy have been presented in

[ANT88a], [ANT88b] and [ANT89a] valid for regular rectangular banyans. These approaches will be recalled in the sequel and their basic algorithms described in an informal way. The reconfiguration is achieved by associating each processing element with *two* different operational phases: during the first one, *the nominal operational phase*, the fault-free element performs the computation associated to its nominal input data, thus producing its nominal output data. In the second one, the processing element acts as a time spare of a (suitably defined) faulty one and performs associated computation (*the spare operational phase*). Redundant links and circuits allow for correct routing of input and output data to and from the time spare element.

Since in these approaches a single processing element can act, in time, as time spare of a single faulty element, all locally data dependent circuitry (e.g., registers, data and/or program memory, coefficients) inside the processor have to be duplicated for allowing correct computation.

The first described approach involves repeated use of fault-free processing elements in adjacent stages (*inter-stage reconfiguration* [ANT88b], [ANT89a]) while the other one uses repeatedly fault-free elements in the same stage where the faults are located (*intra-stage reconfiguration* [ANT88a], [ANT89a]). The nominal structure and topology of the network to be reconfigured is the one in Figure 5.a. In both solutions, performances of the networks are decreased as soon as the first fault is detected: in particular, the throughput is halved, but this degradation in performances remains constant for any number and distribution of faults.

In the case of inter-stage reconfiguration, the redundant links and circuits supporting reconfiguration apply strictly to the topology shown in Figure 5.a: homomorphic computational

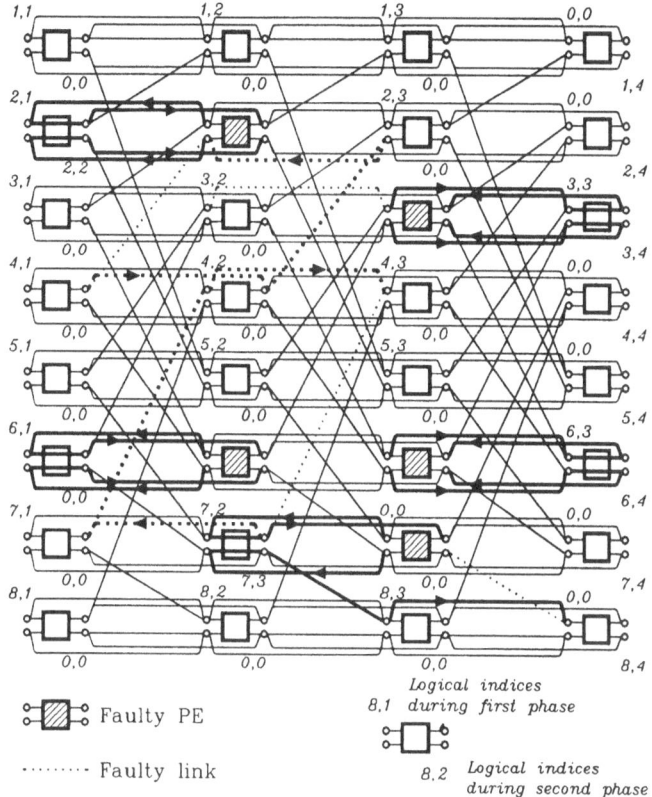

Fig. 6 An example of inter-stage reconfiguration

graphs can be dealt with the same reconfiguration approach, but with a different design of the redundant links and circuits. Differently, in the case of intra-stage reconfiguration, the geometry of the interconnection networks between each stage, and then of the computational graph, is not related to the design of redundant links and circuits allowing for reconfiguration: for this reason, all homomorphic graphs may be straightforward considered. The two structures supporting the alternative reconfiguration approaches are described in the sequel: for each solution, given the fault model, an *equivalent fault reduction* is obtained, and the algorithms are outlined.

4.1 Inter-stage Reconfiguration Using Time Redundancy

The fault-tolerant structure allowing to implement inter-stage reconfiguration algorithms is given in Figure 6 (together with an example of complete reconfiguration, that will be discussed later on): the nominal structure of Figure 5.a has been augmented by a number of switches (circles in the figure) and of added interconnection links.

With the extra connectivity introduced by switches and redundant interconnection links, a faulty processing element is substituted by a fault-free one positioned on the same row and in one of the two adjacent stages. During operation, the fault- free element is, in time, associated to two different pairs of *logical indices*: one pair representing its own indices (*physical indices*), the other one representing the physical indices of the faulty element to be substituted. In the two phases, faulty elements as well as non active fault- free ones are associated with zero indices. Before applying the reconfiguration algorithm, fault collapsing has to be undertaken, so that only two type of fault classes are considered, namely faults in processing elements and in interconnection links. (For example, a fault in a switch makes unusable *all* interconnection links arriving and departing from the switch itself, and the processing element the switch is connected to.)

Considering the redundant structure of Figure 6, it is obvious that the only components *not* allowing reconfiguration - independently from the particular formulation of the algorithm - are the input switches to the first stage (*inputs to the network*) and the output switches of the last stage (*outputs of the network*), representing the hard-core section of the redundant structure.

Following the equivalent fault collapsing, the reconfiguration algorithm is activated. The main characteristics of the reconfiguration algorithm proposed in [ANT89a] may be so summarized:

- The algorithm searches for equivalent faults in components scanning the structure stage by stage, where a stage is defined as the processing elements belonging to it and the nominal interconnection links departing from the processing elements. If necessary, reconfiguration actions are undertaken and the reconfiguration associations will not be modified by possible reconfiguration of other stages.

- Starting from the leftmost stage and the uppermost position and for increasing values of the row index i first ($i = 0, \cdots, N - 1$) (N being the number of inputs) and stage index j ($j = 1, \cdots, log_2 N$) the algorithm checks whether a fault has been associated with the processing element of position ($\lfloor i/2 \rfloor + 1, j$) and/or with the nominal interconnection link ((i, j)($i^*, j+1$)). (The two sets of indices associated with the link represent respectively the origin (i, j), and the destination ($i^*, j + 1$), where i^* identifies the destination row obtained as a *shuffle* of the origin row i.)

- If the processing element and the nominal interconnections are fault-free the row index is updated. Otherwise the reconfiguration takes place.

- In case of fault in a processing element of position ($\lfloor i/2 \rfloor + 1, j$), the algorithm verifies the possibility to define a time spare by first considering an association with the processing element of position (($\lfloor i/2 \rfloor + 1, j - 1$). The association can be created if redundant links allowing for the routing of input and output data of the faulty element are usable for reconfiguration, i.e., fault-free and not yet used. If the association fails, the algorithm searches for the processing element of position ($\lfloor i/2 \rfloor + 1, j + 1$). If also in this case conditions for reconfiguration are not met, an impossible reconfiguration is declared. When the reconfiguration takes place all components involved in reconfiguration are declared as used. Referring to Figure 6, an example of "backward" reconfiguration is given for the faulty processing element ($2, 2$): the processing element ($2, 1$) acts as time spare. In the same figure, the processing element ($3, 3$) is "forward" reconfigured (element ($3, 4$) is the time spare) due to the faulty redundant link not allowing association with element ($3, 2$).

The same approach of backward/forward reconfiguration is followed in the case of a faulty interconnection link. Then, it is necessary to rebuild the path from (i, j) to ($i^*, j + 1$): the reconstruction requires the usability of redundant and nominal interconnection links, as it can be seen again from Figure 6 for the faulty link (($12, 2$)($6, 3$)), which is reconfigured backwards.

In case of fault in a processing element and in a nominal link departing from it, the above described reconfiguration actions are activated in a coherent way (see for example the

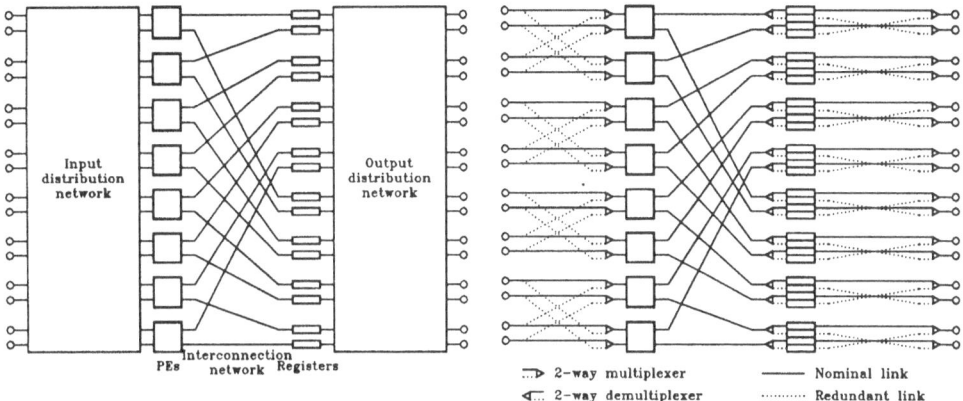

Fig. 7 a) Redundant structure for intra-stage reconfiguration; b) 1-distance network.

reconfiguration of the faulty processing element in position $(7,3)$ and of the nominal interconnection link departing from it $((13,3)(14,4))$ in Figure 6).

- When the whole structure has been examined and reconfiguration actions have been all successful, the complete mapping of logical indices of processing elements onto physical indices is completed, as shown again in Figure 6. Otherwise, fatal failure is declared.

More complex and powerful procedures may be defined. In particular, by allowing an exhaustive search of associations, all possible reconfigurations can be found, still maintaining the concept of backward and forward association.

An alternative redundant structure introduced in [ANT89a] makes use of *switched buses* instead of dedicated redundant links. This alternative reduces the complexity of the interconnection network with only slight modifications in the reconfiguration algorithm. A bidirectional redundant interconnection is used for allowing both backward and forward associations (this in fact reduces possibility in reconfiguration). The comparative evaluation between the dedicated links and the switched buses approaches, given in [ANT89a], shows that dedicated links become more efficient (both for survivability and reliability) in the presence of many faults.

4.2 Intra-stage Reconfiguration Using Time Redundancy

In this second approach, the reconfiguration algorithm searches for time spare(s) of a faulty processing element inside the same stage where the fault has been located. For this reason not only the reconfiguration proceeds stage by stage, but the information necessary for reconfiguration is related only to the stage under reconfiguration. Again, two operational phases are needed for correct computation: in the first one, fault-free elements process their own nominal data, while in the second one they act as time spare.

The fault- tolerant structure supporting this approach is shown in Figure 7.a. Each stage of the nominal computing network (processing elements and interconnection links) has an *input distribution network*, capable of receiving inputs to the stage and of correctly routing them to fault-free processing elements. Registers at the end of nominal interconnection links are necessary to correctly hold processed data, generated either during a nominal computation phase or during a spare phase. An *output distribution network* is needed to present a *fault independent* interface to the following stage.

The actual structure of the distribution networks defines the reconfiguration capabilities of the algorithm: connectivity introduced by the input distribution network identifies the set of processing elements for which a fault-free one can act as a time spare. The structure of the output distribution network is a consequence of that of the input one. Independently from the adopted structures, inputs and outputs of the *redundant stage* are the *hard-core* of the fault-tolerant structure.

A possible example of distribution networks is shown in Figure 7.b. In this particular instance (*1-distance network*), a processing element can act as time spare of the one with row index at distance one (and viceversa). In case of faults, the reconfiguration algorithm associates the faulty elements with the corresponding fault-free ones, if any: during the first operational phases, fault-free elements process nominal data and output results are suitably stored in the upper registers; during the second operational phase, input data referring to faulty elements are routed to the corresponding time spares that perform computation: results are stored in the lower registers. At the end of the two phases, all results are correctly routed, through the output distribution network, at the outputs of the redundant stage.

As in the case of the reconfiguration approach described in the previous subsection, the redundant structure allows to overcome faults in all the components of the structure itself (processing elements, nominal and redundant interconnection links, multiplexers and registers). Here again faults are considered from a functional point of view.

As examined in detail in [ANT89a], the equivalent fault collapsing allows to consider for the reconfiguration algorithm only faults — proper or equivalent — located in processing elements: in particular, a processing element after the fault collapsing procedure is declared as: (1) faulty and unusable, (2) fault-free and not available as time spare, or (3) fault-free and available.

After the collapsing procedure, the reconfiguration procedure may be undertaken: the algorithm scans the structure stage by stage searching for faulty processing elements. The verification of associations between faulty processing elements and correspondent time spares, may be reduced to a complete matching problem.

As detailed in [ANT89a], it is possible to build a bipartite graph in which the source nodes represent *logical* processing elements, while sink nodes represent *physical* processing elements. Edges connecting sources to sinks represent association capabilities of the distribution network: in fact, the structure of the distribution networks can be much more complex of the one shown in Figure 7.b, thus allowing for different associations. (Trivially, each source is connected with the corresponding sink, i.e., a processing element is associated with itself.)

Equivalent fault information are used to modify the graph to take into account the real fault distribution. In particular, if a processing element is declared faulty the corresponding sink node and ingoing edges have to be deleted; if a processing element is declared fault-free and not available as time spare, the edge from the source node, representing it, to the sink node for which unavailability is declared has to be deleted.

After this modification in the graph, if a source node is not connected to any sink node the reconfiguration is not possible. Otherwise, the research for association may start. The reconfiguration of the stage is achieved if each source node is associated with a sink node and if a sink node is used no more than twice (during operation, a processing element can act as time spare of only one other element). The algorithm *exhaustively* searches for associations and then, if the reconfiguration is not achieved, the redundant structure is not reconfigurable.

In [ANT89a], evaluation of survivability considering different instances of distribution networks has been done, and shows a significant increase of survivability as the complexity of the distribution networks increases.

Reliability of redundant structures defined for inter-stage reconfiguration and intra-stage reconfiguration is also evaluated: the inter-stage reconfiguration shows a better trend *vs.* that of intra-stage reconfiguration supported by the redundant structure shown in Figure 7.b (1-distance network). It is interesting to note that the percentage of structure redundancy (i.e., circuit area) necessary for inter-stage reconfiguration is less then the one for 1-distance intra-stage reconfiguration. Intra-stage reconfiguration obtained through more complex distribution networks grants for a better trend in reliability, with the drawback of a significant increase in structure redundancy.

5. CONCLUDING REMARKS

This paper has considered some of the main points of (Multiple) Interconnection Networks (MINs), related in particular with their adoption in supercomputing architectures. The main properties and characteristics of MINs have been recalled, and some typical parallel computer architectures adopting MINs have been summarized in order to show the true technological possibilities of using MINs.

Two main application classes of MINs have been considered: (1) parallel computer systems implemented by connecting together powerful processors and large shared memories — e.g., adopting circuit-switched networks — and (2) dedicated supercomputing structures (e.g., high throughput array structures dedicated to signal and image processing, that directly implement highly parallel algorithms).

For both application classes, the adoption of fault tolerance methods has been discussed. Fault tolerance can be usefully adopted both to overcome production defects and faults arising during systems working life.

This paper also surveys classic approaches to fault tolerance in MINs for parallel computer systems, and some recent results in the less studied field of fault tolerance in dedicated supercomputing structures.

6. REFERENCES

[ADA82]   G.B.Adams, H.J.Siegel: "The Extra Stage Cube: a Fault-Tolerant Interconnection Network for Super Systems", *IEEE Trans. Comp.*, Vol.31, pp. 443-453, 1982.

[ADA84]   G.B.Adams, H.J.Siegel: "Modifications to Improve the Fault Tolerance of the Extra Stage Cube Interconnection Network", *Proc. IEEE Int'l Conf. Parallel Processing*, 1984.

[ADA87]   G.B.Adams, D.P.Agrawal, H.J.Siegel: "Fault-Tolerant Multistage Interconnection Networks (A Survey and Comparison of —)", *IEEE Computer*, June 1987.

[AGR82]   D.P.Agrawal: "Testing and Fault Tolerance of Multistage Interconnection Networks", *IEEE Computer*, April 1982.

[ANT88a] A.Antola, R.Negrini, M.G.Sami, N.Scarabottolo: "Time-Folding: a Solution for Functional and Fault-Tolerance Reconfiguration of Systolic FFTs", *Proc. 3rd Int'l Conf. on Superc. — ICS-3*, Boston, May 1988.

[ANT88b] A.Antola, R.Negrini, M.G.Sami, N.Scarabottolo: "Policies for Fault-Tolerance through Mixed Space- and Time- Redundancy in Semi-Systolic FFTs Arrays", *Proc. IEEE Int'l Conf. on Systolic Arrays*, San Diego, May 1988.

[ANT88c]   A.Antola, R.Negrini, N.Scarabottolo: "Arrays for Discrete Fourier Transforms", *Proc. EUSIPCO 88*, Grenoble, Sept. 1988.

[ANT89a]   A.Antola, R.Negrini, M.G.Sami, N.Scarabottolo: "Fault-Tolerance in FFT Arrays: Time-Redundancy Approaches", Int. Rep. N. 89-16, Dipart. Elettronica, Politecnico di Milano, 1989.

[ANT89b]   A.Antola, M.G.Sami, D.Sciuto: "Testing Approaches for Flow-Graph Derived FFT Arrays" *Proc. Int'l Conf. on Systolic Arrays*, Killarney, May 1989.

[BRO83]   G.Broomell, J.R.Heath: "Classification Categories and Historical Development of Circuit Switching Topologies", *Computing Surveys*, Vol. 15, No.2, June 1983.

[CHE84]   V.Cherkassky, E.Opper, M.Malek: "Reliability and Fault Diagnosis Analysis of Fault- Tolerant Multistage Interconnection Networks", *Proc. IEEE FTCS 14*, June 1984.

[CHO88]   Y.H.Choi, M.Malek: "A Fault-Tolerant FFT Processor" *IEEE Trans. Comp.*, Vol.37, NO.5, May 1988.

[CIM88]   L.Ciminiera, A.Serra: "A Connecting Network with Fault Tolerance Capabilities", *IEEE Trans. Comp.*, Vol.35, NO.6, June 1986.

[FEN81]   T.Y.Feng, C.L.Wu: "Fault-Diagnosis for a Class of Multistage Interconnection Networks", *IEEE Trans. Comp.*, Vol.30, NO.10, October 1981.

[FRA86]   P.D.Franzon: "Yield Modeling for Fault-Tolerant Arrays", *Proc. Int'l Conf. on Systolic Arrays*, Oxford, July 1986.

[GAZ86]   I.Gazit, M.Malek: "Fault Tolerance Capabilities in Multistage Network Based Multicomputer Systems", *IEEE Trans. Comp.*, Vol.35, NO.7, July 1986.

[GAZ88]   I.Gazit, M.Malek: "Fault Tolerance Capabilities in Multistage Network Based Multicomputer Systems", *IEEE Trans. Comp.*, Vol.37, NO.7, July 1988.

[JEN86]   M.Jeng, H.J.Siegel: "A Fault- Tolerant Multistage Interconnection Network for Multiprocessor Systems Using Dynamic Redundancy", *Proc. IEEE Conf. Distrib. Computing Systems*, 1986.

[JEN88]   R.M.Jenevein, T.Mookken: "Traffic Analysis of Rectangular SW-Banyan Networks", *Proc. IEEE Symp. on Computer Architecture*, 1988.

[JOU88]   J.Y.Jou, J.A.Abraham: "Fault-Tolerant FFT Networks" *IEEE Trans. Comp.*, Vol.37, NO.5, May 1988.

[KIM88]   D.W.Kim, G.J.Lipovski, A.Hartmann, R.Jenevein: "Regular CC-Banyan Networks", *Proc. IEEE Symp. on Computer Architecture*, 1988.

[KUM87]   V.P.Kumar, S.M.Reddy: "Augmented Shuffle-Exchange Multistage Interconnection Networks", *IEEE Computer*, June 1987.

[LIP87]   G.J.Lipovski, M.Malek: *Parallel Computing - Theory and Comparisons*, John Wiley and Sons, New York, 1987.

[MAN82]   T.E.Mangir, A.Avizienis: "Fault-Tolerant Design for VLSI: Effect of Interconnection Requirements on Yield Improvement of VLSI Design", *IEEE Trans. Comp.*, Vol.31, NO.7, July 1982.

[NEG88]   R.Negrini, M.G.Sami, R.Stefanelli: *Fault Tolerance through Reconfiguration in VLSI and WSI Arrays*, The MIT Press, Cambridge, 1988.

[RAO88]   S.K.Rao, T.Kailath: "Regular Iterative Algorithms and their Implementation on Processor Arrays", *IEEE Proceedings*, Vol.76, NO.3, March 1988.

[REE87]   D.A.Reed, D.C.Grunwald: "The Performance of Multicomputer Interconnection Networks", *IEEE Computer*, June 1987.

Computer Networking for Interactive Supercomputing

John Lekashman
NASA Ames Research Center
Mail Stop 258-5
Moffett Field, Ca, 94035, USA
lekash@orville.nas.nasa.gov

Abstract

Supercomputers are now integral parts of scientific research. Resources and researchers are scattered over a wide geographic area. The distances involved are sufficiently large that wide area data network access is and will continue to be absolutely necessary.

It is our responsibility to provide an infrastructure such that anywhere someone happens to be, he or she can communicate with whatever people and computational resources that they desire.

The distribution of resources across such wide areas can and is being accomplished with data networks by NASA, DARPA, NSF, and many other US government agencies. Universities and many corporations are also so connected. Widespread cooperation occurs, with exceptional results. Information can and does move between people across continents, on demand.

Here we describe those networking paradigms in use now and in the future at the NASA Numerical Aerodynamic Simulation (NAS) Division. These methods are used very effectively by researchers across the United States. The key ideas which drive these paradigms are: Interactive access to resources is essential, and standardized, effective operating systems and networking protocols are essential. The NAS systems all use UNIX[1] and TCP/IP to fulfill these ideas.

Introduction

The NAS Program

The NAS Program started in 1975. Fluid dynamics researchers needed advances in computer technology for Computational Fluid Dynamics (CFD) modeling. Several technical studies of the use of computers in these applications took place. In 1984, the program became an official New Start, charged with providing leading-edge national computing capability. By the end of 1988, the NAS computer systems served 1246 researchers at 118 locations across the United States.

Leading-edge computing capability causes computer systems to be very costly to build and maintain. It is not practical to site such systems at multiple locations. The requirement for national computing capability therefore causes wide area data networks to become a critical element of the system. In order to reach those 118 sites, communication paths must be constructed to provide methods for researchers to access computational resources on demand.

---

[1] UNIX is a registered trademark of AT&T.

Background Information

The principal computer operating system in use at the NAS Facility is Unix. The networking systems are all based upon the TCP/IP protocol suite, with original implementation from the Berkeley Standard Distribution (BSD 4.3). There is a common thread of implementation, which means that advances can typically apply to all computer systems, from personal workstations to supercomputers.

Local area networks are installed in two performance ranges. Ethernet exists ubiquitously throughout Ames Research Center (ARC). It is available to any researcher at ARC. A small number (40) of computer systems are connected to the higher performance Hyperchannel-50 (tm). This is shown in figure 1.

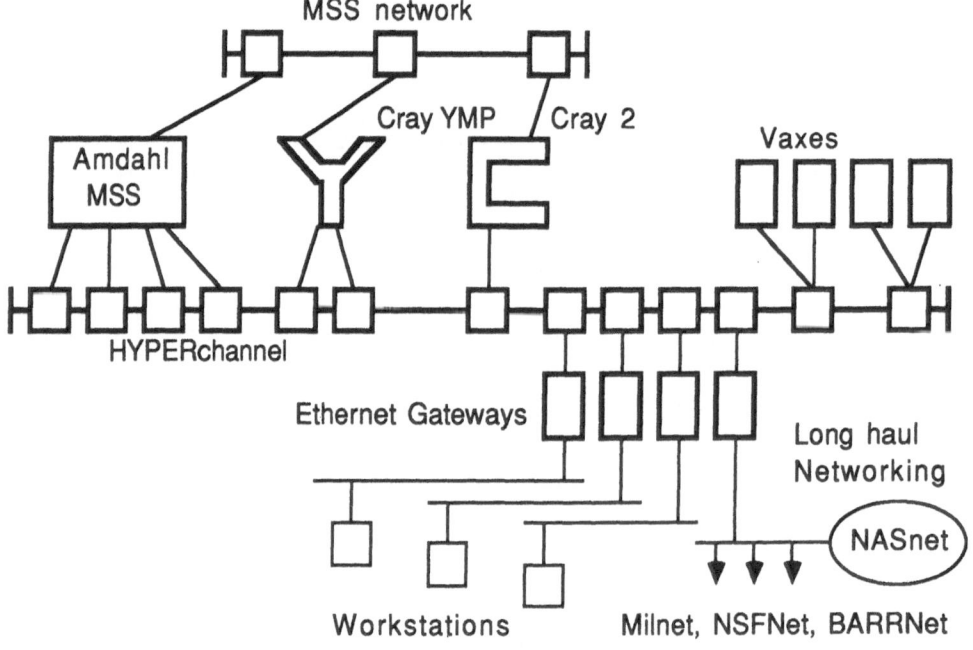

figure 1 - Local Networks

Our wide area networks are typically built out of common carrier communication lines, currently ranging in speed from 56 kb to T1. Communication processors from Vitalink Corporation use these circuits to provide bridging of ethernets between our central supercomputer facility at NASF and user sites across the United States. These networks are shown in figure 2.

Scientific Requirements Driving Network Performance

Visualization is key to performing research. This means that a persons interaction with computational systems must include an effective display of the computational results. We believe the best approach to doing this is with a personal workstation, capable of independent processing and video display. This meshes very well with the requirement for servicing a wide area.

The first facet of data networking is establishing connectivity. As soon as some degree of connectivity is established, the question immediately arises, how good does it have to be?

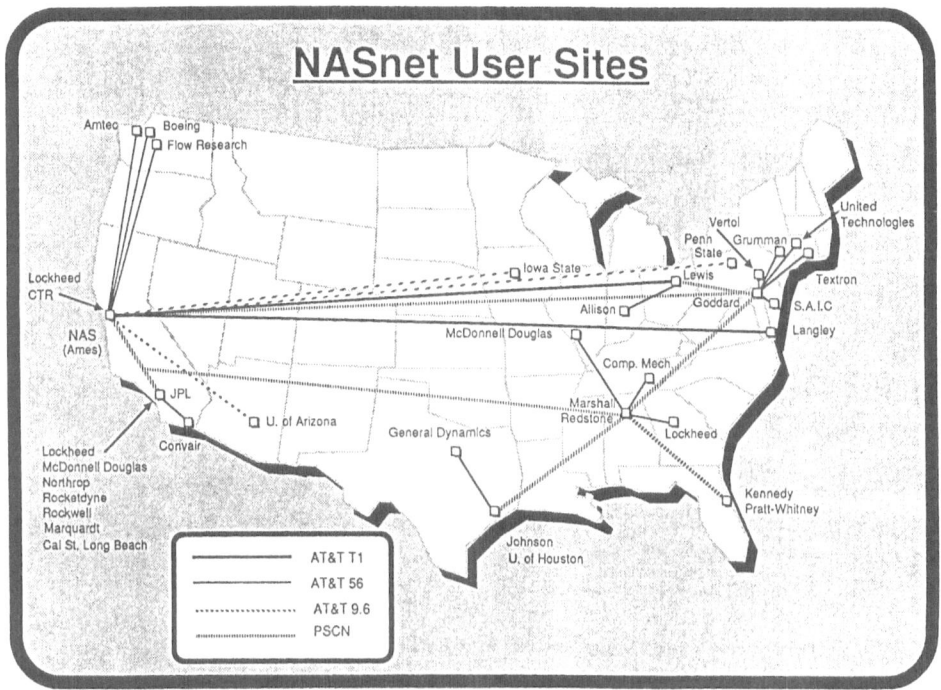

figure 2 - NASnet

System analysis must take place, to see what are the important characteristics of a data network. The following is a list of those important attributes. Each will be expanded upon within the text.

- Interactive Behavior. The systems must respond in a timely manner. The definition of timely manner depends on what the nature of scientific problem being addressed.

- Delivering Throughput. In some cases, the data networks must be able to transfer large amounts of data. Once again, the nature of large is quite dependent upon the scientific problem being addressed.

- Ubiquity. The systems must be present everywhere that we might wish to be connected with at some time. Decisions will be made about resource utilization, and an organization should be in a position to immediately respond to a need. This is only possible if the connectivity is a complete part of the infrastructure.

- Functional Standards. The systems must operate with a well defined, working standard. The data networks are the point of interchange between computer systems. As such, they are the common ground upon which systems must meet.

- Uptime. The systems must be available all the time. Data networks cross time zones, are used by many people with widely varying work habits.

Defining Interactive Behavior

It is very important to note that interactive computing is not the same as timesharing computational resources. Interactive computing means that when someone provides a stimulus, there is a response. It is key that an immediate response occur. Timesharing, on the other hand, describes a method whereby a scarce resource is time sliced among multiple users. This is expected to happen in a transparent fashion. Transparency is achieved because human reaction time is typically much longer than the time it takes for computer system to change context.

The computational resources are there for the convenience of the user base, and should perform in a fashion most advantageous to them. Essentially, the computational environment should be present the smallest possible a barrier to the exploration of physics as we can create. One aspect of this is to encourage interactive use of systems. Timesharing of computer systems takes place for economic reasons, not because it is desirable. Therefore, a computer system must be designed to be usable by single and multiple users.

Interactive computing requirements place a burden upon data networks. This is because the data networks are not expected to present any barrier to investigation. The barrier that does occur is time. This length of time is different for different applications, but the basic idea applies across the spectrum.

There are two principle areas of user visible network performance. These are delivered response time and delivered throughput. How long do we think someone needs to wait for something to happen? Following are some general descriptions of typical computational fluid dynamics problems, and the demands they place about data transfer systems. These demands translate into necessary throughput and delay characteristics for the networks.

Some Computational Fluid Dynamics Problems

Particle Traces

One CFD problem consists of analyzing particle traces. This consists of repeatedly sending a particle trace from a supercomputer to the display. This typically takes about 4K bytes of data per trace. One trace every four seconds is acceptable behavior, although a higher rate would certainly be preferable. To look at a single particle trace, a user really needs to be able to transfer a minimum of 1K byte per second, in order to get effective visualization of the problem. 56 kilobits per second data service accommodates this well. Such service can typically support several users simultaneously working on this type of problem.

Contour Plots

A contour plot analysis is an example of a more advanced type of problem now being solved. Estimates of a typical problem size are as follows:

- 100 particle traces to describe a single contour.
- 10 plots to describe a 3-dimensional problem.

This totals about 4K bytes * 100 * 10 = 4 megabytes, in a single image. In order to solve and display the results of this problem, the data network needs to provide three orders of magnitude more bandwidth than before. This would be about eight megabits per second for a single user. We estimate there will easily be several users simultaneously. At eight megabits per second each, this reaches a requirement for at least T3 service (45 megabits/second) very quickly. Depending upon the actual number of users, the rates necessary to meet the requirements can very easily be higher.

File Transfer

In a flow field problem analysis, a typical problem can have one million nodes. Each node consists of eight floating point numbers:

- density
- momentum : (vector -> 3 points)
- energy
- location (x,y,z)

With 32 bits per floating point number, this is a total of 32 megabytes in a solution file. This amount of data can be transferred over a single T1 communication line fairly quickly, in just over three minutes with complete utilization of the path. However, this data file size is only the beginning.

Time Accurate Problems

Researchers are currently engaged in doing work on time accurate problems. One example is the analysis of a Harrier jet while it is landing. Determining why this is so loud, or why dirt flies into the engine air intakes, could lead to better designed aircraft.

For this analysis we shall assume 1000 steps of the problem during a landing. This means there are now 32 gigabytes of data for the analysis.

It can be useful to move the file from the supercomputer site, where it has been generated, to a local machine for post processing. In order to look at different aspects of the problem, it will be useful to move the file once, and do data analysis locally many times. The computational requirements to display a result of one aspect of a solution are not nearly as large as those required to produce the solution. However, the data network requirements to display it will remain the same each time. Therefore, it can often be very useful to copy the file once, and perform post processing at a site remote from the main supercomputer.

A second motivation is that of fostering the exchange of ideas. Someone at Ames Research Center might want to send a solution they have been working on to someone at Langley Research Center, across the country. This should be able to occur with ease. If it cannot happen in a short time, it is likely happen at all.

In both these cases, a 32 gigabyte file must be copied. How long should this take? If we are going to bother with a network transfer at all, it should be faster than making a tape and shipping it Federal Express. So, an upper bound of 24 hours exists. In day to day use, it should certainly take less than an hour.

$$32 \text{ gigabytes} \times \frac{1 \text{ hour}}{3600 \text{ seconds}} = 8.9 \text{ megabits per second}$$

This rate is fairly close to that required by the above analysis of a contour plot display. Being able to perform an exchange in approximately three minutes, as a single step can now be transferred, would certainly be even more useful.

Threshold of Performance

The nature of a threshold of useful bandwidth is useful to examine. Looking at the above calculations, it can be seen that the single particle trace calculation is solved with fairly low rates. The nature of the next step of problem solving needs several orders of magnitude performance increase. Without adequate bandwidth, there is no reason to use it, and that method of performing research simply will not happen.

An example of this occured several years ago at the NAS program. There was a problem when the Mass Storage System first came on line. It took longer to copy files and retrieve them than to recalculate the data. As a result, the system was not used very much. Now that the transfer rates have significantly increased, the system is heavily used.

Wide Area Networks currently suffer from a similar problem. The current available bandwidths in the 56kb to T1 range is simply not enough to make the next qualitative step in the use of data networks. The activity in networks right now is a quantitative increase in usage, without qualitative changes. Tariff structures for pricing data communication costs are quite simply way out of line. Service should cost less. The solution to this problem will occur in the political realm, not the technical.

Nature of the Solution

Using Widespread and Effective Standards

The diversity of computer systems that people can use continues to grow. Any system that is spread widely throughout a large geographic area requires support for equipment from multiple vendors. The most effective long term solution to this is to use existing standards for data interchange.

There are two principal standards that are used by NAS. The first is the Unix operating system. This is used across all machines. There are a multitude of advantages to this. For the scope of this discussion, the principal advantages are a common code heritage for data networking, and a common user interface for command structures.

The second major standard is the use of the TCP/IP protocol suite. Our program has found great success in the use of this networking code. It is available and in use on every machine that is a part of NAS. Its advantages are myriad. Some of these are:

- Implementations are available. High quality code is produced, in the public domain, suitable for porting to many computer systems. This is key for widespread use. When a new machine architecture arrives, the ability to quickly get operational networking on it is essential.

- It is in widespread use among the academic community. Universities throughout the U.S. are connected into a nationwide Internet. This system provides for rapid dissemination of knowledge.

- We have been able to push the use of this software technology to very high throughput capabilities. We have not encountered an actual limit to the available throughput. Limits are usually encountered due to a poor port to a particular hardware base, or hardware limitations.

- It is sufficiently similar in structure to the emerging OSI networking standads that several interesting effects are occurring. First, software is starting to be produced that enables both protocol stacks to be used. Second, many sites are using this technology until OSI matures completely. Third, many ideas are first proven in the TCP/IP research environment, and then incorporated into the OSI protocol suite.

The NAS Program has made great use of requiring these two standards to be in all computer systems. It removes the need for a large in-house system software development staff. It provides a common ground for operation and requirements. This is particularily important in the Government, due to Federal regulations regarding specification and acquisition of computer systems.

A researcher knows that he or she can use the NAS system, if he or she can find a computer with Unix, that also communicates with the Internet. This sort of infrastructure is essential for rapid advancement. It often removes the need for retraining, and for reconfiguration of Wide Area networks. Both of these can be time-consuming processes, which are not directly related to the aerospace research at hand.

Developing New Software

NAS has engaged in developing some new software, in cases where the current implementations were not quite up to the performance necessary. Recently, we discovered a shortcoming in the current TCP/IP implementations which needs correcting. TCP is a sliding window transport protocol, with positive acknowledgement. For further details on its operation, see RFC 793.[2] Due to this protocol style, in order to effectively use all the bandwidth of a given path, TCP needs to have at least as many data octets in flight as the product of available bandwidth and delay over the connection. The reason for this is that there is a limit on how much data a sliding window protocol will send before receiving an acknowledgement. The amount of data that can be so sent is referred to as the sending window.

The problem we have discovered leads to difficulties in physical networks which have a large value for the product of bandwidth and delay. Large in this case means a value for this product that is greater than 65535 octects.

The solution we came up with was to parallelize the problem. Opening up multiple simultaneous streams for a single data transfer resulted in being able to overcome this. For more discussion on this experiment, see the referenced paper.[3] The key idea is that very little new software was required. The existing software was used as a building block, in order to get more available throughput.

Developing New Hardware Base

The NAS program is involved in several efforts to advance the state of technology in data networks. These can be divided into the local and wide area network parts. Here are described two programs that NAS has undertaken in order to develop these areas.

Local Networks

The NAS program has undertaken three separate efforts, with three different vendor equipment bases, in order to prototype new Local Network technology. The requirements upon the efforts are as follows:

- Provide interfaces to Cray supercomputers.
- Provide interfaces to other computer systems.
- Measure the available throughput between the systems.

As of this writing, hardware has been produced and installed by all three vendors. These vendors are:

---

[2] Postel, J., "Transmission Control Protocol - DARPA Internet Program Protocol Specification", RFC 793, DARPA, September 1981.

[3] Lekashman, J. Type of Service Wide Area Networking, To appear in Supercomputing '89, November 1989.

- Network Systems Corporation.
- Proteon Inc.
- Ultra Network Technologies.

Ultra has completed their testbed and experiments. The maximum rate of data transfer occurred in user memory to user memory between two Cray 2 computer systems. This was 400 megabits per second. This is a very promising data rate. One of the conclusions drawn in the report is that with this particular network, the computer systems are once again the bottleneck in performance. This is a continuing contest, with the CPU manufacturers increasing the amount of data that they can generate, and the networking manufacturers then coming up with networking products that are still faster. This race will continue for years to come. Figure 3 was provided by Ultra, showing one of the configuration of their testbed. For further details on this prototype, a project report is available. [4]

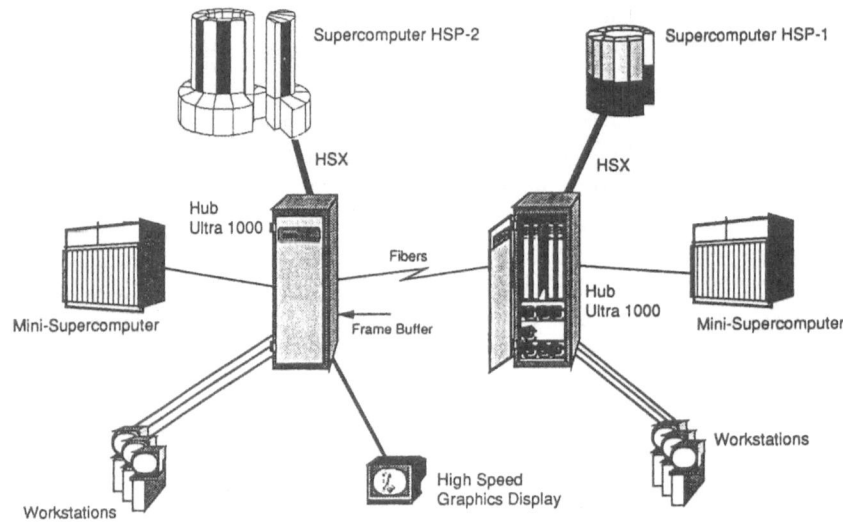

INITIAL HIGH SPEED LAN TOPOLOGY

Feb. 1989

figure 3 - Ultra Prototype Configuration

The other two projects are nearing completion at the time of this writing. Following the testbed activities is acquisition of LAN equipment to provide the next generation of general local network services to the NAS systems. Figure 4 shows a top level view of the requirements to be placed upon the local area networks. Each of the networks shown has a particular functional requirement placed upon it. Our plans are to use the data network best suited to meeting each of the data transfer needs, as described here. This is an example of the engineering effort necessary in order to build effective data networking infrastructure.

---

[4] Clinger, Marke, Very High Speed Network Prototype Development, NASA Ames Research Center, Contract #NAS2-12332 / CTO #9.

• 1. **Campus network.** This network is designed to provide service to host computer systems throughout a campus area. Campus is ill-defined, but can be generally thought of as within a few kilometers.

• 2. **Building network.** This network is designed to provide access to computer systems contained within a particular building. Long distance trade-offs versus bandwidth available do not have to be made for this system.

• 3. **Mass storage network.** This network is under the shortest distance constraint. It is expected to connect only between supercomputers and the Mass Storage System located in the same computer room. This network has only one goal : maximum file transfer throughput capability.

• 4. **Mini-supercomputer network.** This network is designed to provide access to Mass Storage from several mini-supercomputers.

• 5. **Backbone network.** This type of network is necessary at almost every site. Its purpose is to collate together all the other random networks spread throughout, and provide a common point of interconnection.

• 6. **Long Haul network.** This network is designed to provide a point of access for data networks that connect over a Wide Area. This is useful to partition, because there are more operational constraints, and typically lower bandwidth requirements upon this network.

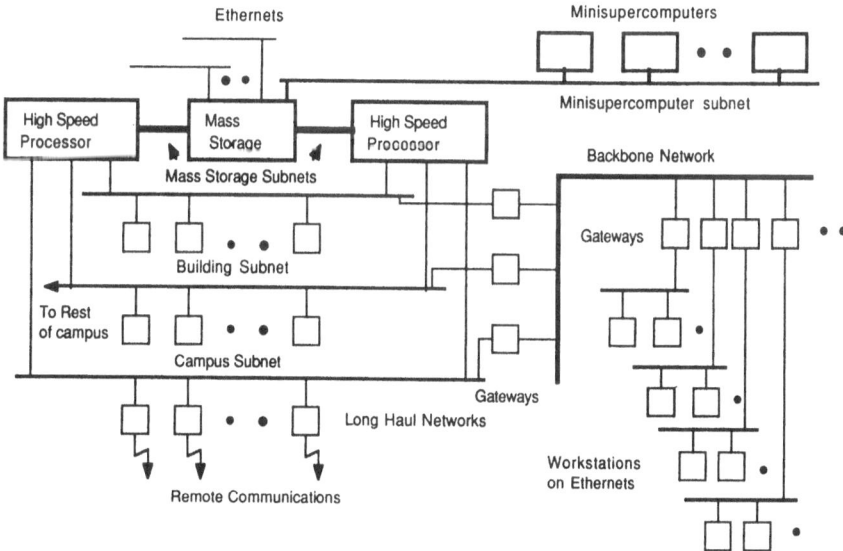

EOC Logical Top Level Network

figure 4 - LAN Logical Topolgy

Given this set of networks, we expect to be able to more closely tailor the choice of hardware to the most appropriate function. This sort of engineering analysis will become more essential over time, given the number of new network technologies and LAN vendors that are coming into existence.

Wide Area Networks

NAS connects to nationwide infrastructure networks. This includes networks such as MIL-net, ARPAnet, and NSFnet. This provides some degree of connectivity. In addition, NAS puts in network infrastructure to those parts of the aerospace community that are not easily reached via one of these national networks, or whose communications requirements are greater than those currently installed. In this fashion, we expect the aerospace community to be able to use the NAS facility for CFD research.

A similar paradigm to development in LAN technology is taking place in next generation of Wide Area Networks. Because of the costs involved, three test beds are occuring in a cooperative program with three different US Government agencies. Information Sciences Institute and Rome Air Development Center for DARPA, the NAS program, for NASA, and Los Alamos National Laboratory for DOE, are each building and evaluating one testbed, each from a different vendor. This method enables the three agencies to push technology in an area of common interest, within cost constraints. The three companies are:

- Bolt, Beranak, and Neumann.
- GTE/Proteon Inc.
- SRI/cisco systems.

These three companies are committed to producing a next generation of high performance network gateway. This project will be a pathfinder for the next generation of Internet gateways. Some of the expected goals of the program are:

- Increased packet throughput in gateways.
- Effective bandwidth allocation among disparate traffic flows.

These objectives are to produce equipment that operates at a higher performance than is currently available commercially. We expect to be able to route and transmit 10,000 datagrams per second through one of these units. We expect to be able to switch 15 megabits per second over long haul lines, per unit. This is the first step toward producing a capability to utilize T3 (45 megabits per second) communication lines. We expect the equipment to evolve in the following generation to be able to fully utilize T3 service.

- Effective use of switched service circuits.
- Effective type of service routing.
- Accounting information.
- Access control capability.

These advances are designed to improve our ability to manage wide area networks. The ability to bring up and tear down circuits dynamically is an element long missing from our IP wide area networks. Type of service routing will enable us to prioritize traffic, and use the circuit service that we feel is most appropriate for a given application. Accounting and access control are becoming more essential as use and abuse of these networks becomes more widespread. We are long past the stage where we know and trust everyone who is connected to our networks, and it is time to know more about what is going on with the systems.

- Multiple commercial products.

This is essential for the work to be a success. There must be a commercial market for the systems, or else we have chosen a wrong path, and the vendors will not be able to produce units at cost effective prices. Also, the existence of multiple vendors will keep the marketplace competitive. This is generally necessary for any high technology area, in order to prevent decay.

The NASA testbed is scheduled for deployment in the following testbed in the first quarter of 1990. This testbed is to verify equipment functionality, and test out various algorithms for using disparate communication paths. it will also provide service toward some of the CFD problems described above. This testbed is shown below in figure 5. We expect it to evolve into a more widespread NASA Aerospace CFD network, as shown in figure 6.

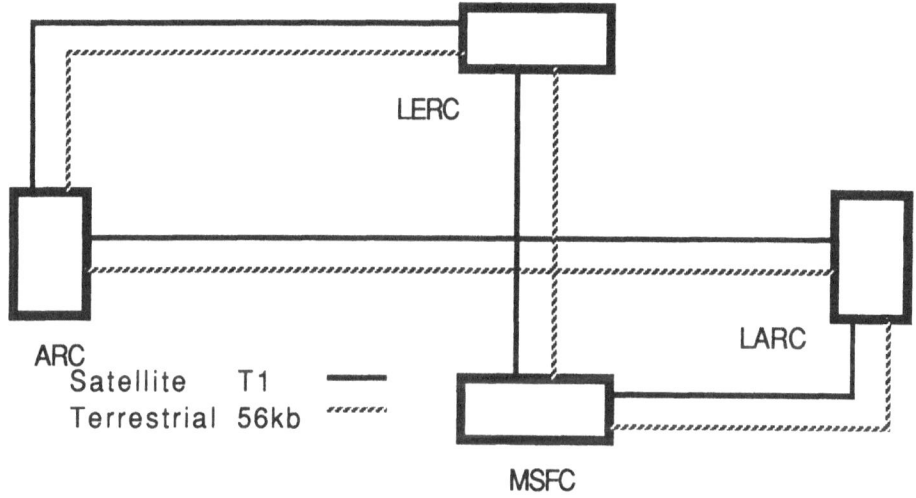

figure 5 - RIG testbed

Conclusion

The first step we take in designing networks is to start with the science to be solved. In our case, this is primarily Computational Fluid Dynamics. Often, as shown here, some calculations can quickly show what sort of implementation is necessary. This needs to be followed by prototype activities, in order to drive the technology in the direction we need it to develop. These activities must take place in a live testbed. Finally, having proven the technology, it is then put into widespread use.

Throughout this is an underlying requirement that design and implementation be done in an open and standard environment. This is absolutely necessary to be able to reuse something. Keeping to these rules results in computational capability that is able to grow and quickly change to meet current needs.

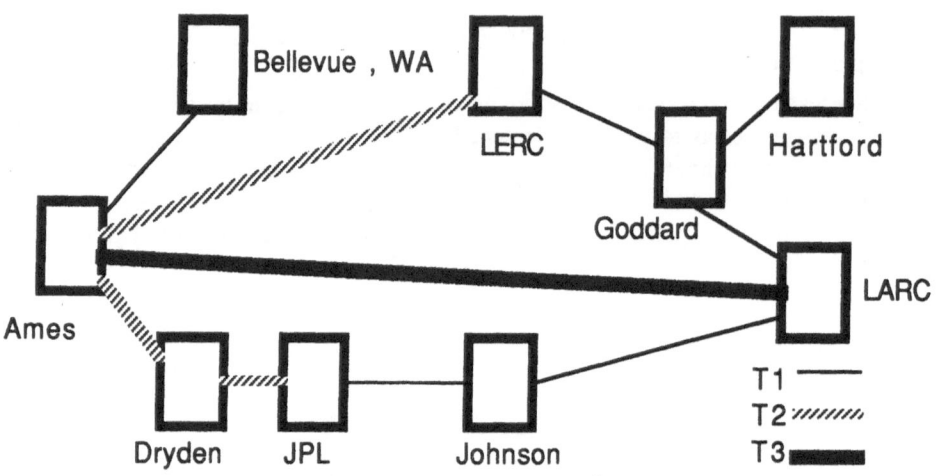

figure 6 - Topology for an Aerospace CFD Network

References

[1] Peterson, V. L., Impact of Computers on Aerodynamics Research and Development, Proceedings of the IEEE, Vol. 72, No 1. January, 1984. pp. 68-79.

[2] Ballhaus, W. F. Jr., Computational Aerodynamics and Supercomputers, NASA Ames Research Center, 1983, U.S. Government work, not protected by U.S. copyright.

[3] Bailey, F. Ron, and Blaylock, B. T., Update of NASA's Numerical Aerodynamic Simulation Program, NAS Internal Paper, unpublished.

[4] Jacobson, V., Braden, R., TCP Extensions for Long-Delay Paths, RFC 1072, October 1988.

PART 9

STORAGE

YOU'RE NOT WAITING, YOU'RE DOING I/O

George A. Michael
Lawrence Livermore National Laboratory
University of California
Post Office Box 808, L-306
Livermore, California 94550

Abstract

We discuss some of the growing problem in computer usage related to storage. As computers continue to get faster and support more memory and larger, more sophisticated applications we are seeing that the I/O and storage services are inadequate. Two rather bad omens are that the disparities are growing, and nothing seems to be being done about it by vendors.

Introduction

I am a voice, crying in the wilderness, " the I/O on Supercomputers is slow, and it's getting worse. " Frankly, it's also getting kind of boring mainly because even if the warning is right, and timely, no one is really interested. I/O balancing is hard to do, and not very glamorous, and the contribution to MFLOPS is not so obvious. Beyond the slowness issue is the one of capacity.

In this paper, we will consider some general factors related to large stores, and their relationships to the computers. We begin with two observations that are generally accepted as true around supercomputer centers.

The first thing is that there isn't enough memory on a supercomputer to allow us to run big applications. Today, there are applications "waiting for" the availability of a 100 Gigaword memory. A simple 3-D problem will easily require at least half that much. We cannot pin all our hopes on someone finding a way to do the computation using less memory.

NATO ASI Series, Vol. F 62
Supercomputing
Edited by J. S. Kowalik
© Springer-Verlag Berlin Heidelberg 1990

The second thing is that the I/O bandwidth is generally too small for moving data from or to this much memory in a suitably short enough time, and even if one could, there is nothing big enough to hold it. Notice that there's more than one user so the needs are for storage for many cycles of many problems.

The Storage Problem

First, one needs some sort of estimate of how much data is to be stored. In **Figure 1** we give an estimate of the upper limit of the number of data bytes that arise in various sources. The term "Experiments" covers laboratory work as well as data returned from satellites, and "Data Collections" covers things as dissimilar as libraries and genome tabulations. The current estimate of the LOC (Library of Congress) covers books and technical reports.

Figure 2 shows data related to the rates at which data is produced in computer modeling. The memory performance numbers do not relate to any specific computer; they are included only to warn about another possible problem.

Figure 3 is nothing more than a litany of desireable features for a large store. We envision a scheme in which each data store is a huge repositiory, and is to be treated as a regional resource.

Figure 4, the (depressing) end of the story merely covers some of the things now being considered as storage devices. They are classed as either disk (2-Dimensional) or tape (1-Dimensional) access devices. Compared to the challenges, as indicated in fig. 2, about the only nice thing one can say for this hardware is that it's consistent; all of it is inadequate. RAID stands for Redundant Arrays of Inexpensive Disks and refers to the work of Profs. D. Patterson and R. Katz at the University of California in Berkeley. They put numbers of disks together and treat the collection as a single source/sink of data, thereby gaining the sum of the transfer rates and capacities. Given such an arrangement, the problems of error and reliability can largely be neutralized. It should also be noted that nothing in the scheme prevents the use of tapes instead of disks. The STC4400 is a silo-like cylinder (approximately 2 metres high by 3 metres in diameter) for storing up to 6000 of the 3480 tape cartridges, any one of which may be

fetched to an associated tape handler by a robot arm that is pivoted on the axis of the cylinder. It is possible to cluster several of these silos together, and to pass a cartridge between them. The designations D1 and D2 are technically related to television recording formats, but they have come to be applied also to the tape cartridges. Their importance is mainly because of their high data capacity and bandwidth, and because they are an accepted standard; clearly a final requirement for the long-term storage of large amounts of data. Using them instead of the standard 3480 cartridges in the STC4400 might be a reasonable next step along the way to producing large data stores.

Conclusion

Storage devices with a broad range of capabilities are needed; with particular emphasis on the high performance stuff. Supercomputer Centers are in the vanguard of these needs, and, they ought to be similarily positioned in the attempts to encourage the development of better devices and systems.

General Sources of Data

	Number of Bytes	Events per year
Experiments	10^{12}	10^4
Computer Modeling	10^{10}	10^3
Data Collections (L.O.C.)	10^{13} 10^{15}	10
User Text	10^4	10^6

Figure 1

Some Estimates

1 Flop "causes" 24 bytes of memory activity

Large Computation Fluid Dynamics (CFD) Monte Carlo codes use

200 - 2,000 floating pt. operations per zone per cycle

for 20,000 - 200,000 zones per run

with 100 - 1,000 cycles (steps)

yields $\sim 10^{12}$ bytes

For these problems a computer memory $\sim 10^9$ B and at a tramsfer rate of

10^7 B/sec it will take $\sim 10^2$ sec/load !

"You're are not waiting, you are doing I/O"

Figure 2

Nature of Mass Storage

Multi User Access

Multi Media and Format

Never Down

Never Full

Never Loses Data

Secure

Modular

Redundant

Figure 3

Some Hardware

	Gigabytes	Megabytes/sec.	Latency (ms)
Big Disks	2 - 3	3 - 20	10 - 20
Raids	20	12 - 40	60
Optical	~10	.05 - 1	500
Tapes			
3480	.2 - .6	~12	~400
STC 4400	x 6,000	--	~--
D1, D2	5-180	~400	2,000

Figure 4

APPENDIX

Report From Trondheim: Trends and Needs
In Supercomputing

Bill Buzbee
Scientific Computing Division
National Center for Atmospheric Research
P. O. Box 3000
Boulder, Colorado   80307

Introduction

The 1989 NATO Advanced Research Workshop on Supercomputing was
held at the Norwegian Institute of Technology in Trondheim, Norway,
June 19-23.   Dr. Janusz Kowalik, Boeing Computer Services in Seattle,
was general chair.   Dr. Bjornar Pettersen, SINTEF, Norway, was
responsible for local arrangements that included not only commodious
conference facilities, but a delightful array of local menus and scenery.

This paper summarizes some of the trends and needs in supercomputing
as distilled by the author from the papers presented at the conference
and related discussions.

Trends

Because six years had elapsed since the previous NATO workshop on
supercomputing, certain trends were easily recognized.   Consider, for
example, parallel processsing.   In 1983, the CRAY X-MP/24 had just been
introduced,   and   no   supercomputer   center   was   routinely   using
multitasking.   Today, centers such as the European Centre for Medium-
range Weather Forecasting (ECMWF), and the National Center for
Atmospheric Research (NCAR), routinely support multitasking as a part
of their daily workload.   Also in 1983, no supercomputer center had an
experimental parallel processor with even tens of processors in it.   At
the 1989 workshop, every supercomputer center represented was
involved with one or more experimental parallel processors with some of
the associated systems containing several thousand processors.   The
trend is self-evident--today parallel processing is much on our minds,

NATO ASI Series, Vol. F 62
Supercomputing
Edited by J. S. Kowalik
© Springer-Verlag Berlin Heidelberg 1990

tomorrow it will be much on our computers. There is simply no other way to achieve the performance levels required by the next generation of simulations. In the 1990s, supercomputers likely will be shared-memory parallel processors, and will contain at least 16 processors. To restate this trend, in the '90s, supercomputing and parallel processing will be synonymous. Incidentally, Harry Jordan, University of Colorado, asserted that massively parallel, shared-memory systems can be built and that the unsubstantiated, widely-held opinion that such cannot be done is a deterrent to progress in this area.

In order to encourage scientists to use parallel processing on supercomputers, several centers offer charging algorithms that favor multitasking, so a trend may be in the making. Further, John Levesque, Pacific-Sierra Research Corporation, conjectured that we may be able to use previously acquired performance data to advantage when using parallel processors. For example, a model could carry with it a history file of previous optimizations for a particular architecture and this file would then be used to aid the system in managing the current simulation. Interestingly, a unique feature the Perfect Benchmark Suite discussed by Joanne Martin, T. J. Watson Research Center, is a "diary" that documents successful optimization strategies on various architectures. (*Shades of another trend?*) Harry Jordan further observed that users have been lead to believe that parallel processing is much more difficult than it really is. We, as a community, need to counter this impression because it partially accounts for user resistance to parallel processing. (*The beginnings of a counter-trend?*)

The future role of super-workstations evoked a lively discussion. There was wide agreement that single-user workstations will soon offer performance that is within an order of magnitude of a single CRAY X-MP processor. Gene Levin's (RIACS at NASA Ames) comparsions of the Ardent and X-MP substantiated this view. Further, the close coupling of high-speed processing and powerful graphics on super-workstations will make them particularly attractive, provided their cost is not too high. Under the caveat of reasonable cost, there was general agreement that some of the simulation being done today on supercomputers will migrate to super-workstations in the 1990s. Several participants noted that this migration has already begun. (*Hence another trend?*) It was also noted that there are many important problems that cannot be solved on current supercomputers that can be solved with the next generation of equipment. These problems are far beyond the capability of super-workstations.

What is needed

Concurrency in algorithms is prerequisite to successful parallelization. Put negatively, if the algorithm does not contain concurrency, then the sophistication of parallelizing compilers and the elegance of parallel architecture are of no avail. Thus, support for research and development of concurrent algorithms merits high priority. We need more theory, more paradigms, and better performance measurement tools for parallel computation. Jack Dongarra, Argonne Natinal Laboratory, reported on the LAPACK project that is developing a collection of linear algebra routines formulated to exploit parallel processing. Similar efforts are needed for other widely used software collections. Finally, and in general, all supercomputer centers will need more expertise in parallel processing and to provide user support therein.

Bruce McCormick, Texas A&M, observed that super-workstations will encourage the use of interactive and dynamically adapted algorthims and this, in turn, affords new opportunity for numerical analysts. Put another way, classical numerical analysis evolved in the era of batch computation, but interactive computation combined with graphical display makes numerical techniques feasible that were not feasible under batch processing. Bruce speculates that such numerical techniques could have far reaching implications for numerical analysis.

Garry Rodrigue, Lawrence Livermore Laboratory, lamented the state of computer documentation in that it is often tedious to find specific details about system usage, hardware behaviors, etc. There is an almost urgent need for better structure and organization of information, along with associated tools, such that users can quickly and easily get the details they need.

Summary

The next five years will be a period of dramatic change in supercomputing as parallel processing and visualization become routine capabilities.

PRESENTERS

Session 1: Introduction and Supercomputing Centers
Chair, R. Bailey

R. Bailey, "R&D In The NAS Program"
B. Buzbee, "Supercomputing--The View From NCAR"
F. Hossfeld, "Supercomputing at KFA Juelich"
S. Karin, "Supercomputing Environment at San Diego Center"
K. Neves, J. Kowalik, "Supercomputing: Key Issues and Challenges"

Session 2: Supercomputer Architecture
Chair, K. Miura

K. Bratbergsengen, "Fast Database Systems"
K. Miura, "Architecture of Fujitsu's Supercomputer"
K. Solchenbach, "Suprenum"
T. Watanabe, "Advanced Architecture and Technology of a Super-
        computing System"

Session 3: Compilers and Programming Tools
Chair, J. Dongarra

J. Dongarra, "LAPACK"
H. Jordan, "Process Scheduling"
S. Kumar, "Maximally Parallel Task System"
J. Levesque, "A Cooperative Approach to Program Optimization"
A. Lichnewsky, "Optimizing CRAY-2 Performance"
D. Sorensen, "Tools For Parallel Programs"

Session 4: User Environments and Visualization
Chair, B. McCormick

E. Levin, "The Role of Superworkstations"
B. McCormick, "Issues in Scientific Visualization"

H. Santo, "The Science of Engineering Visualization: Past, Present, Prospects"
Supercomputing Videos

Session 5: Supercomputer Productivity and Performance
Chair, J. Martin

R. Babb II, "Sara--A Cray Assembly Speed-up Tool"
R. Hockney, "Measurements of Problem Related Performance Parameters"
J. Martin, "Supercomputer Performance"
R. Voigt, "Requirements for Multi-Disciplinary Modeling of Aerospace Vehicles on High Performance Computers"

Session 6: Methods and Algorithms
Chair, G. Rodrigue

N. Deo, "Data Structures for Parallel Computations"
I. Duff, "Large Sparse Linear Systems"
W. Gander, "Solving Linear Equations by Extrapolation"
A. Kiper, "Parallel Algorithms for Eigenvalue Problems"
F. Lootsma, "Parallel Separable Nonlinear Optimization"
K. Miura, "Vectorization and Parallelization of Transport Monte Carlo Methods"
S. Norsett, "Parallel Ode Solvers"
D. Parkinson, "Parallel Algorithms Are Easy But Different"
G. Rodrigue, "The Solution of 3-D Implicit Linear Systems"
D. Waltz, "Information Applications on Connection Machine"

Session 7: Networking

J. Lekashman, "Computer Networking for Interactive Supercomputing"
R. Stefanelli, "Interconnecting Networks for High Speed Parallel Supercomputing"

Session 8: Summary

NATO ASI Series F

NATO ASI Series F

NATO ASI Series F